AF360829

Lipid Delivery System and
Functional Food

Lipid Delivery System and Functional Food

Editors

Qianchun Deng
Ruijie Liu
Xin Xu

MDPI • Basel • Beijing • Wuhan • Barcelona • Belgrade • Manchester • Tokyo • Cluj • Tianjin

Editors
Qianchun Deng
Chinese Academy of Agricultural Sciences
China

Ruijie Liu
Jiangnan University
China

Xin Xu
Yangzhou University
China

Editorial Office
MDPI
St. Alban-Anlage 66
4052 Basel, Switzerland

This is a reprint of articles from the Special Issue published online in the open access journal *Foods* (ISSN 2304-8158) (available at: https://www.mdpi.com/journal/foods/special_issues/Lipid_Delivery_System_Functional_Food).

For citation purposes, cite each article independently as indicated on the article page online and as indicated below:

LastName, A.A.; LastName, B.B.; LastName, C.C. Article Title. *Journal Name* **Year**, *Volume Number*, Page Range.

ISBN 978-3-0365-7498-1 (Hbk)
ISBN 978-3-0365-7499-8 (PDF)

Contents

About the Editors

Qianchun Deng

Ph.D., Professor, doctoral supervisor, National high-level talents special support plan talents, "Agricultural Talents and Leading Talents" of Chinese Academy of Agricultural Sciences, Executive Chief of Oil Quality Chemistry and Processing Utilization Innovation Team of Oil Crop Research Institute of Chinese Academy of Agricultural Sciences, Deputy Director of Key Laboratory of Oil Processing of Ministry of Agriculture, Secretary-General of the National Healthy Oil Industry Technology Innovation Alliance, Vice Chairman of the Clinical Nutrition Special Committee of the Chinese Society for the Promotion of Human Health Science and Technology, Vice Chairman of the Hubei Society for Food Science and Technology, Editorial Board of *"Critical Reviews in Food Science and Nutrition"* and *"Food Science and Human Wellness"*, Associate Editor of *"Chinese Journal of Oil Crop Sciences"*, Youth Editorial Board of *"Food Science"*, and Editorial Board of *"Journal of Light Industry"*. He is mainly engaged in the research of characteristic oilseed processing and plant-based foods. He hosted over 20 National and Provincial level projects and industry-university-research cooperation projects. He authorized 23 national invention patents. He has created more than 20 new products, such as visual fatigue relief health food, cleaning label flaxseed-based plant milk, DHA mixed-protein solid beverage. He has published more than 100 papers as the first or corresponding author in *Comp. Rev. Food Sci. Food Safe*, *Prog. Lipid Res.*, *Food Hydrocoll.*, *Food Science*, etc. The achievements have been industrialized and applied in relevant listed companies and leading industrialization enterprises, with significant economic and social benefits. As the main contributor, he has won five scientific and technological awards, including the second prize of the National Science and Technology Progress Award (ranking second) and the first prize at the Provincial and Ministerial level.

Ruijie Liu

Ph.D., Professor, doctoral supervisor, School of Food Science and technology in Jiangnan University. Member of a council of oil Sub-Association & Cereals and Oils Nutrition Sub-Association of China Cereals and oils Association (CCOA), and editorial board member of China oils and fats. National excellent scientific and technological worker of grain and oil, winner of "Young Scientist Investigator Award" of international fatty acid and lipid Society (ISSFAL) and young science and technology award of CCOA, doctor of entrepreneurship and innovation of "entrepreneurship and innovation plan" of Jiangsu Province, and outstanding young backbone teacher of "blue project" of Jiangsu Province. In recent years, she has received many research projects including 6 General Program fund of National Natural Science Foundation of China, 4 Science and Technology projects (such as the National Key Technology R&D Program), and a lot of cooperation projects from enterprises. She received more than 10 awards/honors, including the Prize for Progress in Science and Technology of Jiangsu Province, Science and Technology Progress Award from China National

Light Industry Council, and China Patent Excellence Award. She has formulated and revised more than 10 national and industry standards for food safety; authorized more than 40 Utility patents and published over 60 SCI/EI journal articles.

Xin Xu

Professor and Doctoral Advisor, School of Food Science and Engineering, Yangzhou University; Visiting Scholar, Cornell University and University of Texas at Austin; Former Mobile Commissioner at the National Natural Science Foundation of China and Ministry of Science and Technology Rural Center; Research Team Leader of Lipid Nutrition and Safety, focused on basic theoretical research and new product development of lipid nutrition and functional food resources; Project Leader of 8 National Natural Science Foundation projects and more than 20 other provincial and enterprise commissioned research projects in recent years; Published over 60 papers in domestic and international academic journals, including *J. Agric. Food Chem.*, *Food Chem.*, *Food Funct.*, and *Food Science* (Chinese journal), among others; Applied for and authorized 10 national invention patents; Peer Review Expert for National Natural Science Foundation, National Food Production Permit Registration Reviewer, and Member of Jiangsu Provincial Food Safety Standards Expert Committee. Research interests include Development and utilization of functional food resources; lipid nutrition and safety; active ingredients of medicinal and edible plants; food biotechnology.

 *foods*

Article

Improvement of Oxidative Stability of Fish Oil-in-Water Emulsions through Partitioning of Sesamol at the Interface

Zhihui Gao [1], Zhongyan Ji [1], Leixi Wang [1], Qianchun Deng [2], Siew Young Quek [3,4], Liang Liu [1] and Xuyan Dong [1,*]

[1] College of Food Science and Engineering, Qingdao Agricultural University, Qingdao 266109, China; liuliangwh@163.com (L.L.)
[2] Oil Crops Research Institute of the Chinese Academy of Agricultural Sciences, Wuhan 430062, China; chunn2@163.com
[3] School of Chemical Sciences, The University of Auckland, Auckland 1142, New Zealand; sy.quek@auckland.ac.nz
[4] Riddet Institute, Palmerston North 4474, New Zealand
* Correspondence: dongxuyan@qau.edu.cn

Abstract: The susceptibility of polyunsaturated fatty acids to oxidation severely limits their application in functional emulsified foods. In this study, the effect of sesamol concentration on the physicochemical properties of WPI-stabilized fish oil emulsions was investigated, focusing on the relationship between sesamol–WPI interactions and interfacial behavior. The results relating to particle size, zeta-potential, microstructure, and appearance showed that 0.09% (w/v) sesamol promoted the formation of small oil droplets and inhibited oil droplet aggregation. Furthermore, the addition of sesamol significantly reduced the formation of hydrogen peroxide, generation of secondary reaction products during storage, and degree of protein oxidation in the emulsions. Molecular docking and isothermal titration calorimetry showed that the interaction between sesamol and β-LG was mainly mediated by hydrogen bonds and hydrophobic interactions. Our results show that sesamol binds to interfacial proteins mainly through hydrogen bonding, and increasing the interfacial sesamol content reduces the interfacial tension and improves the physical and oxidative stability of the emulsion.

Keywords: sesamol; fish oil; O/W emulsion; oxidation; interface

Citation: Gao, Z.; Ji, Z.; Wang, L.; Deng, Q.; Quek, S.Y.; Liu, L.; Dong, X. Improvement of Oxidative Stability of Fish Oil-in-Water Emulsions through Partitioning of Sesamol at the Interface. *Foods* **2023**, *12*, 1287. https://doi.org/10.3390/foods12061287

Academic Editor: Arun K. Bhunia

Received: 5 February 2023
Revised: 11 March 2023
Accepted: 16 March 2023
Published: 17 March 2023

1. Introduction

Polyunsaturated fatty acids (PUFAs), which are rich in docosahexaenoic acid (DHA) and eicosapentaenoic acid (EPA), exhibit beneficial effects on human health, such as anti-inflammatory, anti-cancer, anti-depression, and vascular protective activities, and promote neurological development [1]. PUFAs have great potential as functional food ingredients and thus, have become a research hotspot for domestic and international scholars [2]. However, PUFAs, including DHA and EPA, are highly susceptible to oxidation during processing and storage because of high level unsaturated bonds [3]. The reaction products of lipid oxidation not only adversely affect the flavor, nutrient content, and shelf life of foods but can also even damage the health of consumers [4]. Lipid oxidation severely limits the development and application of PUFAs in functional emulsified foods. Therefore, understanding and preventing lipid oxidation in complex food systems is essential for the construction of PUFAs emulsion systems with good oxidative stability. An emulsion delivery system is a classical lipid carrier that can significantly improve the bioavailability and stability of PUFAs.

Using antioxidants, especially natural antioxidants, is the most direct and effective way to inhibit lipid oxidation in the food industry [5]. Natural polyphenols, comprising chemical structures with antioxidant activity (e.g., catechols and hydroxyl groups), can inhibit lipid oxidation by scavenging free radicals or chelating transition metal ions. Furthermore, natural polyphenols are safe to consume and have no side effects, which has made them popular in research and production [6]. Natural phenolic antioxidants, such as tea polyphenols, quercetin, and tocopherols, improve oxidative stability and extend the shelf life of emulsions [7–9].

Sesamol is a plant-derived monophenolic compound extracted from sesame seeds. Its antioxidant activity is derived from the phenolic group on the benzodioxole ring [10]. Sesamol is used as an antioxidant in food because of its excellent ability to inhibit lipid oxidation. Sesamol not only inhibits single lipid oxidation, but also lipid oxidation in complex systems [11,12]. For example, sesamol significantly improved the oxidative stability of lipids in beeswax organogel systems [13]. Additionally, sesamol inhibits lipid oxidation in sunfower oil-in-water (O/W) emulsion systems [14]. The DPPH radical scavenging rate of the same concentration of sesamol (2.5 μmol/g) was 2.56 times higher than that of 2,6-di-tert-butyl-p-cresol in lard after thermal induction at 180 °C for 80 min [15].

The interfacial region, which separates the oil phase from the aqueous phase, plays a key role in inhibiting lipid oxidation in emulsions [16]. Tocopherols interacting with whey protein isolate (WPI) through hydrophobic and electrostatic interactions, were adsorbed at the interface, where they significantly inhibited lipid oxidation in O/W emulsions of linseed oil [9]. Numerous polyphenols also reside at the oil–water interface via non-covalent bonds with adsorbed proteins [16]. Wang et al. found that sesamol effectively inhibited particle aggregation and lipid oxidation in protein-stabilized flaxseed oil-in-water emulsions, and hypothesized that sesamol molecules could adsorb on the surface of oil droplets and interact with emulsifiers to influence interfacial properties, thereby enhancing the stability of emulsions [17]. However, the interaction between sesamol and emulsifiers was not explored and the partitioning of sesamol in all phases of the emulsion was not clarified. Therefore, we expected to understand the relationship of antioxidant-emulsifier interactions, interfacial partitioning, and emulsion stability by further investigating the interaction and interfacial partitioning of sesamol in WPI-stabilized fish oil emulsions.

In this study, the effect of sesamol on the oxidative stability of WPI-stabilized fish oil O/W emulsions was investigated, focusing on sesamol interfacial partitioning and interactions. Furthermore, a preliminary study of its potential mechanism of action was conducted. The sesamol content in each phase was analyzed separately at different storage periods to elucidate the interfacial distribution of sesamol during lipid oxidation in emulsions. This study provides guidance for the application of sesamol as a plant-based antioxidant in functional foods, thereby providing a theoretical basis for the development of functional emulsified foods enriched with PUFAs.

2. Materials and Methods

2.1. Materials

Sesamol (98%) was purchased from Shanghai Yi'en Chemical Technology Co., Ltd. (Shanghai, China). WPI (92%) was purchased from Shanghai Yuanye Biotechnology Co., Ltd. (Shanghai, China). Fish oil was purchased from DSM Nutritional Products Ltd. (Kaiseraugst, Switzerland). Except for methanol, which was chromatographic grade, all other chemicals were analytical grade chemicals and were purchased from Sinopharm Chemical Reagent Co. (Beijing, China). Double-distilled, deionized water was used for the preparation of all solutions.

2.2. Preparation of Stripped Fish Oil

Stripped fish oil was prepared in order to exclude the interference of the original phenolics in the fish oil. The fish oil was processed based on the method described by Cheng et al. with minor modifications. The chromatographic silica was repeatedly rinsed with double-distilled water at room temperature until it was free of impurities and then activated at 120 °C for 12 h. Equal amounts of fish oil were dissolved in 100 mL of n-hexane, loaded onto the column, and then eluted with 10 times n-hexane (at room temperature). The glassware for collecting fish oil was wrapped with aluminum foil and put in an ice bath to avoid lipid oxidation during collection, and the collected mixture was placed in storage at $-80\ °C$. Before use, n-hexane was removed using a vacuum rotary evaporator (IKA, RV10D, Staufen, Germany) at 37 °C, and the remaining solvent was evaporated with nitrogen [18].

2.3. Preparation of Fish O/W Emulsion

WPI (1% *w/v*) was dissolved in phosphate buffer solution (PBS, 10 mM, pH 7.0) to form the aqueous phase. The oil phase was prepared by adding sesamol (sesamol content in the emulsion was 0, 0.01, 0.03, or 0.09%, w/v) to the fish oil. The 5% (*v/v*) oil phase and 95% (*v/v*) water phase were mixed and processed by a high-speed stirring device (IKA, T25, Staufen, Germany) at 15,000 rpm for 2 min, followed by placing the primary emulsion on an ice bath and subjecting it to 650 W ultrasonication (Nanjing Xianou Instrument Manufacturing Co., Ltd., XO-1000D, Nanjing, China, amplifying bar Φ 6) for 5 min. Sodium azide (0.02% *w/w*) was added to the emulsions to prevent microbial growth.

2.4. Accelerated Storage Experiments

The emulsions were oxidized under the Fenton system for 5 days, with samples collected every 24 h for analysis. The Fenton system was generated from a recovered solution of 10 μM $FeCl_3$, 100 μM ascorbic acid, and 5 mM H_2O_2 [19].

2.5. Measurement of Particle Size and ζ Potential during Storage of Emulsions

The laser particle size analyzer (Microtrac MRB, S3500, Montgomeryville, PA, USA) measured the particle size of the emulsion with refractive indices set to 1.49 (oil) and 1.33 (deionized water) [20]. The results were recorded as volume weighted average particle size ($D_{4,3}$):

$$D_{4,3} = \sum n_i d_i^4 / \sum n_i d_i^3 \tag{1}$$

n_i: number of droplets, d_i: droplet diameter.

The zeta potential was measured using dynamic light scattering (Malvern, NANO ZS90, Malvern, UK). The emulsion samples were diluted 200 times with PBS (10 mM, pH 7) at 25 °C and loaded into DTS1070 measuring dishes. The Hückel model was selected for the measurement calculation [21].

2.6. Observation of the Microstructure and Visual Appearance during Emulsion Storage

The microstructure of the emulsion was photographed using a confocal laser scanning microscope (Zeiss, LSM900, Oberkochen, Germany) equipped with a 60× oil immersion objective [22]. Nile Red (1 mg) was dissolved in 10 mL of ethanol, and 20 μL was added to 1 mL of emulsion to stain the oil phase. An Ar laser was used to excite the Nile red dye fluorescence at 488 nm.

2.7. Analysis of Lipid Oxidation during Emulsion Accelerated Storage

The extent of lipid oxidation was monitored by periodically measuring the amount of hydrogen peroxide and thiobarbituric acid reactive substances (TBARS) formed in the emulsion [23]. The method developed by Shantha and Decker was optimized to determine lipid hydroperoxides [24]. The sample (0.3 mL) was mixed with the extraction solution (isooctane/isopropanol, *v/v*, 3:1; 1.5 mL) and then vortexed for 10 s (three times) with 20 s intervals. The mixture was separated by centrifugation (Cence, Changsha, China) at $2000\times g$ for 2 min (4 °C) and the supernatant (200 μL) was added to MB (methanol–butanol, *v/v*, 2:1; 2.8 mL). To each sample, Fe^{2+} solution (50 μL) was added with ammonium thiocyanate solution (50 μL, 3.94 M), rapidly vortexed, and then reacted for 20 min at room temperature and protected from light. Sample absorbance was analyzed at 510 nm with a UV/Vis spectrophotometer (Shimadzu, UV-2700, Kyoto, Japan). Fe^{2+} solutions were obtained by centrifugation ($2000\times g$, 2 min) of a freshly prepared mixture of $FeSO_4$ (1 mL, 0.144 M)) and $BaCl_2$ (1 mL, 0.132 M, in 0.4 M HCl). The standard curve was constructed using cumene hydroperoxide (CH).

The TBARS quantification method was somewhat refined according to the description of McDonald and Hultin [25]. The emulsion (1 mL) was mixed with thiobarbituric acid (TBA) reagent (2 mL), which contains 150 g/L trichloroacetic acid (TCA), 3.75 g/L TBA, and 0.25 mol/L HCl. The samples were boiled for 15 min, cooled, mixed with chloroform (1 mL), vortexed, and finally centrifuged ($2000 \times g$, 15 min). A UV/Vis spectrophotometer was used to determine the absorbance at 532 nm. The lipid concentration was calculated using TBARS with a standard curve of 1,1,3,3-tetraethoxypropane.

2.8. Protein Oxidation Analysis

The increase in carbonyl groups and the reduction in free sulfhydryl groups were monitored during emulsion storage. The protein carbonyl content of the samples was determined according to the method provided by Levine et al. [26]. The emulsion was mixed with TCA (200 mg/mL) in equal amounts, incubated in ice water for 10 min, and then centrifuged (4 °C, $15,000 \times g$, 10 min). The protein pellet was dissolved in SDS (2 mL, 20 mg/mL, pH 8) solution, and n-hexane (1 mL) was used to separate the oil. A total of 0.1 mL of clear protein solution was taken after centrifugation ($2000 \times g$, 5 min) and added to DNPH (2 mL, 10 mM) solution and allowed to react for 1 hr. TCA (1 mL, 200 mg/mL) was added, vortexed, and centrifuged ($6000 \times g$, 10 min) to recover the WPI pellet, and cleaned three times using ethanol/ethyl acetate (1:1, v/v) solvent. After blowing the residual ethanol/ethyl acetate dry, the WPI pellets were solubilized in guanidine hydrochloride (2 mL, 6.0 M) and incubated at 37 °C for 15 min. Absorbance was collected at 370 nm to analyze the carbonyl content. Results were analyzed using a protein molar extinction coefficient of 22,000 $M^{-1}cm^{-1}$.

The content of free sulfhydryl groups in emulsions was determined according to Beveridge's method [27]. The 100 µL emulsion was thinned to 2 mL with PBS (10 mM, pH 7.0), then blended with 10 mL of Tris-Gly buffer containing urea (8 M, pH 8), and finally 80 µL of Ellman's reagent (5,5′-dithiobis(2-nitrobenzoic acid) (DNTB)) was added and stirred for 30 min at 25 °C and then centrifuged (Cence, Changsha, China, $12,000 \times g$, 10 min). The sulfhydryl content was calculated based on the absorbance at 412 nm and the molar extinction coefficient of 13,600 $M^{-1}cm^{-1}$ using PBS as a blank.

2.9. Isothermal Titration Calorimetry (ITC)

Raw data was measured at room temperature by a MicroCal PEAQ-ITC (version, 1.41, Malvern Instruments Inc., Northampton, MA, USA), and other thermodynamic parameters (K_d, n, ΔG, ΔH, and ΔS) were calculated and analyzed using MicroCal PEAQ-ITC analysis software [28]. The cuvette was injected with 200 µL of PBS (10 mM) containing WPI (0.3 mM), and the reference cuvette was injected with an equal amount of ultrapure water. The WPI solution was stirred continuously (750 rpm), and 4 µL of sesamol (20 mM) solution was injected precisely into it every 120 s. The unit point combination model was selected for data fitting and calculated according to the Van't Hoff equation:

$$\Delta G = -RT \ln K_d \tag{2}$$

$$\Delta G = \Delta H - T\Delta S \tag{3}$$

2.10. Molecular Docking

The non-covalent interactions between sesamol and WPI were analyzed using molecular docking simulation techniques. The most abundant beta-lactoglobulin (β-LG) was selected to represent WPI. The structure of β-LG (PDB ID: 3NPO) was obtained from the Research Collaboratory for Structural Bioinformatics Protein Data Bank (http://www.rcsb.org, accessed on 2 August 2022). The 3D structure of sesamol (Compound CID: 68289) was downloaded from PubChem (https://pubchem.ncbi.nlm.nih.gov/, accessed on 2 August 2022) [29]. Water molecules were removed from the β-LG structure before docking. The β-LG and sesamol were set to rigid and flexible, respectively. AutoDock 4.2 was used to simulate the binding properties of sesamol to β-LG [30]. Finally, the docking result with the lowest energy was selected, visualized, and exported using the Discovery Studio 2020 software (version 4.5.0, Biovea Inc., Omaha, NE, USA).

2.11. Interfacial Tension Measurement

The interfacial activity of sesamol was analyzed using a droplet shape analyzer DSA100 (Krüss GmbH, Hamburg, Germany) [21]. The titration module consists of a DS3210 software-controlled single titration device, a disposable syringe with a Luer lock connector (1 mL), and a standard-fit steel needle (NE45, 1.832 mm). A solution of PBS (10 mM, pH 7) with or without WPI (1 wt%) was first drawn into a syringe, and then the needle was inserted into fish oil containing different proportions of sesamol. A precise drop of suspension is extruded by system control at room temperature, capturing the droplet profile every 5 s for 1 h.

2.12. Determination of Sesamol Distribution in Emulsions

The distribution of sesamol in the emulsion was determined by the method of Cheng et al. [18]. The sesamol emulsion (4 mL) was centrifuged (4 °C, 10,000× g, 1 h), and the aqueous phase was carefully collected with a syringe. PBS (2 mL, 10 mm, pH 7) was added to the remaining emulsified layer, vortexed for 10 min, and centrifuged (4 °C, 3000× g, 5 min) to remove the aqueous phase. This was repeated thrice, then 1 mL of isooctane-isopropanol (3:1) was added, vortexed for 10 min, centrifuged (4 °C, 3000× g, 5 min) to remove the lower aqueous phase and the intermediate emulsifier layer, and the organic phase (oil phase) was carefully collected using a syringe. The sample (emulsion, oil, or aqueous phase) of 0.2 mL was vortexed in a 5 mL centrifuge tube with 2 mL of methanol (3 min) and then placed in an ultrasonic water bath (5 min). The supernatant was collected, and the process was repeated thrice. The supernatant was transferred to a dark glass vial to evaporate the solvent and fix the volume to 1 mL. The sample (1 mL) was injected into the sample vial with an organic 0.22 μm filter. The sesamol content was analyzed by HPLC LC-20A (Shimadzu Corporation, Kyoto, Japan) and calculated using a standard curve made by dissolving sesamol in methanol. The sesamol distribution ratio in each phase (aqueous phase, oil phase, interfacial layer) is based on the following formula:

$$\text{Partitioning ratio in the aqueous phase } (\%) \approx \frac{C_a \times V_a}{C_e \times V_e} \times 100\% \tag{4}$$

$$\text{Partitioning ratio in the oil phase } (\%) \approx \frac{C_o \times V_o}{C_e \times V_e} \times 100\% \tag{5}$$

$$\text{Partitioning ratio in the interface layer } (\%) \approx 100 - (4) - (5) \tag{6}$$

C_e, C_a, and C_o are the concentrations of sesamol in the emulsion, aqueous, and oil phases. V_e is the volume of 1 mL of emulsion, V_a and V_o are the volumes of the aqueous phase (0.95 mL) and oil phase (0.05 mL) in 1 mL of emulsion.

2.13. Statistical Analysis

All results, obtained via triplicate experiments, are presented as the mean ± standard deviation. Statistical data analysis was performed using SPSS (V20, SPSS Inc., Chicago, IL, USA). Differences between samples were assessed using analysis of variance (ANOVA). $p < 0.05$ was considered statistically significant.

3. Results and Discussion

3.1. Effect of Sesamol on the Physical Properties of Emulsions

The physical stability of emulsions with and without sesamol was evaluated by monitoring changes in emulsion particle size, charge, and microstructure during storage under the Fenton system for 5 days.

3.1.1. Effect of Sesamol on the Droplet Size of WPI-Stabilized Emulsions during Storage

All the emulsions initially contained relatively small droplets ($D_{4,3} < 1.7$ μm; Figure 1A). Nevertheless, the sesamol-added emulsions had lower $D_{4,3}$ values than the control sample (0.00% sesamol emulsions), and the average droplet size tended to decrease with the increase in sesamol content. The presence of sesamol promoted the production of small oil droplets during the preparation of emulsions. In addition, the oil droplet size of the control emulsion decreased on day 2 of storage and remained at a lower value from day 2 to day 4 (Figure 1A). The larger the oil droplet size, the faster the rate of agglomeration. The newly prepared control emulsion had the largest droplet size and formed a creamy or oil layer on top of the emulsion first because of the flocculation or agglomeration of large droplets. The proportion of small size droplets remaining in the emulsion was much greater, so the light scattering measurement reflects only the size of the small droplets remaining in the emulsion [17]. The addition of 0.09% sesamol not only promoted the formation of small oil droplets during emulsion preparation but also inhibited the degree of droplet aggregation during the first four days of storage. A similar study found that catechins interact with rice bran proteins through hydrogen bonding and hydrophobic interactions and enhanced the ability of WPI as an emulsifier to form and stabilize emulsions [31].

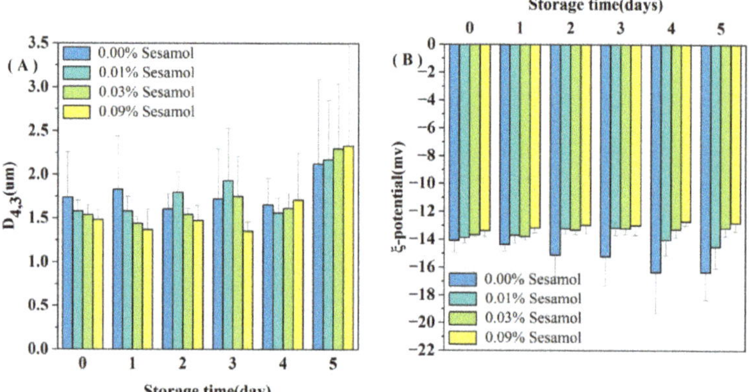

Figure 1. Changes in the mean droplet size $D_{4,3}$ (**A**) and ζ-potential (**B**) of emulsions containing different levels of sesamol during storage (30mL of emulsion in a 50mL glass bottle with screw cap stored at room temperature and protected from light).

3.1.2. Effect of Sesamol on the ζ-*Potential* of WPI-Stabilized Emulsions during Storage

All the emulsions had a slight negative charge (less than 15 in absolute value; Figure 1B). The pH 7 of the emulsion is higher than the WPI isoelectric point, so the charges of the droplets are all negative [32]. The smaller absolute values of charge may be because ultrasound during emulsion preparation promoted the binding of WPI to polyphenols and reduced the exposure of negatively charged groups on WPI [33]. More sesamol masks more WPI negatively charged groups; thus, the magnitude of the initial ζ-potential depended on the amount of sesamol added: 0.00% > 0.01% > 0.03% > 0.09%. Our observations are consistent with Yi et al., who found that the addition of the antioxidant black rice anthocyanins to the aqueous phase of walnut O/W nanoemulsions resulted in a decrease in the absolute value of the ζ-potential [34]. We observed a decreased trend in the zeta potential of the control emulsion during storage (Figure 1B), indicating that the interfacial composition and structure of the emulsion changed during storage.

3.1.3. Effect of Sesamol on the Visual Appearance and Microstructure of WPI -Stabilized Emulsions during Storage

The visual appearance of the emulsion was recorded after 5 days of storage under the Fenton system (Figure 2A). The freshly prepared emulsion (0 days) was uniformly creamy, indicating that the oil droplets were evenly distributed, free of flocculation and agglomeration, and were not altered by the presence of sesamol. After 5 days of storage, the overall color of the control emulsion gradually turned slightly yellow, and the color change in the top layer was apparent. The products of the Schiff base reaction during the oxidation of the emulsion changed its color to yellow [35], so the color change proved that the formation of the cream layer originated from the oxidation of lipids. Emulsions with 0.01% sesamol were observed to have insignificant color changes, and 0.03 and 0.09% sesamol emulsions maintained a homogeneous milky appearance throughout storage. Meanwhile, the changes in the microstructure of the emulsions were evidently reflected by CLSM observation(Figure 2B). The oil droplets (green) of all emulsions showed aggregation with the increase in storage time, and especially those in the control emulsion showed the most apparent aggregation with the largest oil droplets. As the sesamol content increased, the inhibition of the size and number of aggregated oil droplets and the delay in the appearance of oil droplet aggregation increased. On day 5, all emulsions were observed to have droplet flocculation and coalescence, with the smallest droplets observed for the 0.09% sesamol emulsion as well as the least coalescence.

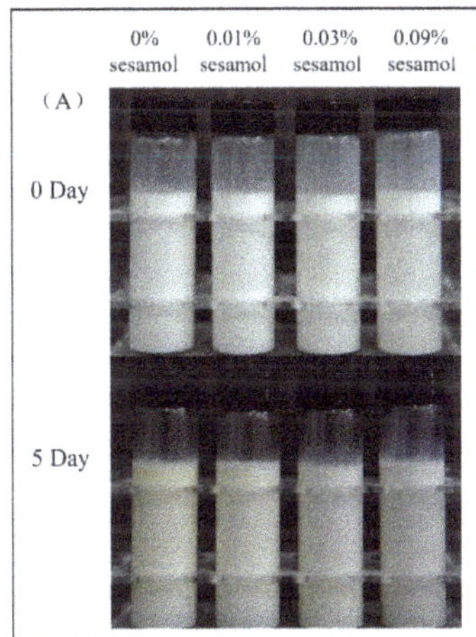

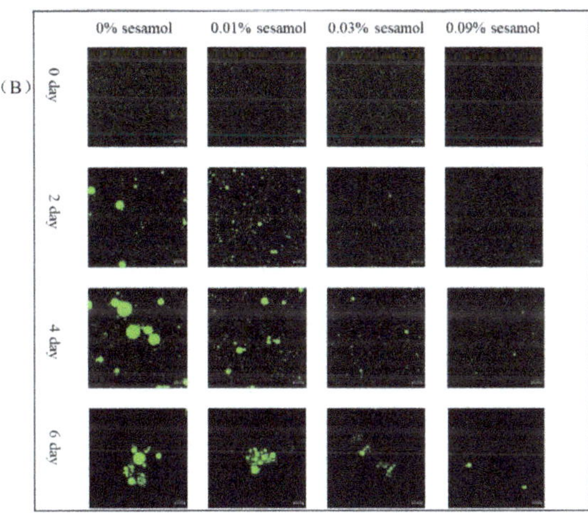

Figure 2. The effects of sesamol on the visual appearance (**A**) (from left to right: 0% sesamol, 0.01% sesamol, 0.03% sesamol and 0.09% sesamol) and emulsions' confocal micrographs (**B**) (30 mL of emulsion in a 50 mL glass bottle with screw cap stored at room temperature and protected from light).

3.2. Effect of Sesamol on the Lipid Oxidation of WPI-Stabilized Emulsions during Storage

The presence of sesamol inhibited the oxidation of lipids in WPI-stabilized emulsions, especially the formation of hydrogen peroxide was reduced by 46.44 (0.01% sesamol), 63.88 (0.03%), and 72.6% (0.09%) after 5 days of storage(Figure 3A). The ability of sesamol to inhibit lipid oxidation was positively correlated with the amount added, with high levels (0.09%) of sesamol having the best antioxidant effect, thereby delaying the accumulation of lipid hydroperoxides in the emulsion by 4 days. The effect of sesamol in inhibiting the accumulation of TBARS in emulsions was similar to that of lipid hydroperoxides. The addition of sesamol significantly reduced the generation of secondary reaction products during storage ($p < 0.05$), further confirming the antioxidant capacity of sesamol(Figure 3B). Similar results have been found with the addition of natural polyphenols to other fish oil-fortified emulsions. For example, lipid oxidation inhibition by resveratrol was observed in fish oil emulsions [36]. The excellent antioxidant properties of sesamol originate from its unique structure and chemical reactivity and the phenolic group on its benzodioxyl group has the same powerful antioxidant scavenging ability as the phenolic hydroxyl group [37]. Sesamol can not only effectively scavenge various oxidative radicals to interrupt the lipid oxidation chain inhibiting oxidation through its powerful electron-donating ability, but can also inhibit the catalyzing lipid oxidation by chelating transition metals [38]. In contrast, the benzodioxole group of sesamol generates another antioxidant (1,2-dihydroxy benzene) when it scavenges hydroxyl radicals; 1,2-dihydroxy benzene can continue to scavenge oxidative free radicals to exert antioxidant effects, and this reaction of sesamol gives it a sustainable antioxidant capacity [10]. In addition, the relationship between emulsifiers and antioxidants on the interfacial layer has a great influence on the stability of lipid oxidation.

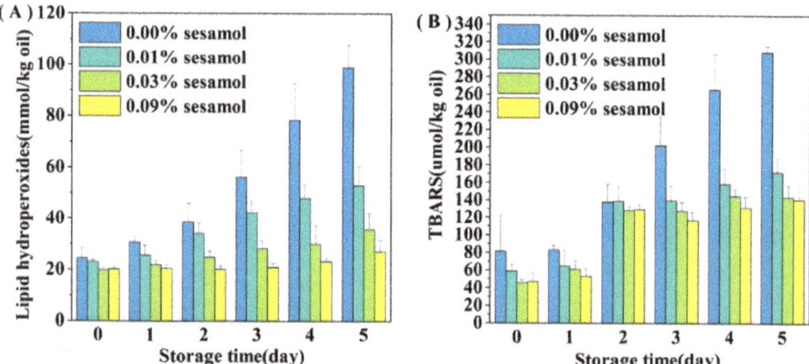

Figure 3. Effect of sesamol on lipid hydrogen peroxide (**A**) and TBARS (**B**) concentrations in WPI-stabilized fish oil emulsions during storage (30 mL of emulsion in a 50 mL glass bottle with screw cap stored at room temperature and protected from light).

3.3. *Effect of Sesamol on the Protein Oxidation of WPI-Stabilized Emulsions during Storage*

For the initial emulsion, the carbonyl content in the sesamol emulsion was lower than that of the control emulsion. The results indicated that sesamol effectively inhibited protein oxidation during the emulsion preparation. Carbonyl groups are generated by protein functional groups through reactions with lipid oxidation products or catalyzed by excess metal ions [39]. The accumulation of carbonyl groups in all emulsions represented a deepening of protein oxidation. However, the degree of protein oxidation in the emulsions decreased significantly with the addition of sesamol (Figure 4A), indicating that sesamol inhibits protein oxidation in emulsions. All the above results were consistent with the lipid oxidation results. Moreover, natural phenols can inhibit the oxidation of proteins in emulsions. For example, tea polyphenols effectively inhibited protein oxidation in whey protein-stabilized emulsions [40]. The protein peptide backbone is attacked by reactive oxygen species to lose hydrogen atoms to form protein radicals, which react with oxygen to form peroxyl radicals, followed by a series of protein oxidation reactions [41]. Sesamol can inhibit protein oxidation through its strong hydrogen supply capacity [38].

Interestingly, the initial sulfhydryl content of the control emulsion was lower than that of the sesamol emulsion(Figure 4B), and this may be due to the fact that sesamol promotes the disruption of protein disulfide bonds (S-S) and the generation of new sulfhydryl groups in the presence of ultrasound [42]. In contrast, sesamol promotes the unfolding of protein structure and the exposure of internal thiol groups [43]. The rate of sulfhydryl loss increased with the amount of sesamol during storage, indicating that the presence of sesamol increased the efficiency of protein sulfhydryl scavenging radicals to inhibit lipid oxidation in emulsions. Previous studies have also shown that protein can replace lipid oxidation as a lipid antioxidant [44]. Although our results suggest that sesamol is an effective antioxidant against lipid and protein oxidation at all the levels tested, the level of sesamol needs to be controlled to avoid negative effects on proteins. At 2 days before emulsion storage, the addition of 0.09% sesamol inhibited the growth trend of lipid hydroperoxides and thiobarbituric acid reactants and only reduced the growth of protein oxidation products. Compared with the control emulsion, the addition of 0.09% sesamol inhibited the formation of 72.6% lipid hydroperoxides and 54.33% thiobarbituric acid reactants in the emulsion after 5 days of storage; however, the reduction of protein sulfhydryl groups and the production of carbonyl groups were little inhibited. It indicates that sesamol mainly inhibits lipid oxidation in emulsions rather than protein oxidation.

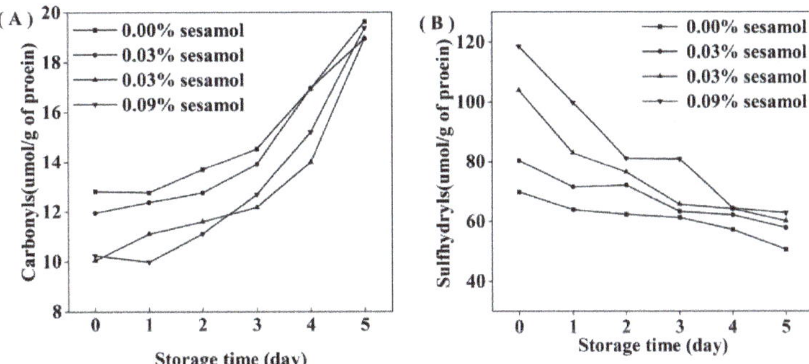

Figure 4. Effect of different sesamol concentrations on protein sulfhydryl (**A**) and carbonyl (**B**) groups in emulsions during storage (30 mL of cmulsion in a 50 mL glass bottle with screw cap stored at room temperature and protected from light).

3.4. Determination of Sesamol Binding to WPI by Isothermal Titration Calorimetry (ITC)

The interaction of sesamol with WPI in the interface is important and related to its antioxidant capacity in the emulsion. Therefore, ITC was used to thermodynamically characterize sesamol and WPI to investigate their non-covalent interactions in this experiment. The heat generated by the interaction of sesamol with WPI was recorded as a function of time, and an integral calculation was performed to obtain the enthalpy curve. The relevant thermodynamic parameters were obtained by model fitting the enthalpy to molar ratio curves (Figure 5). The results show that sesamol bonded spontaneously to WPI and the reaction was exothermic ($\Delta G < 0$, $\Delta H < 0$). The larger equilibrium dissociation constants (KD) and smaller stoichiometric binding numbers (n) for sesamol and WPI binding were 1.75×10^{-3} M and 1.6×10^{-2} M, respectively, exhibiting weak affinity and nonspecific binding. The non-covalent interactions between polyphenols and proteins may be driven by hydrophobic, hydrogen bonding, electrostatic interactions, and Van der Waals forces [45]. Since $\Delta H < 0$, $\Delta s < 0$, and the smaller Gibbs free energy ($\Delta G = -3.76$ kcal/mol is closely related to the larger enthalpy and entropy changes, the reaction of sesamol with WPI may mainly be driven by hydrogen bonding and Van der Waals forces [46]. The same results were found in the ITC studies of the ascorbic acid–bovine serum albumin and a-tocopherol–bovine serum albumin systems, where the favorable enthalpy and unfavorable entropy indicated that the main driving forces of the binding reaction were hydrogen bonding forces and Van der Waals forces [47]. The negatively charged O and N atoms on the protein polypeptide chain can form hydrogen bonds with the positively charged hydrogen atoms on the polyphenol phenolic hydroxyl group [48]. In summary, the results indicate that sesamol and WPI interact spontaneously at the interface through non-covalent interactions and aggregate with each other at the interfacial layer to stabilize the emulsion.

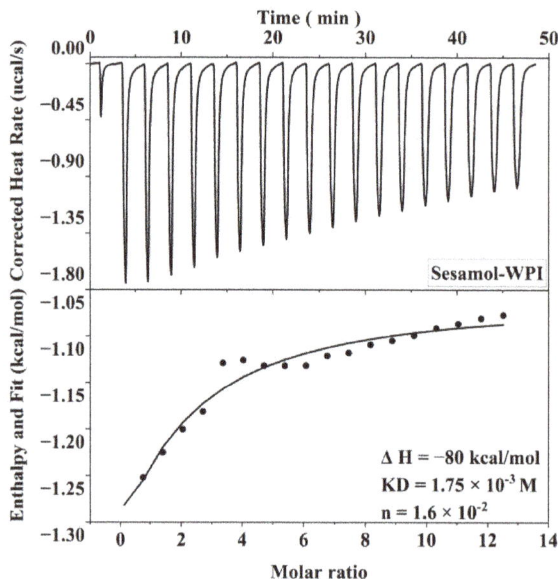

Figure 5. Correction heat rate versus time and enthalpy change corresponding to 20 mM sesamol titration of 0.3 mM WPI.

3.5. Molecular Docking

Molecular docking simulations were used to understand the binding sites and types of interaction forces between WPI and sesamol. Figure 6 shows the best simulation with the lowest binding energy among multiple simulations of sesamol docking with WPI (β-LG) molecules. The 3D docking results show that the binding site of sesamol to β-LG tended to bind to the surface cavity of β-LG rather than the hydrophobic cavity, which may be related to the less hydrophobic nature of sesamol and is consistent with the unfavorable entropic change of ITC. The docking results indicate that the interaction between sesamol and WPI (β-LG) is mediated by not only hydrogen bonding and Van der Waals forces but also hydrophobic interactions. The docking results are consistent with the results of ITC analysis, i.e., hydrogen bonding plays an important role in the binding. As shown in the lower right 2D diagram of Figure 6, there is a hydrogen bonding interaction between Leu-10 of WPI (β-LG) and the methoxy of sesamol, whereas sesamol has Van der Waals forces with five amino acids (Thr-6, Met-7, Lys-8, Gly-9, and Ile-78), hydrophobic and aromatic amino acids (Pro-79 and Ala-80) have hydrophobic interactions (pi-alkyl) with the benzene ring of sesamol. Similar studies have found that hydrogen bonding and hydrophobic interactions are also key interaction forces in the reaction, both between hydroxylated PAHs and peroxidase [29], and between grape skin extract and wheat gliadin [30].

3.6. Effect of Sesamol on the Interfacial Activity of WPI Solution with Oil Phase

Generally, the lower the interfacial tension, the higher the aggregation stability of the emulsion [49]. Therefore, we measured the interfacial tension between fish oil and WPI solutions with different concentrations of sesamol at room temperature (Figure 7). WPI adsorption to the oil-water interface changed from fast to slow, and the interfacial tension decreased with the increase in adsorbed WPI at the oil–water interface. This effect resulted from the ability of WPI to adsorb and form interconnected, viscoelastic films at the oil-water interface [50,51]. The addition of sesamol further reduced the interfacial tension, by 19.4% especially at high levels (0.09%). The lower interfacial tension facilitates the formation of small, stable droplets [52]. This is consistent with the previous observation of minimum droplet size and optimal physical stability of high-level sesamol emulsions. The additional reduction in interfacial tension can be attributed to the WPI–sesamol interaction. Hydrogen bonding, Van der Waals forces, and hydrophobic interactions supported and competed with each other to promote the proximity of sesamol to the WPI surface cavity binding site and its aggregation near the binding site [53]. In addition, the further cross-linking of sesamol with the adsorbed layer WPI may form a synergistic mechanism. Dimitris et al. hypothesized that chlorogenic acid at the interface can form hydrogen bonds with multiple adjacent protein molecules, inducing protein unfolding to form a more efficient interfacial coverage [54].

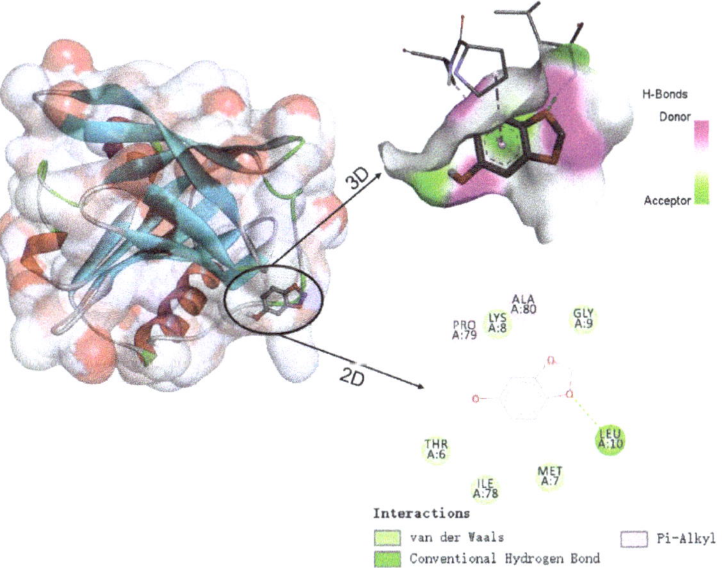

Figure 6. Schematic diagram of the 3D docking model and 2D interaction between β-lactoglobulin (β-LG) and sesamol.

3.7. Interfacial Partitioning Analysis of Sesamol in Emulsions

Emulsions containing different concentrations of sesamol showed similar inter-tissue partitioning behavior (Figure 8). The sesamol in the newly prepared emulsions was mainly located in the aqueous phase (54–58%), followed by the interfacial layer (30–33%) and the oil phase (11–12%). In previous studies, sesamol was entirely located in the aqueous phase in freshly prepared emulsions because it was bound to unabsorbed proteins [17]. The partitioning of sesamol in the aqueous phase in this study was lower than in previous studies, which may be related to the treatment with ultrasound during our emulsion preparation. Ultrasound may have promoted the partition of sesamol into the interfacial layer and oil phases.

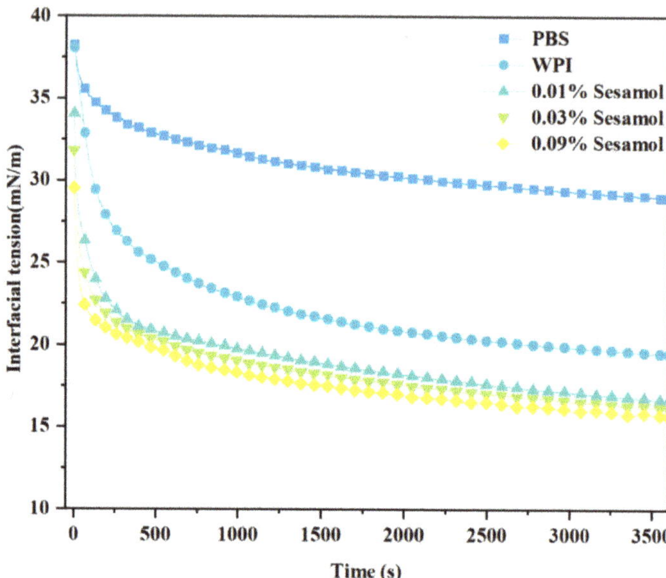

Figure 7. Effect of sesamol on the interfacial tension between the oil phase and WPI solution with time.

After 5 days of storage, the total amount of sesamol in emulsion was reduced to one-third of the original level, which was associated with chemical degradation during oxidation [22]. Sesamol content in the oil phase increased significantly during the first 2 days of emulsion storage, indicating that sesamol in the aqueous phase and interfacial layer diffused into the oil phase, increasing the partition ratio of sesamol in the oil phase [55]. After redistribution by diffusion, the partition ratio of the oil phase sesamol reached a maximum, while the partition ratio of the interfacial layer sesamol reached the lowest level. On the third day of storage, the sesamol content in both the oil and aqueous phases decreased by 50%, while the sesamol content in the interfacial layer increased, indicating that some of the sesamol in the oil and aqueous phases diffused into the interfacial layer in addition to oxidative degradation. In summary, the distribution of sesamol in the emulsion during storage is dynamic. We hypothesize that the micelle or vesicle structure in the emulsion facilitates the dynamic diffusion of sesamol between the phases [56].

Accurately determining the interfacial distribution of antioxidants in emulsions is not a simple task. Separating different phases by centrifugation can disrupt the equilibrium of the interfacial region, and the interfacial distribution of antioxidants detected on this basis may not reflect the actual situation. However, there is no method to determine the interfacial distribution of antioxidants by directly measuring the content of antioxidants in each phase. The relatively scientific approach is based on the reaction between the 4-hexadecyl diazenium ion ($16\text{-}ArN_2{}^+(BF_4{}^-)$) molecular probe and the antioxidant, and the kinetic equations are used to calculate the partition constants and interfacial molar values of the antioxidant between different interfaces in the emulsion [57–60]. In the future, we will validate the accuracy of the centrifugation method using molecular probe methods.

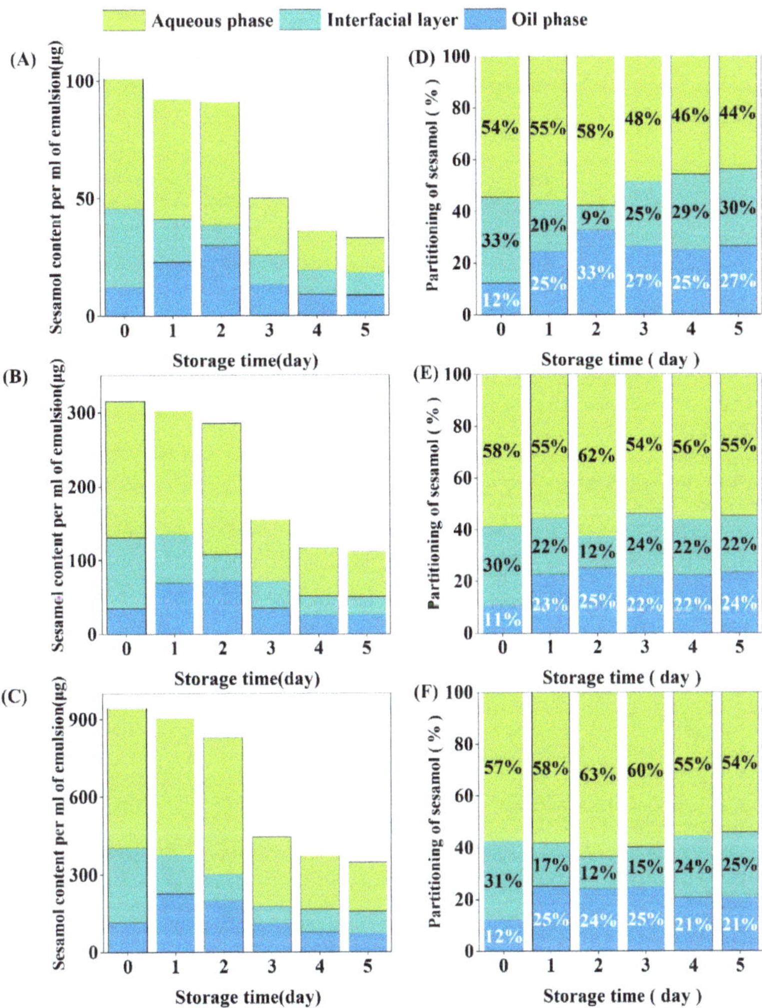

Figure 8. Interfacial partitioning of sesamol in emulsions. Content of sesamol in the aqueous phase, oil phase, and interfacial layer per ml of emulsion ((**A**): 0.01% sesamol, (**B**): 0.03% sesamol, (**C**): 0.09% sesamol). Percentage partitioning of sesamol in the aqueous phase, oil phase and interfacial layer ((**D**): 0.01% sesamol, (**E**): 0.03% sesamol, (**F**): 0.09% sesamol).

4. Conclusions

In conclusion, WPI-stabilized emulsions supplemented with sesamol have superior physical and chemical stability during storage. The presence of sesamol in emulsions has a desirable positive effect, especially in emulsion with a sesamol concentration of 0.09%. ITC and molecular docking results indicated that sesamol is spontaneously bound to WPI mainly through hydrogen bonding, Van der Waals forces, and hydrophobic interactions. In addition, the presence of sesamol reduced the interfacial tension, indicating sesamol-WPI adsorption and interactions at the interface. Sesamol on the interfacial layer exerts antioxidant capacity in a timely and efficient manner and is replenished by diffusion from the oil and aqueous phases when consumed. In summary, sesamol has rich potential for the development of DHA-fortified emulsion delivery systems and the food industry.

Our results have important implications for antioxidant studies of PUFA-rich emulsion delivery systems.

Author Contributions: Methodology, Z.J.; investigation, Q.D.; data curation, L.W.; writing—original draft, Z.G.; writing—review and editing; S.Y.Q., L.L. and X.D. All authors have read and agreed to the published version of the manuscript.

Funding: This work was supported by the Natural Science Foundation of Shandong Province (Grant No ZR2022MC172 and ZR2020MC206), Scientific Research Foundation for High-Level Talents of Qingdao Agricultural University (Grant No 663/1120085 and 663/1120084).

Data Availability Statement: All related data and methods are presented in this paper. Additional inquiries should be addressed to the corresponding author.

Conflicts of Interest: The authors declare no conflict of interest.

References

1. Shahidi, F.; Ambigaipalan, P. Omega-3 Polyunsaturated Fatty Acids and Their Health Benefits. *Annu. Rev. Food Sci. Technol.* **2018**, *9*, 345–381. [CrossRef]
2. Venugopalan, V.K.; Gopakumar, L.R.; Kumaran, A.K.; Chatterjee, N.S.; Soman, V.; Peeralil, S.; Mathew, S.; McClements, D.J.; Nagarajarao, R.C. Encapsulation and Protection of Omega-3-Rich Fish Oils Using Food-Grade Delivery Systems. *Foods* **2021**, *10*, 1566. [CrossRef]
3. Qiu, C.Y.; Zhao, M.M.; Decker, E.A.; McClements, D.J. Influence of anionic dietary fibers (xanthan gum and pectin) on oxidative stability and lipid digestibility of wheat protein-stabilized fish oil-in-water emulsion. *Food Res. Int.* **2015**, *74*, 131–139. [CrossRef]
4. Berton-Carabin, C.C.; Ropers, M.-H.; Genot, C. Lipid Oxidation in Oil-in-Water Emulsions: Involvement of the Interfacial Layer. *Compr. Rev. Food Sci. Food Saf.* **2014**, *13*, 945–977. [CrossRef]
5. Jie, Y.; Chen, F. Progress in the Application of Food-Grade Emulsions. *Foods* **2022**, *11*, 2883. [CrossRef] [PubMed]
6. Carocho, M.; Barreiro, M.F.; Morales, P.; Ferreira, I. Adding Molecules to Food, Pros and Cons: A Review on Synthetic and Natural Food Additives. *Compr. Rev. Food Sci. Food Saf.* **2014**, *13*, 377–399. [CrossRef] [PubMed]
7. Von Staszewski, M.; Ruiz-Henestrosa, V.M.P.; Pilosof, A.M. Green tea polyphenols-β-lactoglobulin nanocomplexes: Interfacial behavior, emulsification and oxidation stability of fish oil. *Food Hydrocoll.* **2014**, *35*, 505–511. [CrossRef]
8. Hoshino, C.; Tagawa, Y.; Wada, S.; Oh, J.H.; Park, D.K.; Nagao, A.; Terao, J. Antioxidant Activity of Quercetin against Metmyoglobin-induced Oxidation of Fish Oil-bile Salt Emulsion. *Biosci. Biotechnol. Biochem.* **1997**, *61*, 1634–1640. [CrossRef]
9. Wang, L.; Yu, X.; Geng, F.; Cheng, C.; Yang, J.; Deng, Q. Effects of tocopherols on the stability of flaxseed oil-in-water emulsions stabilized by different emulsifiers: Interfacial partitioning and interaction. *Food Chem.* **2022**, *374*, 131691. [CrossRef] [PubMed]
10. Joshi, R.; Kumar, M.S.; Satyamoorthy, K.; Unnikrisnan, M.K.; Mukherjee, T. Free Radical Reactions and Antioxidant Activities of Sesamol: Pulse Radiolytic and Biochemical Studies. *J. Agric. Food Chem.* **2005**, *53*, 2696–2703. [CrossRef]
11. Hadeel, S.Y.; Khalida, S.A.; Walsh, M. Antioxidant activity of sesame seed lignans in sunflower and flaxseed oils. *Food Res.* **2019**, *4*, 612–622. [CrossRef] [PubMed]
12. Toorani, M.R.; Farhoosh, R.; Golmakani, M.; Sharif, A. Antioxidant activity and mechanism of action of sesamol in triacylglycerols and fatty acid methyl esters of sesame, olive, and canola oils. *LWT Food Sci. Technol.* **2019**, *103*, 271–278. [CrossRef]
13. Hong, S.; Jo, S.; Kim, M.J.; Park, S.; Lee, S.; Lee, J.; Lee, J. Addition of Sesamol Increases the Oxidative Stability of Beeswax Organogels and Beef Tallow Matrix Under UV Light Irradiation and Thermal Oxidation. *J. Food Sci.* **2019**, *84*, 971–979. [CrossRef] [PubMed]
14. Keramat, M.; Golmakani, M.-T.; Niakousari, M. Effect of polyglycerol polyricinoleate on the inhibitory mechanism of sesamol during bulk oil oxidation. *Sci. Rep.-UK* **2022**, *12*, 11946. [CrossRef] [PubMed]
15. Yeo, J.D.; Jeong, M.K.; Park, C.U.; Lee, J. Comparing antioxidant effectiveness of natural and synthetic free radical scavengers in thermally-oxidized lard using DPPH method. *J. Food Sci.* **2010**, *75*, C258–C262. [CrossRef]
16. McClements, D.J.; Decker, E. Interfacial Antioxidants: A Review of Natural and Synthetic Emulsifiers and Coemulsifiers That Can Inhibit Lipid Oxidation. *J. Agric. Food Chem.* **2018**, *66*, 20–35. [CrossRef]
17. Wang, X.; Yu, K.; Cheng, C.; Peng, D.; Yu, X.; Chen, H.; Chen, Y.; Julian McClements, D.; Deng, Q. Effect of sesamol on the physical and chemical stability of plant-based flaxseed oil-in-water emulsions stabilized by proteins or phospholipids. *Food Funct.* **2021**, *12*, 2090–2101. [CrossRef]
18. Cheng, C.; Yu, X.; McClements, D.J.; Huang, Q.; Tang, H.; Yu, K.; Xiang, X.; Chen, P.; Wang, X.; Deng, Q. Effect of flaxseed polyphenols on physical stability and oxidative stability of flaxseed oil-in-water nanoemulsions. *Food Chem.* **2019**, *301*, 125207. [CrossRef]
19. Yang, J.; Xiong, Y.L. Comparative time-course of lipid and myofibrillar protein oxidation in different biphasic systems under hydroxyl radical stress. *Food Chem.* **2018**, *243*, 231–238. [CrossRef]

20. Cui, S.; Yang, Z.; McClements, D.J.; Xu, X.; Qiao, X.; Zhou, L.; Sun, Q.; Jiao, B.; Wang, Q.; Dai, L. Stability mechanism of Pickering emulsions co-stabilized by protein nanoparticles and small molecular emulsifiers by two-step emulsification with different adding sequences: From microscopic to macroscopic scales. *Food Hydrocoll.* **2023**, *137*, 108372. [CrossRef]
21. Bu, X.; Wang, X.; Dai, L.; Ji, N.; Xiong, L.; Sun, Q. The combination of starch nanoparticles and Tween 80 results in enhanced emulsion stability. *Int. J. Biol. Macromol.* **2020**, *163*, 2048–2059. [CrossRef] [PubMed]
22. Tian, L.; Zhang, S.; Yi, J.; Zhu, Z.; Cui, L.; Decker, E.A.; McClements, D.J. Antioxidant and prooxidant activities of tea polyphenols in oil-in-water emulsions depend on the level used and the location of proteins. *Food Chem.* **2022**, *375*, 131672. [CrossRef] [PubMed]
23. Liang, L.; Chen, F.; Wang, X.; Jin, Q.; Decker, E.A.; McClements, D.J. Physical and Oxidative Stability of Flaxseed Oil-in-Water Emulsions Fabricated from Sunflower Lecithins: Impact of Blending Lecithins with Different Phospholipid Profiles. *J. Agric. Food Chem.* **2017**, *65*, 4755–4765. [CrossRef] [PubMed]
24. Shantha, N.C.; Decker, E.A. Rapid, sensitive, iron-based spectrophotometric methods for determination of peroxide values of food lipids. *J. AOAC Int.* **1994**, *77*, 421–424. [CrossRef] [PubMed]
25. Mcdonald, R.E.; Hultin, H.O. Some Characteristics of the Enzymic Lipid Peroxidation System in the Microsomal Fraction of Flounder Skeletal Muscle. *J. Food Sci.* **1987**, *52*, 15–21. [CrossRef]
26. Levine, R.L.; Garland, D.; Oliver, C.N.; Amici, A.; Climent, I.; Lenz, A.-G.; Ahn, B.-W.; Shaltiel, S.; Stadtman, E.R. Determination of carbonyl content in oxidatively modified proteins. *Methods Enzymol.* **1990**, *186*, 464–478.
27. Beveridge, T.; Toma, S.J.; Nakai, S. Determination of SH- and SS-groups in some food proteins using Ellman's reagent. *J. Food Sci.* **1974**, *39*, 49–51. [CrossRef]
28. Wang, Y.; Sun, Y.; Li, M.; Xiong, L.; Xu, X.; Ji, N.; Dai, L.; Sun, Q. The formation of a protein corona and the interaction with α-amylase by chitin nanowhiskers in simulated saliva fluid. *Food Hydrocoll.* **2020**, *102*, 105615. [CrossRef]
29. Zhang, J.; Chen, L.; Zhu, Y.; Zhang, Y. Study on the molecular interactions of hydroxylated polycyclic aromatic hydrocarbons with catalase using multi-spectral methods combined with molecular docking. *Food Chem.* **2020**, *309*, 125743. [CrossRef]
30. Guo, Z.; Huang, Y.; Huang, J.; Li, S.; Zhu, Z.; Deng, Q.; Cheng, S. Formation of protein-anthocyanin complex induced by grape skin extracts interacting with wheat gliadins: Multi-spectroscopy and molecular docking analysis. *Food Chem.* **2022**, *385*, 132702. [CrossRef]
31. Li, D.; Zhao, Y.; Wang, X.; Tang, H.; Wu, N.; Wu, F.; Yu, D.; Elfalleh, W. Effects of (+)-catechin on a rice bran protein oil-in-water emulsion: Droplet size, zeta-potential, emulsifying properties, and rheological behavior. *Food Hydrocoll.* **2020**, *98*, 105306. [CrossRef]
32. Hu, M.; McClements, D.J.; Decker, E.A. Lipid oxidation in corn oil-in-water emulsions stabilized by casein, whey protein isolate, and soy protein isolate. *J. Agric. Food Chem.* **2003**, *51*, 1696–1700. [CrossRef] [PubMed]
33. Rodríguez, S.D.; von Staszewski, M.; Pilosof, A.M.R. Green tea polyphenols-whey proteins nanoparticles: Bulk, interfacial and foaming behavior. *Food Hydrocoll.* **2015**, *50*, 108–115. [CrossRef]
34. Yi, J.; Qiu, M.; Liu, N.; Tian, L.; Zhu, X.; Decker, E.A.; McClements, D.J. Inhibition of Lipid and Protein Oxidation in Whey-Protein-Stabilized Emulsions Using a Natural Antioxidant: Black Rice Anthocyanins. *J. Agric. Food Chem.* **2020**, *68*, 10149–10156. [CrossRef] [PubMed]
35. Chen, E.; Wu, S.; McClements, D.J.; Li, B.; Li, Y. Influence of pH and cinnamaldehyde on the physical stability and lipolysis of whey protein isolate-stabilized emulsions. *Food Hydrocoll.* **2017**, *69*, 103–110. [CrossRef]
36. Xu, M.; Lian, Z.; Chen, X.; Yao, X.; Lu, C.; Niu, X.; Xu, M.; Zhu, Q. Effects of resveratrol on lipid and protein co-oxidation in fish oil-enriched whey protein isolate emulsions. *Food Chem.* **2021**, *365*, 130525. [CrossRef] [PubMed]
37. Jayaraj, P.; Narasimhulu, C.A.; Rajagopalan, S.; Parthasarathy, S.; Desikan, R. Sesamol: A powerful functional food ingredient from sesame oil for cardioprotection. *Food Funct.* **2020**, *11*, 1198–1210. [CrossRef]
38. Palheta, I.C.; Borges, R.S. Sesamol is a related antioxidant to the vitamin E. *Chem. Data Collect.* **2017**, *11–12*, 77–83. [CrossRef]
39. Farber, J.M.; Levine, R.L. Sequence of a peptide susceptible to mixed function oxidation. Probable cation binding site in glutamine synthetase. *J. Biol. Chem.* **1986**, *261*, 4574–4578. [CrossRef]
40. Tian, L.; Zhang, S.; Yi, J.; Zhu, Z.; Cui, L.; Decker, E.A.; McClements, D.J. Factors impacting the antioxidant/prooxidant activity of tea polyphenols on lipids and proteins in oil-in-water emulsions. *LWT Food Sci. Technol.* **2022**, *156*, 113024. [CrossRef]
41. Lund, M.N.; Heinonen, M.; Baron, C.P.; Estevez, M. Protein oxidation in muscle foods: A review. *Mol. Nutr. Food Res.* **2011**, *55*, 83–95. [CrossRef] [PubMed]
42. Berton, C.; Ropers, M.H.; Guibert, D.; Sole, V.; Genot, C. Modifications of interfacial proteins in oil-in-water emulsions prior to and during lipid oxidation. *J. Agric. Food Chem.* **2012**, *60*, 8659–8671. [CrossRef] [PubMed]
43. Zhang, S.; Tian, L.; Yi, J.; Zhu, Z.; Dong, X.; Decker, E.A. Impact of high-intensity ultrasound on the chemical and physical stability of oil-in-water emulsions stabilized by almond protein isolate. *LWT Food Sci. Technol.* **2021**, *149*, 111972. [CrossRef]
44. Viljanen, K.; Kylli, P.; Hubbermann, E.M.; Schwarz, K.; Heinonen, M. Anthocyanin Antioxidant Activity and Partition Behavior in Whey Protein Emulsion. *J. Agric. Food Chem.* **2005**, *53*, 2022–2027. [CrossRef] [PubMed]
45. Liu, J.; Yong, H.; Yao, X.; Hu, H.; Yun, D.; Xiao, L. Recent advances in phenolic-protein conjugates: Synthesis, characterization, biological activities and potential applications. *RSC Adv.* **2019**, *9*, 35825–35840. [CrossRef] [PubMed]
46. Ross, P.D.; Subramanian, S. Thermodynamics of protein association reactions forces contributing to stability. *Biochemistry* **1981**, *20*, 3096–3102. [CrossRef]

47. Li, X.; Wang, G.; Chen, D.; Lu, Y. Binding of ascorbic acid and alpha-tocopherol to bovine serum albumin: A comparative study. *Mol. Biosyst.* **2014**, *10*, 326–337. [CrossRef]
48. Wang, N.; Han, X.; Li, J.; Wang, Y.; Yu, W.; Wang, R.; Chang, J. Comparative study of the bindings between 3-phenyl-1H-indazole and five proteins by isothermal titration calorimetry, spectroscopy and docking methods. *J. Biomol. Struct. Dyn.* **2019**, *37*, 4580–4589. [CrossRef]
49. Gu, L.; Su, Y.; Zhang, M.; Chang, C.; Li, J.; McClements, D.J.; Yang, Y. Protection of beta-carotene from chemical degradation in emulsion-based delivery systems using antioxidant interfacial complexes: Catechin-egg white protein conjugates. *Food Res. Int.* **2017**, *96*, 84–93. [CrossRef]
50. Richter, M.J.; Schulz, A.; Subkowski, T.; Boker, A. Adsorption and rheological behavior of an amphiphilic protein at oil/water interfaces. *J. Colloid Interface Sci.* **2016**, *479*, 199–206. [CrossRef]
51. Schröder, A.; Berton-Carabin, C.; Venema, P.; Cornacchia, L. Interfacial properties of whey protein and whey protein hydrolysates and their influence on O/W emulsion stability. *Food Hydrocoll.* **2017**, *73*, 129–140. [CrossRef]
52. Amine, C.; Dreher, J.; Helgason, T.; Tadros, T. Investigation of emulsifying properties and emulsion stability of plant and milk proteins using interfacial tension and interfacial elasticity. *Food Hydrocoll.* **2014**, *39*, 180–186. [CrossRef]
53. Wu, D.; Mei, S.; Duan, R.; Geng, F.; Wu, W.; Li, X.; Cheng, L.; Wang, C. How black tea pigment theaflavin dyes chicken eggs: Binding affinity study of theaflavin with ovalbumin. *Food Chem.* **2020**, *303*, 125407. [CrossRef] [PubMed]
54. Karefyllakis, D.; Altunkaya, S.; Berton-Carabin, C.C.; van der Goot, A.J.; Nikiforidis, C.V. Physical bonding between sunflower proteins and phenols: Impact on interfacial properties. *Food Hydrocoll.* **2017**, *73*, 326–334. [CrossRef]
55. Geetha, T.; Singh, N.; Deol, P.K.; Kaur, I.P. Biopharmaceutical profiling of sesamol physiochemical characterization. *RSC Adv.* **2015**, *5*, 4083–4091. [CrossRef]
56. Kiralan, S.S.; Dogu-Baykut, E.; Kittipongpittaya, K.; McClements, D.J.; Decker, E.A. Increased antioxidant efficacy of tocopherols by surfactant solubilization in oil-in-water emulsions. *J. Agric. Food Chem.* **2014**, *62*, 10561–10566. [CrossRef]
57. Silva, R.; Losada-Barreiro, S.; Paiva-Martins, F.; Bravo-Díaz, C. Partitioning and antioxidative effect of protocatechuates in soybean oil emulsions: Relevance of emulsifier concentration. *Eur. J. Lipid Sci. Technol.* **2017**, *119*, 1600274. [CrossRef]
58. Ferreira, I.; Costa, M.; Losada-Barreiro, S.; Paiva-Martins, F.; Bravo-Diaz, C. Modulating the interfacial concentration of gallates to improve the oxidative stability of fish oil-in-water emulsions. *Food Res. Int.* **2018**, *112*, 192–198. [CrossRef]
59. Afzal, S.; Lone, M.S.; Nazir, N.; Dar, A.A. pH Changes in the Micelle-Water Interface of Surface-Active Ionic Liquids Dictate the Stability of Encapsulated Curcumin: An Insight Through a Unique Interfacial Reaction between Arenediazonium Ions and t-Butyl Hydroquinone. *ACS Omega* **2021**, *6*, 14985–15000. [CrossRef]
60. Costa, M.; Losada-Barreiro, S.; Paiva-Martins, F.; Bravo-Diaz, C. Polyphenolic Antioxidants in Lipid Emulsions: Partitioning Effects and Interfacial Phenomena. *Foods* **2021**, *10*, 539. [CrossRef]

Article

Fabrication and Characterization of Botanical-Based Double-Layered Emulsion: Protection of DHA and Astaxanthin Based on Interface Remodeling

Mengjia Sun [1], Hongjian Chen [1,†], Fang Geng [2], Qi Zhou [1], Qian Hao [3], Shan Zhang [1], Yashu Chen [1,*] and Qianchun Deng [1,*]

1   Key Laboratory of Oilseeds Processing, Hubei Key Laboratory of Lipid Chemistry and Nutrition, Oil Crops Research Institute of the Chinese Academy of Agricultural Sciences, Ministry of Agriculture, Wuhan 430062, China
2   Key Laboratory of Coarse Cereal Processing (Ministry of Agriculture and Rural Affairs), School of Food and Biological Engineering, Chengdu University, No. 2025 Chengluo Avenue, Chengdu 610106, China
3   College of Biological Engineering and Food, Hubei University of Technology, Wuhan 430068, China
*   Correspondence: yashuchen@sina.com (Y.C.); dengqianchun@caas.cn (Q.D.); Tel.: +86-18696198198 (Q.D.)
†   These authors contributed equally to this work.

Abstract: Both DHA and astaxanthin, with multiple conjugated double bonds, are considered as health-promoting molecules. However, their utilizations into food systems are restricted due to their poor water solubility and high oxidizability, plus their certain off-smell. In this study, the interactions between perilla protein isolate (PPI) and flaxseed gum (FG) were firstly investigated using multiple spectroscopies, suggesting that hydrophobic, electrostatic force and hydrogen bonds played important roles. Additionally, double-layer emulsion was constructed by layer-by-layer deposition technology and exhibited preferable effects on masking the fishy smell of algae oil. Calcium ions also showed an improving effect on the elasticity modulus of O/W emulsions and was managed to significantly protect the stability of co-delivered astaxanthin and DHA, without additional antioxidants during storage for 21 days. The vegan system produced in this study may, therefore, be suitable for effective delivery of both ω-3 fatty acid and carotenoids for their further incorporation into food systems, such as plant-based yoghourt, etc.

Keywords: DHA; flaxseed gum; perilla protein isolate; astaxanthin; double-layered emulsion

Citation: Sun, M.; Chen, H.; Geng, F.; Zhou, Q.; Hao, Q.; Zhang, S.; Chen, Y.; Deng, Q. Fabrication and Characterization of Botanical-Based Double-Layered Emulsion: Protection of DHA and Astaxanthin Based on Interface Remodeling. *Foods* **2022**, *11*, 3557. https://doi.org/10.3390/foods11223557

Academic Editor: Isabel Borrás-Linares

Received: 4 September 2022
Accepted: 18 October 2022
Published: 8 November 2022

Publisher's Note: MDPI stays neutral with regard to jurisdictional claims in published maps and institutional affiliations.

1. Introduction

Algal oil, rich in docosahexaenoic acid (DHA), is believed to contribute to the development of vision and the brain in newborns, inhibiting inflammation and reducing the risk of cardiovascular disease, obesity, diabetes and hypertension [1]. However, high sensitivity to oxidation, poor taste and low water solubility of DHA algal oil are the main obstacles to its application in the food industry [2]. Astaxanthin (AST), as a typical group of lutein carotenoids, possesses strong antioxidant activity and can be used to reduce the risk of cancer, cardiovascular disease, and Helicobacter pylori infection, etc. [3]. AST cannot be synthesized in mammals and, therefore, needs to be obtained from the diet. According to previous research, supplementation of DHA and AST at the same time had a significant improving effect on the baseline redox metabolism of Wistar rats, while single DHA did not, suggesting an advantage of dual delivery [4]. Similar to DHA, AST is highly susceptible to oxidation, sensitive to pH, temperature, lights and oxygen [5]. In addition, high hydrophobicity and low bioavailability of AST also make it difficult to apply as a nutritional enhancer [6].

According to previous studies, the O/W emulsion delivery system is a promising way to encapsulate, protect and jointly deliver hydrophobic functional active substances. These active ingredients are embedded in the hydrophobic core of the emulsion system droplets, which might improve their stability and sensory acceptability, extend their shelf life and even further improve their bioavailability and accumulation at the target site [7,8].

Emulsion, as a typical representative of efficient delivery systems, comes in a variety of forms. Among them, in contrast to single-layer emulsions, double-layer emulsions manufactured by layer-by-layer deposition technology are designed to wrap the oil droplets inside them, with a thicker and denser interfacial film. In a general way, biopolymers, such as charged emulsifier protein and polysaccharides, could combine through non-covalent interaction, including electrostatic interaction, hydrophobic interaction and so on. These aforesaid interactions at the interface could enhance the thickness of the interface layer, hence, effectively protecting the bioactive substances inside from extreme environmental degradation and restricting the diffusion of odor molecules. The emulsions stabilized by protein–polysaccharide double-layered interface were proved to exhibit better stability at high ionic strength, a broad range of pH values and thermal treatment than single-layered emulsions (e.g., emulsions with protein alone as emulsifier) [9–11]. Since the addition of plant polyphenols as antioxidants in food might induce a bitter taste, double-layered emulsion using plant protein and polysaccharides might be a promising solution for delivery of oxidizing nutraceuticals.

The macromolecules could interact with each other through multiple mechanisms, including dot–charge electrostatic interaction, hydrophobic interaction, Van der Waals' force and hydrogen–bond interaction, etc. The uneven distributed charge and complicated spatial structures of both protein and polysaccharides make their interaction alterable at the oil–water interface. Thus, it is crucial to focus on their unique and fundamental interaction in order to build high-efficient delivery systems. For the past few years, plant proteins have aroused wide concern as a partial substitute of animal protein because of their low carbon, high functionality and reasonable price. Perilla protein isolate (PPI) is considered as byproduct of perilla seed after oil manufacture, which possesses high emulsification activity and nutritional value. As a new source of healthy vegetable oil, perilla is rich in α-linolenic acid (50–75% of total fatty acids) and known as homology of medicine and food, with important physiological functions from ancient China. Therefore, the exploitation and utilization of PPI are considered as both sustainable and profitable, which could increase the utilization rate and economic worth of perilla; it is also low carbon and environment friendly [12]. The amino acid composition of PPI is balanced, with relatively high content in perilla (~35%); thus, PPI is a new type of promising plant protein resource. Flaxseed gum (FG), a natural polysaccharide extracted from flaxseeds, is widely used in foods due to its functional properties, such as thickening, swelling, water retention, weak gel formation and emulsification. In addition, flaxseed gum is also considered as a special active component, which possesses several health benefits, such as promoting gut health, preventing obesity, modulating lipid metabolism, et al. [13]. Thus, it is believed to be both study worthy and pragmatic to build an efficient vegan delivery system with high nutritional perilla protein and flaxseed gum to deliver both algae-derived DHA-oil and AST.

Additionally, Calcium ions (Ca^{2+}) can be added to liquid foods as both microenvironmental regulators and nutrient reinforcement. Ca^{2+} added into emulsions would, on one hand, change the repulsive electrostatic forces of oil droplets through modulating the ionic environment and, on the other hand, influence the flow consistency index by interacting with polysaccharide chains [14–16]. Thus, the Ca^{2+} was added into the double-layered emulsion, the influence of which, on the long-term storage stability of emulsions, was also investigated. Overall, the aim of this research was to investigate the possible interactive activities of perilla protein isolate and flaxseed gum and their impact on interface enhancement. Beyond that, the delivery stability of DHA oil and AST-encapsulated emulsions with PPI–FG was thoroughly measured. The headspace solid-phase microextraction coupled with gas chromatography and mass spectroscopy was applied to investigate their interfacial thickening effect. Multiple spectroscopy measurements were performed to investigate molecular interactions between PPI and FG, as well as their influence on the stability of emulsions. We hope that the results of this research contribute to the manufacture of superior vegan delivery systems for oxidable nutraceuticals.

2. Materials and Methods

2.1. Materials

DHA algae oil was supplied by Cabio Biological Engineering Co., Ltd. (Wuhan, China). Perilla Protein Isolate was purchased from Ruizi Biological Technology Co., Ltd. (Shanxi, 83.6 wt.% on dry matter basis). Flaxseed (Baiya NO.2) was provided by Gansu Academy of Agricultural Sciences of China. Astaxanthin (>97%) was purchased from Sigma-Aldrich Co., Ltd. (St. Louis, MO, USA). Other chemicals and reagents were analytical grade.

Extraction of Flaxseed Gum

Initially, flaxseeds were eluted with deionized water to remove dust and mixed again in a 9:1 water–seed ratio. According to the previously described method [17], the solution was stirred for 2 h with a magnetic stirrer at 3000 rpm in a water bath maintained at 60 °C and centrifuged at 4500 rpm for 10 min followed by separation from sedimented flaxseeds. The viscous liquid dissolved with FG was collected by adding ethanol (95%, 10:1), stored at 4 °C overnight and then centrifuged at 9000 rpm for 15 min to collect the lower layer. The flaxseed gum powder is collected after freeze-drying and grinding

2.2. Methods

2.2.1. Preparation of Biopolymer Solutions

Different concentrations of FG (0.01, 0.05, 0.1, 0.2, 0.3, 0.4 wt.%) and PPI (0.25 wt.%) were dissolved in 5 mM phosphate buffer to prepare an aqueous solution of biopolymers. The aqueous solutions were left overnight to completely hydrate. Adjust the solution to the desired pH (2–8) by using HCl and NaOH (0.1 and 1 M) [18].

2.2.2. SDS-PAGE

The composition of perilla protein isolates was determined according to the SDS-PAGE method of Sun et al. [19]

2.2.3. Zeta-Potential of Solutions

The ζ potential value of the solution was measured using an electrophoresis instrument (Zetasizer Nano ZS, Malvern Instruments, Malvern City, UK). Each sample was measured three times during the measurement and its average value was selected.

2.3. Fluorescence Spectroscopy

Measure endogenous fluorescence spectra of PPI/PPI–FG solution (pH = 5). Excitation wavelength = 290 nm, emission wavelength = 300 to 500 nm and slit width = 5 nm in fluorescence spectrometer (F-4500, Hitachi, Tokyo, Japan) [20].

2.4. Fourier Transform Infrared Spectroscopy

With reference to the method in [19], the structural properties of PPI–FG(PF) complexes were determined by means of Fourier transform infrared spectroscopy (FTIR) (TENSOR 27, Brucker, Billerica, MA, USA).

2.5. Emulsion Preparation and Characterization

2.5.1. Preparation of Emulsions

Single-layer emulsion was prepared by dispersing the perilla protein isolate into phosphate buffer (5 mM, pH 7) containing 0.044 wt.% NaN_3, followed by stirring for at least 6 h and stored overnight at 4 °C to ensure complete hydration. The oil phase was prepared by dispersing astaxanthin (0.5 mg/mL) in algal oil, followed by heating (50 °C, 1 h) and sonicating (30 min) until completely dissolved. The mixture of protein solution and oil phase was operated at 10,000 rpm using a high-speed mixer (IKA, T25, Königswinter, Germany) for 2 min. The obtained primary emulsion (10 wt.% oil and 0.5 wt.% PPI) was further homogenized at 10,000 psi by microfluidizer (model M-110L, microfluidizer, Newton, MA, USA), followed by adjusting the pH to 5 to obtain the single-layer emulsion.

Double-layer emulsion was prepared by mixing the single-layer emulsion and flaxseed gum solution by magnetic stirring for 30 min. Calcium chloride (0–0.5 wt.%, phosphate buffer 5 mM, pH 5) was slowly added to obtain double-layer emulsions containing PPI/FG (0.01–0.4 wt.%) -Ca^{2+} (0–0.5 wt.%)

2.5.2. Particle size and Zeta Potential

A laser diffraction instrument (Malvern Mastersizer 3000, Malvern Instruments, Worcs, UK) was used for the determination of emulsion particle size and particle size distribution [21].

The emulsion Zeta potential is determined using the microelectrophoresis device (Zetasizer Nano ZS, Malvern Instruments, Malvern, UK). The results were tested three times and the average was taken for analysis [22].

2.5.3. Gravity Separation Measurement

The physical stability of the emulsion was characterized by instrumental multi-light scattering (MLS) (Turbiscan LAB, Formulaction, Toulouse, France). By measuring the relationship between the backscatter and altitude of near-infrared light, microscopic instability phenomena such as aggregation and flocculation of emulsion during storage are further monitored [23]. The whole process of measurement is maintained at 25 °C and each sample is scanned at 30 s intervals from top to bottom for 30 min. The Turbiscan Stability Index (TSI) was calculated by Turbisoft 2.1 software.

2.5.4. CLSM and Cryo-SEM Analysis

The emulsion microstructure was observed using confocal laser scanning microscopy (Nikon D-Eclipse C1 80i, Nikon, Melville, NY, USA), all stained with Nile red solution (1 mg/mL ethanol dissolved). Prior to the assay, images were taken at excitation and emission wavelengths of 543 and 605 nm, recording storage data analysis [24].

The microstructure of emulsions was further observed by cryo-scanning electron microscopy. The sample pre-treatment process uses a frozen supersphere freezer (at −110 °C) to obtain a cross-section of a freshly prepared sample. The sample was sputtered with platinum (30 s) and imaged at an accelerated voltage of 3 kV at −125 °C in the ZEISS Auriga field emission SEM [25].

2.5.5. Rheological Properties of Emulsions

The dynamic shear rheometers are used for the determination of the rheological properties of emulsions (AR 2000 Rheometer, TA Instruments, West Sussex, UK). At temperatures of 25 °C, the modulus of storage (G') and loss (G'') obtained are measured in a frequency range of 1–100 rad/s [26].

2.6. Aroma Analyses of Emulsions

The determination of volatile compounds of emulsions is analyzed by using headspace solid-phase microextraction combined with gas chromatography and mass spectrometry (Agilent 7890A-5975C). The specific assay method was analyzed with reference to the literature method of Sun et al. [19].

2.7. Long Storage Stability

The prepared emulsion is loaded into a glass test tube and completely sealed, stored in a low-light environment at 4 °C for 21 days and part of the sample is removed regularly for observational test analysis. The mean particle diameter, zeta-potential, TSI, CLSM, Hydroperoxide and TBARS help determine the astaxanthin content in the sample. Changes in the appearance of the emulsion during storage are recorded by taking pictures.

2.7.1. Chemical Stability (Hydroperoxide and TBARS)

The primary and secondary oxidation products, including hydroperoxides and thiobarbituric acid reactive substances (TBARS), were identified as indicators of lipid oxidation of emulsions during storage [24]. Hydrogen peroxide was detected at 510 nm and TBARS values were determined at 532 nm using a UV/VIS spectrophotometer (DU 800, Beckman Coulter, Brea, CA, USA).

2.7.2. Astaxanthin Retention Rate

The concentration of AST was determined with slight modification according to the published method [27]. The emulsion (50 μL) is extracted in a 4.95 mL (dichloromethane: methanol = 2:1 v) solvent, mixed well and centrifuged at 5000 rpm (30 min), pipetting the supernatant. Astaxanthin absorbance in samples was measured at 480 nm using a UV/VIS spectrophotometer (DU 800, Beckman Coulter, USA).

2.8. Statistical Analysis

All samples were measured three times to take the average for analysis. Data were processed using drawing software (Origin Software 8.5, Origin lab Corporation, Northampton, MA, USA). ANOVA and significance analysis were performed using statistical analysis software (SPSS 24, SPSS Inc., Chicago, IL, USA).

3. Results and Discussion

3.1. PPI–FG Solution Interactios Analysis

The properties of the aqueous solution of perilla protein isolate, flaxseed gum and their mixtures were determined first, providing some preliminary analytical basis for exploring the potential mechanism of their interaction. The composition of the perilla protein isolate was firstly analyzed using SDS-PAGE (Figure 1A). From the analysis of the spectrum, six clearly visible bands of perilla protein isolate were obtained. Combined with the results reported in the literature, it was found that the protein isolate of perilla protein was composed of multiple protein subunits. According to Figure 1, eight bands in the reduced state suggested that there were subunit structures connected by disulfide bonds within the protein molecule.

Figure 1B shows the existence state of the perilla protein isolate solution in the PBS buffer in a pH range of 3–8. Combined with the result of the charge characteristics of the solution, the analysis shows that the solution has the lowest solubility in a range of pH 4 to 5 and the ζ-potential value showed that the net charge is close to zero. Due to the electrostatic interaction caused by the charge asymmetry of the protein, the protein can be observed to produce more obvious precipitates in appearance. The solubility is the largest at pH 8, with a ζ-potential value around -40 mV.

When the pH value is 5, the turbidity measurement, charge characteristics, as well as the appearance of the perilla protein isolate–flaxseed gum composite solution are shown in Figure 1C–E. The initial perilla protein isolate solution is at pH = 5, which is near the isoelectric point (pI), and there is hardly enough electrostatic repulsion to overcome the attractive force (including hydrophobic force and van der Waals attraction, etc.), leading to a poor solubility and a high turbidity. According to the published literature [19], FG, as a natural anionic polysaccharide, showed a ζ-potential value in a pH 2–8 range from -2 mV to -19 mV and the potential is -18 mV at pH 5. When the protein solution was mixed with low-concentration (0.01–0.05%) FG solution (volume ratio, 1:1), the net charge of the protein–polysaccharide complex increased and the turbidity decreased significantly. This is due to the particle aggregation of the low-concentration FG and perilla protein molecules through absorption and bridging effect, involved with electrostatic interaction, attractive potential energy, etc. With an increase in the concentration of FG (0.1–0.4%), the zeta potential value and turbidity of protein changed, suggesting a different interaction with polysaccharide molecules. The PPI–FG composite solution increased the charge from -3 mV to 3 mV at a pH of 5. It may be that the solubility of the flaxseed protein in the PPI and FG decreased near the isoelectric point, resulting in a precipitated particle with a net positive charge, resulting in a certain measurement error in the result [28].

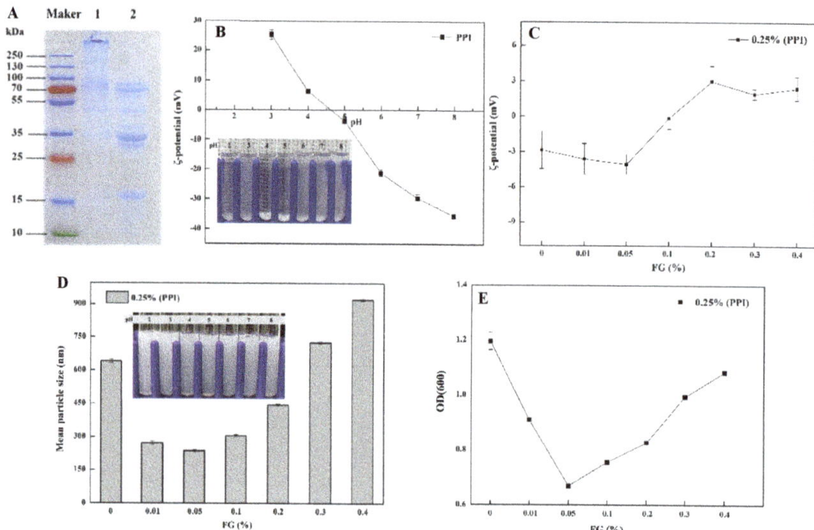

Figure 1. SDS-PAGE analysis of perilla protein isolate (**A**). Line 2 is reducing perilla protein isolate (with 2-ME). pH value to the potential and appearance of PPI/PPI–FG in aqueous solution: (**B**) PPI; (**C**) PPI–FG (1:1); impact of FG concentration on the average droplet size and appearance of PPI–FG systems (**D**) and turbidity (**E**) (pH = 5, PPI–FG (1:1), 5 mM phosphate buffer).

3.2. Fluorescence and FTIR Analysis

The fluorescence intensity of tryptophan group of PPI–FG (0–0.4%) solution was measured at pH = 5 (Figure 2). The λmax of PPI is around 345 nm and after adding FG, an obvious red shift (λmax = 360 nm) was observed, indicating that when PPI interacts with FG, the hydrophilic microenvironment around Trp is enhanced [29]. In addition, the addition of a low concentration of FG (0.01%, 0.05 wt.%) compared to PPI solutions results in a gradual decrease in fluorescence intensity. This effect can be attributed to the fluorescence quenching caused by the interaction between the protein molecule of the PPI and the polysaccharide molecule of FG, resulting in a decrease in fluorescence intensity. When the PPI is combined with a low concentration of FG (less than 0.05%), the resulting precipitate may also lead to a decrease in fluorescence intensity [30]. In addition, when the concentration was between 0.1% and 0.2%, FG, as a hydrophilic colloid, would be distributed around tryptophan after being dissolved in water, which could enhance the polarity of the environment and produce a shielding effect, reducing the fluorescence intensity of tryptophan [12]. Since FG itself contains an amount of protein, when the concentration of FG further increased (up to 0.4%), the protein content in the FG solution increased correspondingly, resulting in a significant increase in the fluorescence intensity of the PPI–FG complex.

The interaction between FG and PPI molecules is inferred by FTIR spectroscopic analysis (Figure 2). PPI has a strong CH stretching band at 2957 cm^{-1}, $-$OH contraction vibration band at 3288 cm^{-1}. At 1300–1700 cm^{-1}, there are C=O, NH and CN tensile/curved bands to form amide bands, respectively. Because of the overlap of OH stretching (3500–2900 cm^{-1}) and CH (2900–2950 cm^{-1}) vibrations produced by the anionic carboxyl group in FG, the spectrum of pure FG has a broad peak at 3493 cm^{-1}. The peaks at 1581 and 1471 cm^{-1} corresponded to the symmetrical vibrations of amide I (C=O and C-N stretching) and carboxyl groups, respectively [31]. According to Figure 2B, the peaks of amide I and II moved from 1529 and 1657 cm^{-1} in PPI to 1543 and 1659 cm^{-1} in PPI–FG. The reason for this change can be inferred from the electrostatic interaction of anionic FG and cationic PPI under acidic conditions. Similar experimental results were found in the literature on relevant protein polysaccharide solutions, including flaxseed protein–flaxseed gum, gelatin–alginate and whey protein–arabic gum, etc. [32]. Compared with PPI, the -OH vibration peak in the PPI–FG complex changed from 3288 to 3304 cm^{-1}, suggesting that hydrogen bonds were formed between the PPI and FG. In addition, the amide II peak shifted from 1529 cm^{-1} to 1543 cm^{-1}, which indicated that there was a hydrophobic interaction between PPI and FG [33].

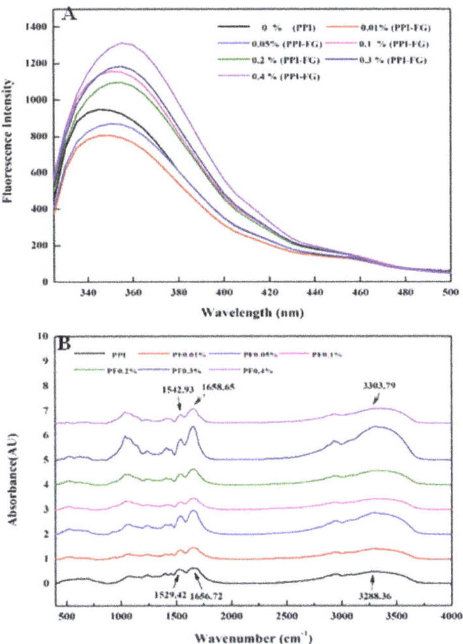

Figure 2. Spectroscopic analysis (PPI/FG solutions pH 5, FG = 0–0.4 wt.%); (**A**) fluorescence spectroscopy; (**B**) FTIR spectroscopy.

3.3. Interaction of PPI and FG in Emulsions

3.3.1. Particle Size and Zeta-Potential

Under the condition of pH = 5, without adding FG, the average particle size of the emulsion is relatively large (D(4,3) = 36 ± 0.75 μm). Because the pH value is close to the isoelectric point of PPI, the droplets in the emulsion had a greater degree of aggregation (Figure 3).

When the added FG concentration was 0.01–0.05 wt.%, the average particle size and particle size distribution of the emulsion showed a certain downward trend, which may be that the protein and polysaccharide interaction degree was weak and the resulting insoluble complex leads to a decrease in the particle size measurement [28]. As the FG concentration increased from 0.1 wt.% to 0.5 wt.%, the average particle size of the emulsion showed a significant downward trend and the higher concentration of polysaccharides could be more evenly adsorbed to the surface of the protein-coated oil droplets. This result is attributed to the adsorption of FG on the surface of the emulsion droplets, which prevented flocculation and coalescing. As a stabilizer, FG can migrate to the oil–water interface through molecular interactions (spatial potential resistance and hydrogen bonding) and alter the final droplet particle size distribution [34].

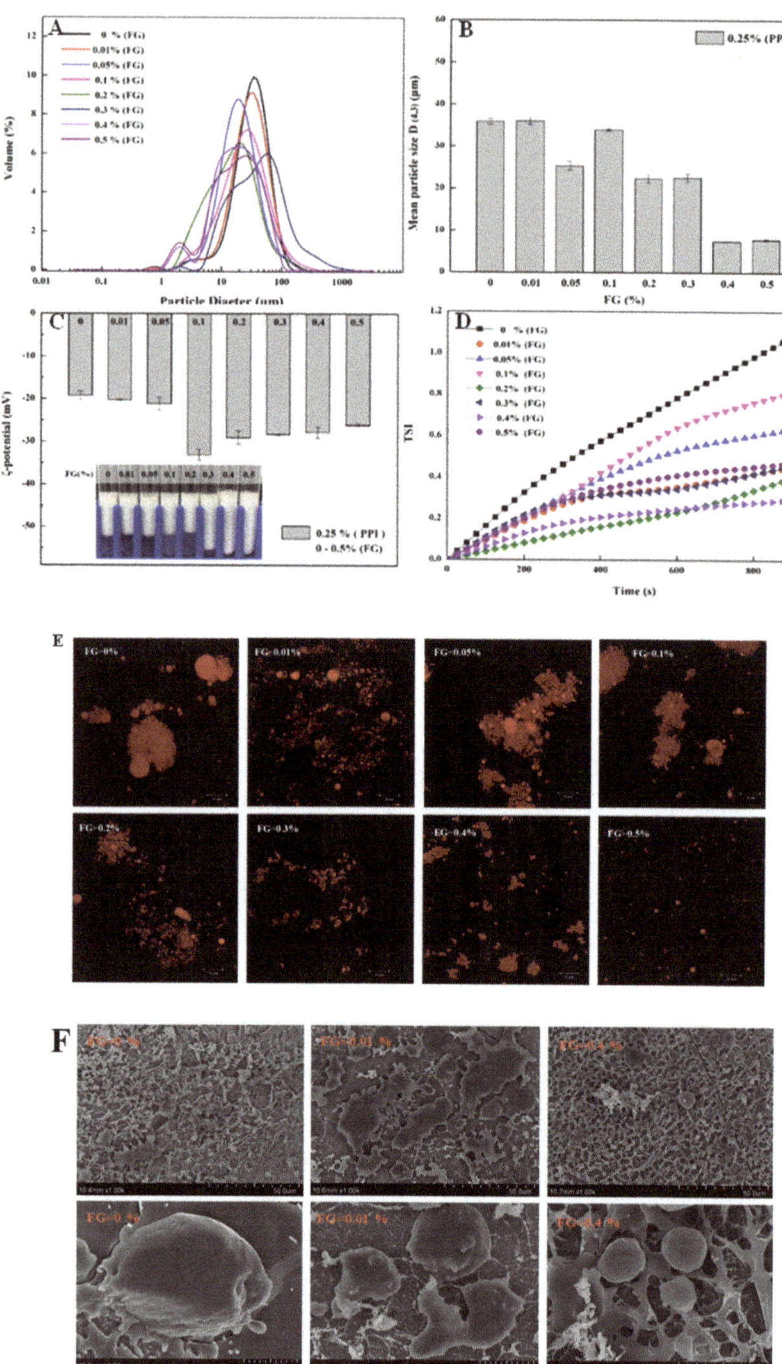

Figure 3. Effect of FG concentration on related properties: (**A**) particle size distribution, (**B**) average droplet size, (**C**) ζ-potential and appearance diagram, (**D**) Turbiscan stability index (TSI), (**E**) CLSM and (**F**) SEM images.

According to Figure 3C, the ζ potential value was reduced to -28 mV when the FG concentration was increased to 0.1 wt.% at pH 5. Combined results of Zeta potential and emulsion appearance indicated that the FG–PPI interacted in emulsion, which inhibited the aggregation of oil droplets in the system. Combined with the morphological figure of the emulsion, it is concluded that when the FG concentration reached 0.4 wt.%, the polysaccharide formed an adsorption "saturation state" of the protein at the interface [11].

3.3.2. Emulsion Microstructure

The appearance of the droplets showed different appearances with different concentrations of FG. The microstructure of the emulsion by CLSM showed that the PPI single-layer emulsion had poor solubility near the isoelectric point under the condition of pH = 5, resulting in a greater degree of aggregation (Figure 3E). When the FG concentration is between 0.01 wt.% and 0.3 wt.%, the protein–polysaccharide bridging flocculation phenomenon occurred. As the concentration of FG increased, the degree of aggregation of oil droplets in the emulsion slowly decreased and the microstructure of the droplet remained consistent with the particle size of the emulsion, showing a slow decreasing trend. When the concentration of FG added is around 0.35–0.4 wt.%, the adsorption of protein and polysaccharide at the interface reached a saturated state and the droplet distribution is uniform, with a smaller droplet size. The results obtained by cryo-scanning electron microscope (Figure 3F) also showed bridging flocculation at a low concentration of FG and a uniform dispersion of droplets at a high concentration of FG.

3.3.3. Gravitational Separation

TSI can provide a quantitative measurement for the resistance of emulsion to phase separation: the higher the TSI value, the more significant the phase separation [35]. Figure 3D shows that adding FG to PPI-coated oil droplets had a significant impact on the resistance to gravity-induced phase separation in the DHA algae oil emulsion. As the concentration of FG increased, the TSI value of the emulsion gradually decreased and 0.4% addition of FG showed the lowest TSI value. This phenomenon can be attributed to many factors. Firstly, when FG was coated on the protein emulsion droplets, the thickness of the interfacial layer and the electrostatic repulsion increased, resisting phase separation. Secondly, when the emulsion contained a higher FG concentration, the viscosity and flow resistance of the aqueous phase were enhanced [36].

3.4. Analysis of Volatile Substances in Algal Oil Emulsions

The GC-MS chromatogram analysis of headspace volatile compounds (Table 1) was used to determine the ability of the FG-PPI two-layer emulsion system to inhibit the release of fishy odor. Compared with pure algae oil, the amount and intensity of volatile substances presented in the headspace of the emulsions were significantly reduced (Figure 4). In addition, the presence of volatile substances in the two-layer emulsion was less than that in the single-layer emulsion, which indicated that the presence of FG contributed to forming an interfacial membrane to inhibit lipid oxidation or reduce the tendency of lipids to release volatile substances into the headspace. Another reason may be that FG molecules or protein–polysaccharide bilayer interface membrane can bind flavor molecules to reduce their diffusion from droplets. Previous studies have shown that the two main lipid oxidation products of algal oil that caused its unpleasant smell were heptanal and (E, Z)-3,5-octadiene-2-one [37]. Therefore, the research results showed that the FG–PPI double-layer emulsion was more effective in reducing the fishy odor in algae oil

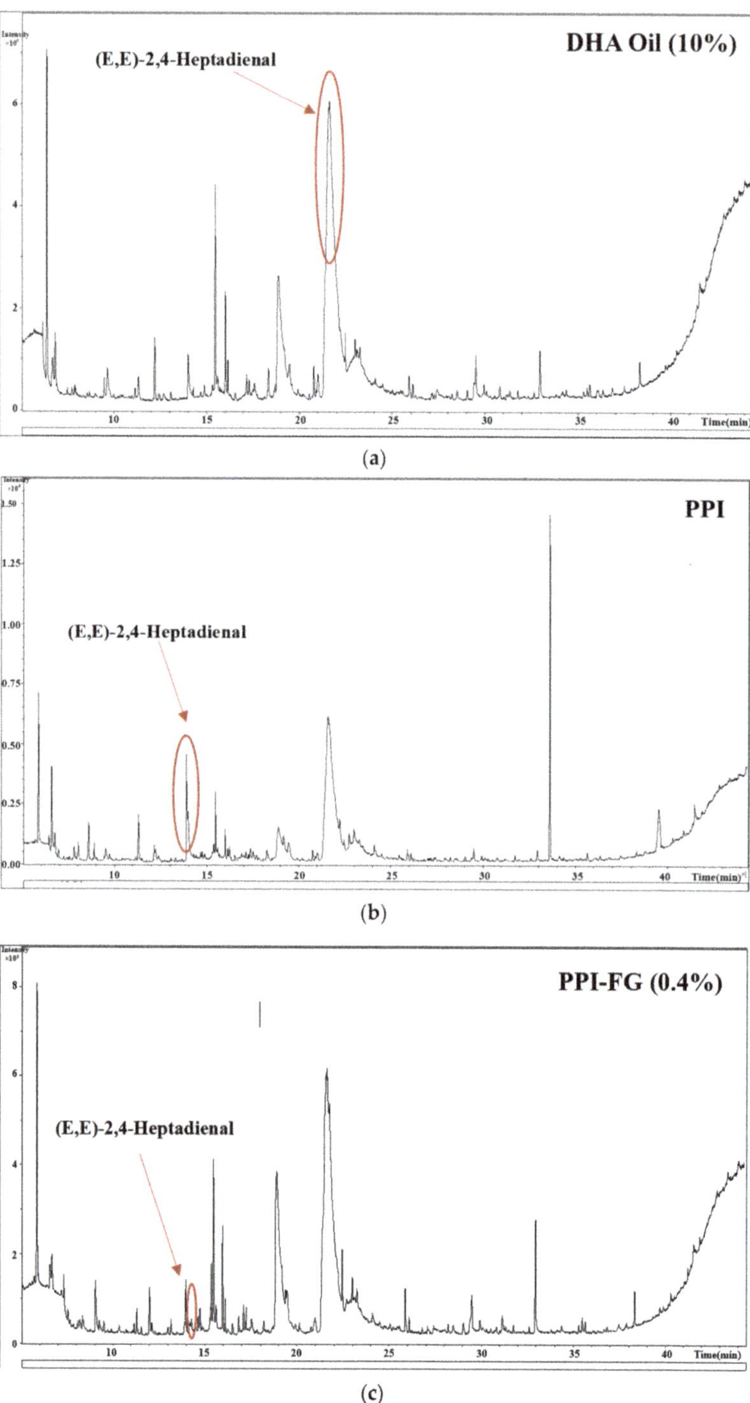

Figure 4. Algal oil emulsion volatile substance gas chromatogram: (**a**) algae oil, (**b**) PPI emulsion, (**c**) PPI–FG (0.4%) emulsion.

Table 1. SPME-GC-MS detects characteristic volatile compounds in different emulsion systems.

Compounds	DHA Oil (Control)	0% FG	0.01%FG	0.2% FG	0.4% FG
Heptanal	0.7875 (100%)	0	0	0	0
Octanal	1.6673 (100%)	0	0.3342 (20%)	0	0
Nonanal	4.0145 (100%)	0.8035 (20%)	0.9553 (23.8%)	0.4746 (11.8%)	0.3441 (8.6%)
(E,E)-2,4-Heptadienal	1.7717 (100%)	0.8148 (46%)	0.6963 (39.3%)	0.3505 (19.8%)	0.1856 (10.5%)
Benzaldehyde	0.5213 (100%)	0.215 (41.4%)	0.1825 (35%)	0.1715 (32.9%)	0
3,5-Octadien-2-one		1.209	0.7031	0.2826	0.1888

DHA oil: algal oil; 0% FG: PPI single-layer emulsion; 0.01%FG: 0.01%FG-PPI Double-layer emulsion; 0.2% FG: 0.2%FG-PPI Double-layer emulsion; 0.4% FG: 0.4%FG-PPI Double-layer emulsion. Percentage (%) expressed as a proportion of per volatile compound (odor) in algal oil systems (100%). The percentage of volatile compound content in different emulsion systems indicates their effect on odor masking.

3.5. Physical and Chemical Stability during Long Storage

3.5.1. Physical Stability

According to Figure 5A, the concentration of calcium ion added into FG-PPI emulsion showed an obvious effect on dynamic rheological properties. The storage modulus (G') of all emulsion systems is higher than the loss modulus (G''), indicating that they mainly have elastic behavior characteristics. The viscoelasticity of FG-PPI emulsion increased with an increase in calcium ion concentration, until the calcium ion concentration reached 0.4 wt.% and the (G') and (G'') viscoelasticity of the emulsion were relatively maximum. The underlying mechanism could be that calcium ions bound to free carboxyl groups in FG to form a calcium bridge, which helped to form a more stable gel network structure. Meanwhile, calcium ions caused different electrostatic interactions, increasing entanglements between FG polysaccharide molecules, resulting in an increase in emulsion flow resistance and a more pronounced trend towards an increase in modulus [38].

According to Figure 5C, the particle size of double-layered emulsions with different calcium ion concentrations showed different variation trends during storage for 21 days, indicating their different stabilities. During storage, the particle size of the 0.4 wt.% calcium ion double-layer emulsion showed no obvious changes, suggesting that an appropriate calcium ion concentration could significantly promote the physical stability of FG–PPI double-layered emulsion systems. Additionally, the TSI value of this group was the smallest and the change was the least during 21 days of storage (Figure 5B). According to Figure 5D, except for double-layer emulsion added with 0.4 wt.% calcium ion, zeta-potential values of the other emulsion system had undergone obvious changes during storage. The changes in interfacial compositions induced by chemical degradation of compounds within the emulsion system would severely influence the zeta-potential values [39], which suggested that double-layer emulsion added with 0.4 wt.% calcium ion might have the highest stability.

According to Figure 5A, the appearance of emulsions also showed the same trends. In general, the double-layered emulsion could better maintain the stable appearance of the emulsion and slow down the rate of phase separation of the emulsion droplets. The 0.4 wt.% calcium-ion-added double-layer emulsion system had the strongest anti-gravity separation ability, indicating that the calcium ion and FG were finely cross-linked through electrostatic interactions. The Appearance topography image analysis of different emulsion systems (Figure 5B,G) showed that the droplets of the 0.4 wt.% calcium-ion-added double-layer emulsion system had a uniform distribution. After storage for 21 days, a certain degree of aggregation occurred in almost all the emulsion systems, while the 0.4 wt.% calcium-ion-added double-layer emulsion system showed better maintenance of both the internal structure and the appearance.

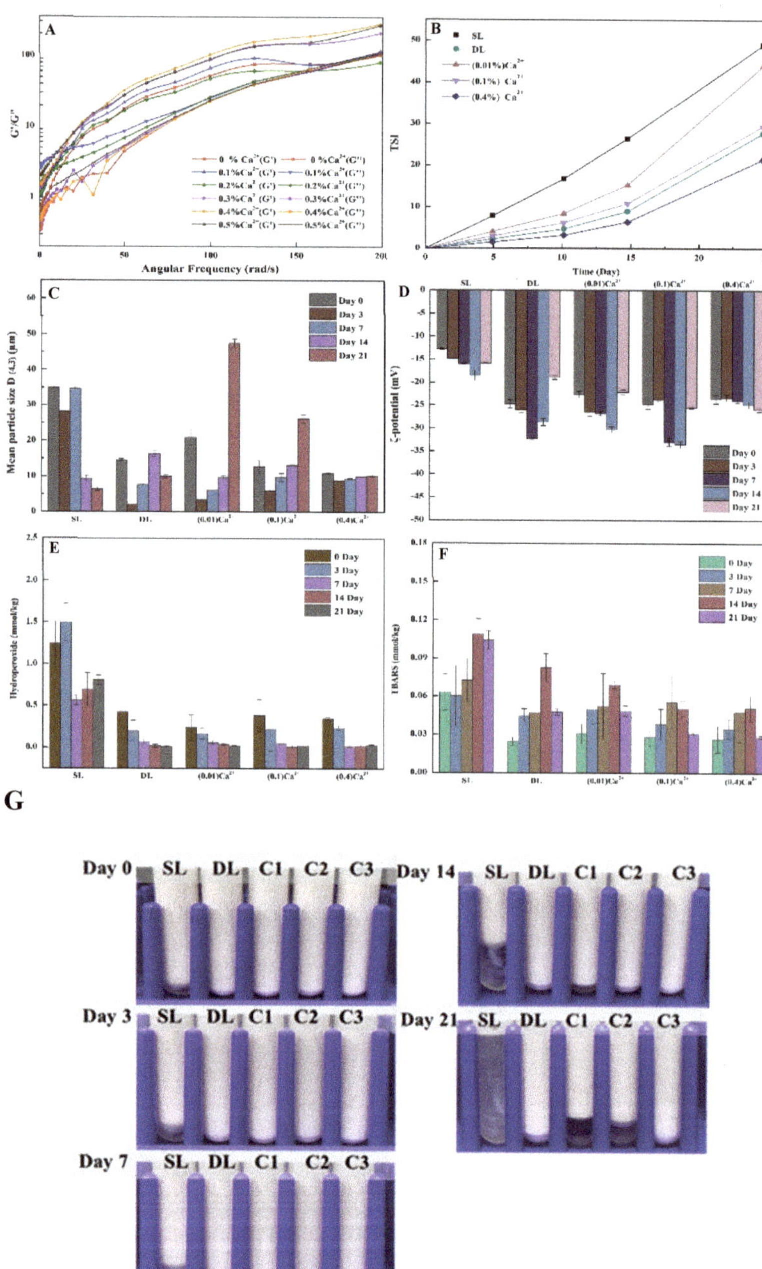

Figure 5. Effects of differences in calcium ion concentration of emulsions: (**A**) shear modulus versus frequency; stability of astaxanthin-loaded emulsion stored for 21 days: (**B**) TSI; (**C**) average droplet size; (**D**) ζ-potential; (**E**) hydroperoxide; (**F**) TBARS and appearance diagram (pH 5, 0.25 wt.%, PPI, 0.4 wt.% FG, 0.01, 0.1, 0.4 wt.% Ca^{2+}); (**G**) Appearance of emulsions loaded with astaxanthin during 21 days of storage (Day 0, 3, 7, 14, 21); SL, single-layer PPI emulsions; DL, PPI–FG (0.4%) double-layer emulsions system. ± standard deviations ($p < 0.05$).

3.5.2. Chemical Stability and Astaxanthin Retention Rate

The chemical stability of different emulsion systems loaded with astaxanthin varied greatly. According to Figure 5E,F, the oxidation degree of double-layered emulsion was obviously lower than that of single-layered emulsion. Both the hydroperoxide (0.025 mmol/kg) and TBARS (0.03 mmol/kg) values of the 0.4 wt.% calcium-ion-added double-layer emulsion system were the lowest, which suggested the highest chemical stability.

According to Table 2, the loss ratio of astaxanthin of the emulsion system was consistent with the above-obtained results of physicochemical stability. It showed that the calcium-ion-added double-layer emulsion system could effectively protect astaxanthin from oxidative decomposition. The results of the study found that when the calcium ion concentration was low (0.01 wt.%), the oxidation of astaxanthin would be accelerated during storage. It was speculated that the low-concentration calcium ion could form a "calcium bridge" with the macromolecules during the storage process. It caused the emulsion to produce flocculation or aggregation, which reduced the physical and chemical stability of the emulsion and caused a more intensive degradation of astaxanthin.

Table 2. Analysis of astaxanthin retention rate and color change in algae oil emulsion after storage.

Emulsion Systems	Retention (%)
PPI	28.1 ± 0.65
FG-PPI	51.9 ± 0.23
$0.01Ca^{2+}$-FG-PPI	31.5 ± 0.43
$0.1Ca^{2+}$-FG-PPI	57.5 ± 0.24
$0.4Ca^{2+}$-FG-PPI	65.0 ± 0.26

PPI: PPI single-layer emulsion; FG-PPI: 0.4% FG-PPI Double-layer emulsion; 0.01% Ca^{2+}-FG-PPI:0.01% Ca^{2+}-FG-PPI Double-layer emulsion; 0.1% Ca^{2+}-FG-PPI:0.1% Ca^{2+}-FG-PPI Double-layer emulsion; 0.4% Ca^{2+}-FG-PPI:0.4% Ca^{2+}-FG-PPI Double-layer emulsion.

4. Conclusions

In this study, a relatively stable and plant-based delivery system for both DHA algae oil and astaxanthin was fabricated and characterized. The molecular interaction between perilla protein isolate and flaxseed gum was investigated, coupled with their influences on the covering of the off-flavor of DHA algae oil. A denser interface and more even distribution of oil droplets with network structure of FG helped to maintain the stability and emulsions and prevent the overflow of the fishy smell. Calcium ion further strengthens the physio-chemical stability of the FG–PPI bilayer emulsion system by interacting with the polysaccharide chains. Both the DHA and astaxanthin were confirmed to be protected during storage by the calcium-ion-added double-layer emulsion system from oxidative decomposition. These findings are quite important for the fabrication of plant-based omega-3 delivery systems and their application in multiple food matrices, including plant-based yoghurt (acidic food matrix), etc.

Author Contributions: Performed the experiments and Writing—original draft, M.S., Resources, H.C., Q.H., S.Z.; Software, F.G.; Methodology, Q.Z.; Data curation and Validation, review and editing, Y.C.; Funding acquisition and Supervision, Q.D. All authors have read and agreed to the published version of the manuscript.

Funding: Key research projects of Hubei Province and Wuhan Scientific and Technical Payoffs Transformation Project (2020BCA086, 2020BBA045); National Natural Science Foundation of China (32001740); General project of Natural Science Foundation of Hubei Province, China (ZRMS2020002499); Earmarked Fund for China Agriculture Research System (CARS-14).

Institutional Review Board Statement: Not applicable.

Informed Consent Statement: Not applicable.

Data Availability Statement: Data are contained within the article.

Acknowledgments: The work was supported by Key research projects of Hubei Province and Wuhan Scientific and Technical Payoffs Trans-formation Project (2020BCA086, 2020BBA045), National Natural Science Foundation of China (32001740), General project of Natural Science Foundation of Hubei Province, China (ZRMS2020002499), Earmarked Fund for China Agriculture Research System (CARS-14) and these are greatly appreciated.

References

1. Castejón, N.; Señoráns, F.J. Enzymatic modification to produce health-promoting lipids from fish oil, algae and other new omega-3 sources: A review. *New Biotechnol.* **2020**, *57*, 45–54. [CrossRef] [PubMed]
2. Vellido-Perez, J.A.; Ochando-Pulido, J.M.; Brito-de la Fuente, E.; Martinez-Ferez, A. Novel emulsions–based technological approaches for the protection of omega–3 polyunsaturated fatty acids against oxidation processes–A comprehensive review. *Food Struct.* **2021**, *27*, 100175. [CrossRef]
3. Sánchez, C.O.; Zavaleta, E.B.; García, G.U.; Solano, G.L.; Díaz, M.R. Krill oil microencapsulation: Antioxidant activity, astaxanthin retention, encapsulation efficiency, fatty acids profile, in vitro bioaccessibility and storage stability. *LWT* **2021**, *147*, 111476. [CrossRef]
4. Polotow, T.G.; Poppe, S.C.; Vardaris, C.V.; Ganini, D.; Guariroba, M.; Mattei, R.; Hatanaka, E.; Martins, M.F.; Bondan, E.F.; Barros, M.P. Redox status and neuro inflammation indexes in cerebellum and motor cortex of wistar rats supplemented with natural sources of Omega-3 fatty acids and astaxanthin: Fish oil, krill oil, and algal biomass. *Mar. Drugs* **2015**, *13*, 6117–6137. [CrossRef] [PubMed]
5. Ambati, R.R.; Siew Moi, P.; Ravi, S.; Aswathanarayana, R.G. Astaxanthin: Sources, extraction, stability, biological activities and its commercial applications—A review. *Mar. Drugs* **2014**, *12*, 128–152. [CrossRef]
6. Aneesh, P.A.; Ajeeshkumar, K.K.; Lekshmi, R.K.; Anandan, R.; Ravishankar, C.N.; Mathew, S. Bioactivities of astaxanthin from natural sources, augmenting its biomedical potential: A review. *Trends Food Sci. Technol.* **2022**, *125*, 81–90. [CrossRef]
7. Boonlao, N.; Ruktanonchai, U.R.; Anal, A.K. Enhancing bioaccessibility and bioavailability of carotenoids using emulsion-based delivery systems. *Colloids Surf. B Biointerfaces* **2022**, *209*, 112211. [CrossRef]
8. Santos, M.; Okuro, P.K.; Fonseca, L.R.; Cunha, R.L. Protein-based colloidal structures tailoring techno- and bio-functionality of emulsions. *Food Hydrocoll.* **2022**, *125*, 107384. [CrossRef]
9. Guo, Q.; Bayram, I.; Shu, X.; Su, J.; Liao, W.; Wang, Y.; Gao, Y. Improvement of stability and bioaccessibility of β-carotene by curcumin in pea protein isolate-based complexes-stabilized emulsions: Effect of protein complexation by pectin and small molecular surfactants. *Food Chem.* **2022**, *367*, 130726. [CrossRef]
10. Xu, D.; Aihemaiti, Z.; Cao, Y.; Teng, C.; Li, X. Physicochemical stability, microrheological properties and microstructure of lutein emulsions stabilized by multilayer membranes consisting of whey protein isolate, flaxseed gum and chitosan. *Food Chem.* **2016**, *202*, 156–164. [CrossRef]
11. Wang, S.; Yang, J.; Shao, G.; Liu, J.; Wang, J.; Yang, L.; Jiang, L. pH-induced conformational changes and interfacial dilatational rheology of soy protein isolated/soy hull polysaccharide complex and its effects on emulsion stabilization. *Food Hydrocoll.* **2020**, *109*, 106075. [CrossRef]
12. Zhao, Q.; Xie, T.; Hong, X.; Zhou, Y.; Fan, L.; Liu, Y.; Li, J. Modification of functional properties of perilla protein isolate by high-intensity ultrasonic treatment and the stability of o/w emulsion. *Food Chem.* **2022**, *368*, 130848. [CrossRef]
13. Luo, J.; Li, Y.; Mai, Y.; Gao, L.; Ou, S.; Wang, Y.; Liu, L.; Peng, X. Flaxseed gum reduces body weight by regulating gut microbiota. *J. Funct. Foods* **2018**, *47*, 136–142. [CrossRef]
14. Tokle, T.; McClements, D.J. Physicochemical properties of lactoferrin stabilized oil-in-water emulsions: Effects of pH, salt and heating. *Food Hydrocoll.* **2011**, *25*, 976–982. [CrossRef]
15. Singh, H.; Sarkar, A. Behaviour of protein-stabilised emulsions under various physiological conditions. *Adv. Colloid Interface Sci.* **2011**, *165*, 47–57. [CrossRef]
16. Fan, C.; Han, S.; Liu, F.; Liu, Y.; Wang, L.; Pan, S. Influence of calcium lactate and pH on emulsification of low-methoxylated citrus pectin in a Pickering emulsion. *J. Disper. Sci. Technol.* **2017**, *38*, 1175–1182. [CrossRef]
17. Qian, K.Y.; Cui, S.W.; Wu, Y.; Goff, H.D. Flaxseed gum from flaxseed hulls: Extraction, fractionation, and characterization. *Food Hydrocoll.* **2012**, *28*, 275–283. [CrossRef]
18. Zhao, Q.; Long, Z.; Kong, J.; Liu, T.; Sun-Waterhouse, D.; Zhao, M. Sodium caseinate/flaxseed gum interactions at oil–water interface: Effect on protein adsorption and functions in oil-in-water emulsion. *Food Hydrocoll.* **2015**, *43*, 137–145. [CrossRef]
19. Sun, M.; Li, X.; McClements, D.J.; Xiao, M.; Chen, H.; Zhou, Q.; Xu, S.; Chen, Y.; Deng, Q. Reducing off-flavors in plant-based omega-3 oil emulsions using interfacial engineering: Coating algae oil droplets with pea protein/flaxseed gum. *Food Hydrocoll.* **2022**, *122*, 107069. [CrossRef]
20. Xu, X.; Luo, L.; Liu, C.; McClements, D.J. Utilization of anionic polysaccharides to improve the stability of rice glutelin emulsions: Impact of polysaccharide type, pH, salt, and temperature. *Food Hydrocoll.* **2017**, *64*, 112–122. [CrossRef]

21. Jiang, T.; Charcosset, C. Encapsulation of curcumin within oil-in-water emulsions prepared by premix membrane emulsification: Impact of droplet size and carrier oil type on physicochemical stability and in vitro bioaccessibility. *Food Chem.* **2022**, *375*, 131825. [CrossRef] [PubMed]
22. Chen, M.; Lu, J.; Liu, F.; Nsor-Atindana, J.; Xu, F.; Goff, H.D.; Ma, J.; Zhong, F. Study on the emulsifying stability and interfacial adsorption of pea proteins. *Food Hydrocoll.* **2019**, *88*, 247–255. [CrossRef]
23. Cheng, C.; Yu, X.; Huang, F.; Peng, D.; Chen, H.; Chen, Y.; Huang, Q.; Deng, Q. Effect of different structural flaxseed lignans on the stability of flaxseed oil-in-water emulsion: An interfacial perspective. *Food Chem.* **2021**, *357*, 129522. [CrossRef] [PubMed]
24. Cheng, C.; Yu, X.; McClements, D.J.; Huang, Q.; Tang, H.; Yu, K.; Deng, Q. Effect of flaxseed polyphenols on physical stability and oxidative stability of flaxseed oil-in-water nanoemulsions. *Food Chem.* **2019**, *301*, 125207. [CrossRef]
25. Ntone, E.; Bitter, J.H. Not sequentially but simultaneously: Facile extraction of proteins and oleosomes from oilseeds. *Food Hydrocoll.* **2020**, *102*, 105598. [CrossRef]
26. Liu, W.Y.; Feng, M.Q.; Wang, M.; Wang, P.; Sun, J.; Xu, X.L.; Zhou, G.-H. Influence of flaxseed gum and NaCl concentrations on the stability of oil-in-water emulsions. *Food Hydrocoll.* **2018**, *79*, 371–381. [CrossRef]
27. Khalid, N.; Shu, G.; Holland, B.J.; Kobayashi, I.; Nakajima, M.; Barrow, C.J. Formulation and characterization of O/W nanoemulsions encapsulating high concentration of astaxanthin. *Food Res. Int.* **2017**, *102*, 364–371. [CrossRef]
28. Kaushik, P.; Dowling, K.; Barrow, C.J.; Adhikari, B. Complex coacervation between flaxseed protein isolate and flaxseed gum. *Food Res. Int.* **2015**, *72*, 91–97. [CrossRef]
29. Wang, S.; Qu, D.; Zhao, G.; Yang, L.; Zhu, L.; Song, H.; Liu, H. Characterization of the structure and properties of the isolating interfacial layer of oil–water emulsions stabilized by soy hull polysaccharide: Effect of pH changes. *Food Chem.* **2022**, *370*, 131029. [CrossRef]
30. Chen, S.; Sun, C.; Wang, Y.; Han, Y.; Dai, L.; Abliz, A.; Gao, Y. Quercetagetin-loaded composite nanoparticles based on zein and hyaluronic acid: Formation, characterization, and physicochemical stability. *J. Agric. Food Chem.* **2018**, *66*, 7441–7450. [CrossRef]
31. Huang, G.Q.; Sun, Y.T.; Xiao, J.X.; Yang, J. Complex coacervation of soybean protein isolate and chitosan. *Food Chem.* **2012**, *135*, 534–539. [CrossRef]
32. Timilsena, Y.P.; Wang, B.; Adhikari, R.; Adhikari, B. Preparation and characterization of chia seed protein isolate–chia seed gum complex coacervates. *Food Hydrocoll.* **2016**, *52*, 554–563. [CrossRef]
33. Chen, S.; Li, Q.; McClements, D.J.; Han, Y.; Dai, L.; Mao, L.; Gao, Y. Co-delivery of curcumin and piperine in zein-carrageenan core-shell nanoparticles: Formation, structure, stability and in vitro gastrointestinal digestion. *Food Hydrocoll.* **2020**, *99*, 105334. [CrossRef]
34. Wang, M.; Feng, M.Q.; Jia, K.; Sun, J.; Xu, X.L.; Zhou, G.H. Effects of flaxseed gum concentrations and pH values on the stability of oil-in-water emulsions. *Food Hydrocoll.* **2017**, *67*, 54–62. [CrossRef]
35. Cerimedo MS, Á.; Iriart, C.H.; Candal, R.J.; Herrera, M.L. Stability of emulsions formulated with high concentrations of sodium caseinate and trehalose. *Food Res. Int.* **2010**, *43*, 1482–1493. [CrossRef]
36. Yang, H.; Li, Q.; Xu, Z.; Ge, Y.; Zhang, D.; Li, J.; Sun, T. Preparation of three-layer flaxseed gum/chitosan/flaxseed gum composite coatings with sustained-release properties and their excellent protective effect on myofibril protein of rainbow trout. *Int. J. Biol. Macromol.* **2022**, *194*, 510–520. [CrossRef]
37. Chen, X.W.; Chen, Y.J.; Wang, J.M.; Guo, J.; Yin, S.W.; Yang, X.Q. Phytosterol structured algae oil nanoemulsions and powders: Improving antioxidant and flavor properties. *Food Funct.* **2016**, *7*, 3694–3702. [CrossRef]
38. Wang, H.; Ke, L.; Ding, Y.; Rao, P.; Xu, T.; Han, H.; Shang, X. Effect of calcium ions on rheological properties and structure of Lycium barbarum L. polysaccharide and its gelation mechanism. *Food Hydrocoll.* **2022**, *122*, 107079. [CrossRef]
39. Chen, Y.S.; Zhang, R.J.; Xie, B.J.; Sun, Z.D.; McClements, D.J. Lotus seedpod proanthocyanidin-whey protein complexes: Impact on physical and chemical stability of β-carotene-nanoemulsions. *Food Res. Int.* **2020**, *127*, 108738. [CrossRef]

Article

Encapsulation of Functional Plant Oil by Spray Drying: Physicochemical Characterization and Enhanced Anti-Colitis Activity

Hao Zhang [1], Zhenxia Xu [1], Zhixian Qiao [2], Xu Wang [3], Hu Tang [1], Chen Yang [1,*] and Fenghong Huang [1,4]

[1] Oil Crops Research Institute of the Chinese Academy of Agricultural Sciences, Oil Crops and Lipids Process Technology National & Local Joint Engineering Laboratory, Key Laboratory of Oilseeds Processing of Ministry of Agriculture and Rural Affairs, Hubei Key Laboratory of Lipid Chemistry and Nutrition, Wuhan 430062, China

[2] Institute of Hydrobiology, Chinese Academy of Sciences, Wuhan 430072, China

[3] Hubei Collaborative Innovation Center for Animal Nutrition and Feed Safety, Huazhong Agricultural University, Wuhan 430070, China

[4] Institute of Food & Nutrition Science and Technology, Shandong Academy of Agricultural Science, Jinan 250100, China

*   Correspondence: yangchen@caas.cn; Tel.: +86-27-86711526

Abstract: In this study, an encapsulation system was developed for functional plant oil delivery. Through a series of orthogonal experiments and single factor experiments, the raw material compositions, emulsification conditions, and spray drying conditions for the preparation of flaxseed oil and safflower seed oil powders were optimized, and the final encapsulation efficiency was as high as 99% with approximately 50% oil loading. The storage stability experiments showed that oil powder's stability could maintain its physicochemical properties over six months. Oral supplementation of the spray-dried flaxseed oil powder exhibited a significant and better effect than flaxseed oil on alleviating colitis in C57BL/6J mice. It suppressed the pro-inflammatory cell factors, including IL-6 and TNF-α, and repaired gut microbial dysbiosis by increasing the microbial diversity and promoting the proliferation of probiotic taxa such as *Allobaculum*. This work suggests that spray-dried flaxseed oil powder has great potential as a nutraceutical food, with spray drying being a good alternative technique to improve its bioactivity.

Keywords: flaxseed oil; encapsulation; oil powder; colitis; gut microbiota

Citation: Zhang, H.; Xu, Z.; Qiao, Z.; Wang, X.; Tang, H.; Yang, C.; Huang, F. Encapsulation of Functional Plant Oil by Spray Drying: Physicochemical Characterization and Enhanced Anti-Colitis Activity. *Foods* **2022**, *11*, 2993. https://doi.org/10.3390/foods11192993

Academic Editor: Jean-Christophe Jacquier

Received: 1 August 2022
Accepted: 15 September 2022
Published: 26 September 2022

Publisher's Note: MDPI stays neutral with regard to jurisdictional claims in published maps and institutional affiliations.

1. Introduction

Plant-based functional oil, such as flaxseed oil and safflower seed oil, contains a rich source of polyunsaturated fatty acids (PUFAs), which have been proven beneficial to human health when consumed in the diet. Previous research has shown that PUFAs can decrease many disease risks (such as coronary heart disease, high blood pressure, and Alzheimer's disease) and improve mental health and brain function [1–4]. However, the unsaturated fatty acids within these oil types are prone to oxidation which may result in the formation of harmful products and have disadvantages such as low availability and unpleasant odor, which limit their applications [5].

Transforming functional liquid oil into solid oil powders has become a very attractive process in the food and pharmaceutical industries, as this produces encapsulated products that can provide a physical barrier between the oil and the external environment that can avoid deterioration and unpleasant odor. In addition, the obtained powder forms are more acceptable in the market due to their improved odor and extended shelf-life [6,7].

Previous studies have shown a lot of encapsulation combinations and methods for oil encapsulation, including coacervation, ionic gelation, freeze drying, cross-linking etc., but most of them only focus on the physicochemical properties of the encapsulated products. Limited information exists relating to the confirmation of the health effects of these products [8,9]. The experimental induction of colitis has been used not only to study gut inflammatory processes but also to evaluate the effects on intestinal barrier integrity and homeostasis for various encapsulated products. As representative plant-based oil resources rich in PUFAs, flaxseed oil and safflower seed oil have been widely explored due to their high levels of n-3 PUFA-α-linolenic acid (50–60%) and n-6 PUFA-linoleic acid (>70%), respectively [3,10]. Investigation of the encapsulation of these two oils, as well as the evaluation of the physiochemical properties and the health benefits of the obtained oil powders, can help us learn more about the advantages of encapsulation and contribute to the development of relative nutraceuticals including n-3 PUFA, n-6 PUFA, or their blends.

This study systematically studied the fabrication of an encapsulation system for flaxseed oil and safflower seed oil. Modified starch and maltooligosaccharide were used as wall material combinations, and orthogonal experiments and single factor experiments were designed to obtain the optimal raw material compositions, emulsification, and spray drying conditions. Additionally, the physicochemical properties of these powdered oil products were evaluated. Furthermore, the potential effect of flaxseed oil powder on maintaining the intestinal epithelial barrier in dextran sodium sulfate (DSS)-induced colitis mice were examined. The research will provide further insight into the method of fabricating spray-dried oil powders and their beneficial roles in alleviating colitis.

2. Materials and Methods

2.1. Materials

Flaxseed oil was purchased from Hongjingyuan Oil Co. Ltd. (Xilingol, China). Safflower seed oil was purchased from Zhongliang Food Marketing Co., Ltd. (Alashankou, China). Fatty acid compositions of flaxseed oil and safflower seed oil are shown in Table S1. Modified starch (HI-CAP 100) was purchased from Ingredion Incorporated (New York, , NY, USA). Maltooligosaccharide with a dextrose equivalent (DE) of 18–28 was purchased from Baolingbao Biotechnology Co., Ltd. (Dezhou, China). Mono- and diglycerides of fatty acids were purchased from Jia Li Shi Additives (Hai An) Co., LTD (Haian, China). Petroleum ether with bp range from 30 to 60 °C was used in this study. Other reagents were of analytical grade and were obtained from Sinopharm Chemical Reagent Co. Ltd. (Shanghai, China).

2.2. Preparation of Emulsion

The oil emulsion was composed of an aqueous phase and an oil phase. The aqueous phase was prepared by dissolving modified starch and maltooligosaccharide in distilled water, and the oil phase was obtained by dispersing mono- and diglycerides of fatty acids in the oil. A coarse emulsion was then prepared by gradually pouring the oil phase into the aqueous phase with continuous shearing at 11,000 rpm for 15 min with a high-shear mixer IKA-T25 (IKA Instruments Ltd., Staufen, Germany). The secondary emulsion was finally obtained by passing this coarse emulsion through a high-pressure homogenizer ATS AH-Basic (ATS Industrial Co. Ltd., Toronto, Canada) at designed homogenization pressure and homogenization cycles.

2.3. Particle Size Distribution of Emulsion

The particle size distribution of emulsion was measured by a dynamic light scattering instrument Mastersizer 2000 (Malvern Instruments, Malvern, UK) at 25 °C. D [0.1], D [0.5], and D [0.9] represent 10%, 50%, and 90% below their sizes. D [2,3] is the surface weighted mean diameter, and D [3,4] is the volume-weighted mean diameter. Span value was obtained by dividing the difference of D [0.1] and D [0.9] by D [0.5] [11].

2.4. Preparation of Oil Powders

To prepare oil powders, the emulsion was spray-dried in a QZR-5 spray drier (Linzhou Spray Dryer Co., Wuxi, China). The emulsion was fed into the main chamber through a peristaltic pump with adjustable feeding speed (LongerPump BT100-2J).

2.5. Orthogonal Experiment Design

In this study, various types of factors: including raw material compositions, emulsification conditions, and spray drying conditions, can affect the characteristics of the final products when transforming the liquid oil into powdered oil by spray drying. To obtain the products that have targeted total oil content (50%) with high encapsulation efficiency, these conditions were optimized. Based on previous research and our preliminary experiments, the effects of raw materials (including wall ratio, core/wall ratio, solid concentration, and emulsifier content) were firstly evaluated to determine the optimal raw material compositions. Flaxseed oil was used in this section, and the selected raw material factors and respective levels are shown in Table 1. These four different factors were optimized using the orthogonal $L_{16} (4)^4$ experiment. A total of 16 experimental trials were formulated (Table 2) using the same process parameters: homogenization pressure (20 MPa), homogenization cycles (2 cycles), and inlet air temperature (170 °C).

Table 1. Selected raw material factors and respective levels.

Level	Factors			
	Wall Ratio	Core/Wall Ratio	Solid Concentration	Emulsifier Content
1	0.6:1	1:1	26%	0.5%
2	1:1	1:0.9	34%	1%
3	1:0.6	1:0.8	42%	2%
4	1:0.2	1:0.7	50%	5%

Table 2. Design of orthogonal $L_{16} (4)^4$ experiments.

Run	Factors			
	Wall Ratio	Core/Wall Ratio	Solid Concentration	Emulsifier Content
1	0.6:1	1:1	26%	0.5%
2	0.6:1	1:0.9	34%	1%
3	0.6:1	1:0.8	42%	2%
4	0.6:1	1:0.7	50%	5%
5	1:1	1:1	34%	2%
6	1:1	1:0.9	26%	5%
7	1:1	1:0.8	50%	0.5%
8	1:1	1:0.7	42%	1%
9	1:0.6	1:1	42%	5%
10	1:0.6	1:0.9	50%	2%
11	1:0.6	1:0.8	26%	1%
12	1:0.6	1:0.7	34%	0.5%
13	1:0.2	1:1	50%	1%
14	1:0.2	1:0.9	42%	0.5%
15	1:0.2	1:0.8	34%	5%
16	1:0.2	1:0.7	26%	2%

2.6. Characterization of the Oil Powders

2.6.1. Total Oil Content

The total oil content was measured according to an acid hydrolysis method described in the Chinese National Standard GB 5009.6—2016 [12]. Briefly, 2–3 g of powder samples were accurately weighed (precise to 0.0001 g) and added to a test tube. Then, 8 mL of water were added and well mixed, followed by the addition of 10 mL of hydrochloric acid.

The test tube flask was then placed in a water bath at $70 \pm 5\,^{\circ}\text{C}$ for 40–50 min, during which it was stirred with a glass rod every 5–10 min. The test tube was then taken out and followed by the addition of 10 mL alcohol. The mixture was transferred into a 100 mL mixing cylinder with a stopper; 25 mL of ethyl ether was added several times to wash the test tube, and then the mixture was transferred into the mixing cylinder. After closing the stopper, the mixing cylinder was shaken for 1 min, carefully opening the stopper to release the gas and closing it again to stand for 12 min. The stopper was then opened, and the oil attached to the bottleneck of the mixing cylinder was washed with ethyl ether. Keeping the cylinder standing for 20 min until clear liquid was observed in the upper layer, the supernatant was removed and placed in a weighing dish (previously dried in an oven at $105\,^{\circ}\text{C}$ for 1 h). Another 5 mL of ethyl ether were added to the mixing cylinder to repeat the process. The solvents in the weighing dish were then evaporated, and the dish was dried to constant weight in an oven at $105\,^{\circ}\text{C}$. The total oil content was calculated according to the following Equation (1):

$$\text{Total oil} = \frac{m_1 - m_0}{m} \times 100\%,\tag{1}$$

where m_1 is the weight of the weighing dish and oil, m_0 is the weight of the weighing dish, and m is the weight of the powdered samples.

2.6.2. Surface Oil Content

The surface oil content of the encapsulated powders was measured according to the method described in the Chinese Aquaculture Industry Standard SC/T 3505—2006 with a minor modification [13]. In total, 5.0 g of powdered samples were weighed and added to a 250 mL volumetric flask. Then, 30 mL of petroleum ether (bp 30–60 $^{\circ}\text{C}$) were added, and the flask was slightly shaken by hand for 10 s. The suspension was filtered into a weighing dish which was pre-dried in an oven at $105\,^{\circ}\text{C}$. The powder residue was washed using 20 mL of petroleum ether, then shaken and filtered into the same weighing dish. This process was repeated twice. The weighing dish was then placed in a water bath at $60\,^{\circ}\text{C}$ to evaporate the organic solvent and dried at $105\,^{\circ}\text{C}$ until it reached a constant weight. The surface oil content was calculated using the following Equation (2):

$$\text{Surface oil} = \frac{W_2 - W_1}{W} \times 100\%,\tag{2}$$

where W_1 is the weight of the weighing dish, W_2 is the weight of the weighing dish and oil, and W is the weight of powdered samples.

2.6.3. Encapsulation Efficiency

The encapsulation efficiency was calculated according to the following Equation (3):

$$\text{Encapsulation efficiency} = \frac{\text{Total oil} - \text{Surface oil}}{\text{Total oil}} \times 100\%.\tag{3}$$

2.6.4. Moisture Content

The moisture content of the spray-dried powders was measured according to the hypobaric drying method described in the Chinese National Standard GB/T 5009.3—2016 [14]. In total, 2~10 g powders were added into a weighing bottle with constant weight and weighed (precision to 0.0001 g). The samples were then dried at $60 \pm 5\,^{\circ}\text{C}$ under a vacuum drying oven with a pressure of ~45 kPa. The moisture content was calculated via Equation (4):

$$\text{Moisture content} = \frac{m_2 - m_3}{m_2 - m_4} \times 100\%,\tag{4}$$

where m_2 is the weight of the weighing bottle and powdered samples before drying, m_3 is the weight of the weighing bottle and powdered samples after drying, and m_4 is the weight of the weighing bottle.

2.6.5. Acidity Value Analysis

The acidity value was determined according to the cold solvent indicator titration method described in the Chinese National Standard GB/T 5009.229—2016 [15]. The encapsulated oil was extracted from the powders prior to the test. The extracted oil samples were weighed and added to a 250 mL conical flask. In total, 50 mL of ether-isopropanol (1:1) solvent mixture were added to the flask, followed by the addition of three drops of phenolphthalein indicator solution. The mixture was then titrated with 0.1 M potassium hydroxide standard solution. Blank tests were performed without the addition of extracted oil. The acidity value was calculated according to the following Equation (5):

$$\text{Acidity value} = \frac{(V - V_0) \times c \times 56.1}{m} \times 100\%, \tag{5}$$

where V and V_0 is the consumption of potassium hydroxide standard solution in the main test and in the blank test (mL), respectively, c refers to the molar concentration (molarity) of the potassium hydroxide standard solution (mol/L), 56.1 is the molar mass of potassium hydroxide (g/mol), and m is the weight of the extracted oil.

2.6.6. Peroxide Value Analysis

The peroxide value was measured with a titration method described in the Chinese National Standard GB/T 5009.227—2016 [16]. The encapsulated oil was extracted from the powders prior to the test. The extracted oil samples were weighed (precision to 0.0001 g) and added to a 250 mL iodine flask. In total, 30 mL of glacial acetic acid-chloroform (3:2) solvent mixture were added into the flask, shaken slightly for 0.5 min, and placed under darkness for 3 min. Then, 100 mL of deionized water were added and shaken. The mixture was then titrated with sodium thiosulfate solution. The peroxide value was calculated according to the following Equation (6):

$$\text{Peroxide value} = \frac{(V - V_0) \times c \times 0.1269}{m} \times 100\%, \tag{6}$$

where V and V_0 are the consumption of sodium thiosulfate solution in the main test and the blank test (mL), respectively, c is the molar concentration (molarity) of the sodium thiosulfate solution (mol/L), 0.1269 is the mass (g) of iodine titrated with 1 mL sodium thiosulfate solution (1 mol/L), and m is the weight of the extracted oil.

2.6.7. Powder Morphology

The morphologies of the powdered samples were observed using scanning electron microscopy (SEM, TESCAN Vega3). The samples were coated with gold prior to the tests. The particle sizes of the powdered samples were obtained by analyzing the SEM images with ImageJ software.

2.7. Storage Stability of the Spray-Dried Oil Powders

To evaluate the storage stability of the spray-dried oil powders, the newly prepared oil powders were sealed in aluminum laminated polyethylene (ALPE) pouches. The pouches were placed in a constant temperature and humidity incubator at storage conditions of 30 °C and 60% relative humidity (RH). Oil powders were withdrawn at different time intervals (30, 60, 90, 120, 150, and 180 days, respectively) for analyses, and the moisture content, surface oil content, total oil content, acidity value, and peroxide value were measured as described in the preceding text.

2.8. Animal Experiment

Thirty-two male C57BL/6J mice at 4–5 weeks old (22–23 g) were split at random into four groups ($n = 8$): NC group, DSS group, FS group, and FSP group. Mice in the NC group (normal control group) were administered with 0.2 mL/day of sterile water for 21 days. Mice in the DSS group (colitis model group) were administered DSS (2%) drinking water for seven days, followed by administration with distilled water for 14 days. Mice in the FS group and FSP group received DSS (2%) oral solution for seven days and were then treated with flaxseed oil and dried flaxseed oil powders for 14 days, respectively. The flaxseed oil or flaxseed oil powder was dissolved in 0.1% flaxseed gum solution [17] for subsequent oral administration. The dose at 500 mg/kg·Bw/day that has high therapeutic efficacy in colitis was referred to in our preliminary experiments. This animal experiment was approved by the Animal Ethics Committee of Huazhong Agricultural University (Permission HZAUMO-2021-0170).

Colitis-related parameters, including body weight loss and colon length, were measured after the mice were sacrificed. Haematoxylin and eosin (HE) stain was carried out for the distal colon tissues fixed in 4% paraformaldehyde solution, and microscopic observation at a magnification of 40X was performed for histologic assessment. Serum cytokines levels (IL-1β, IL-6, TNF-α, and IL-10) were detected by ELISA (elabscience Biotechnology Co., Ltd., Wuhan, China) following the supplier's recommendations. Fecal DNA was extracted using the DNA Stool Mini Kit and sequenced using the Illumina Miseq platform (2×150 pair-end) by Shanghai Paiseno Biological Technology Co., LTD. Bioinformatic analysis of the microbial component and structure in gut microbiota was implemented according to Yang et al. [18].

2.9. Statistical Analysis

The experiments were conducted in triplicate. The data are reported as means $\pm$ standard deviations. Difference significance in the gut microbiota across subgroups was processed using One-way ANOVA. p-value < 0.05 was set as significant.

3. Results and Discussion

3.1. Effect of Raw Material Compositions on Spray-Dried Flaxseed Oil Powders

During the encapsulation of oil with a spray drying method, various factors such as raw material compositions, emulsification conditions, and spray drying conditions can affect the properties of the powered oil products. Four different factors, including wall ratio, core/wall ratio, solid concentration, and emulsifier content, were optimized using the orthogonal L_{16} $(4)^4$ experiments. A total of 16 experimental trials were spray dried, and the encapsulation efficiency and encapsulation yield results were determined as indicators.

The encapsulation efficiency results are shown in Figure 1a, with a range from 28.8% to 94.95%. There were two experiments that showed encapsulation efficiency higher than 90%; they were run 10 (94.95%) and run 13 (94.09%), respectively. The corresponding analyses are exhibited in Table S2, showing that the optimal levels of raw material compositions are A4 (wall ratio, 1:0.2), B2 (core/wall ratio, 1:0.9), C3 (solid concentration, 42%), and D2 (emulsifier content, 1%). The encapsulation yields of these experiments were then evaluated (Figure 1b, Table S3), which showed that the optimal level of raw material compositions was A4 (wall ratio, 1:0.2), B1 (core/wall ratio, 1:1), C3 (solid concentration, 42%), and D2 (emulsifier content, 1%). In addition, only the result of run 13 exceeded 90% (Figure 1b). Combining these results and analyses, the raw material compositions at wall ratio (1:0.2), core/wall ratio (1:1), solid concentration (50%), and emulsifier content (1%) were selected for further experiments. The total oil and surface oil contents of the oil microcapsules under this condition were 49.05% and 2.90%, respectively. The encapsulation efficiency is comparable to other spray-dried flaxseed oil microcapsules which have similar or lower oil loading. Farzaneh Mohseni encapsulated flaxseed oil using the combination of oxidized tannic acid-gelatin and flaxseed mucilage and obtained microcapsules with 50% oil loading and 94.2% encapsulation efficiency, which are higher than that obtained in a similar work conducted by Pratibha Kaushik [19,20].

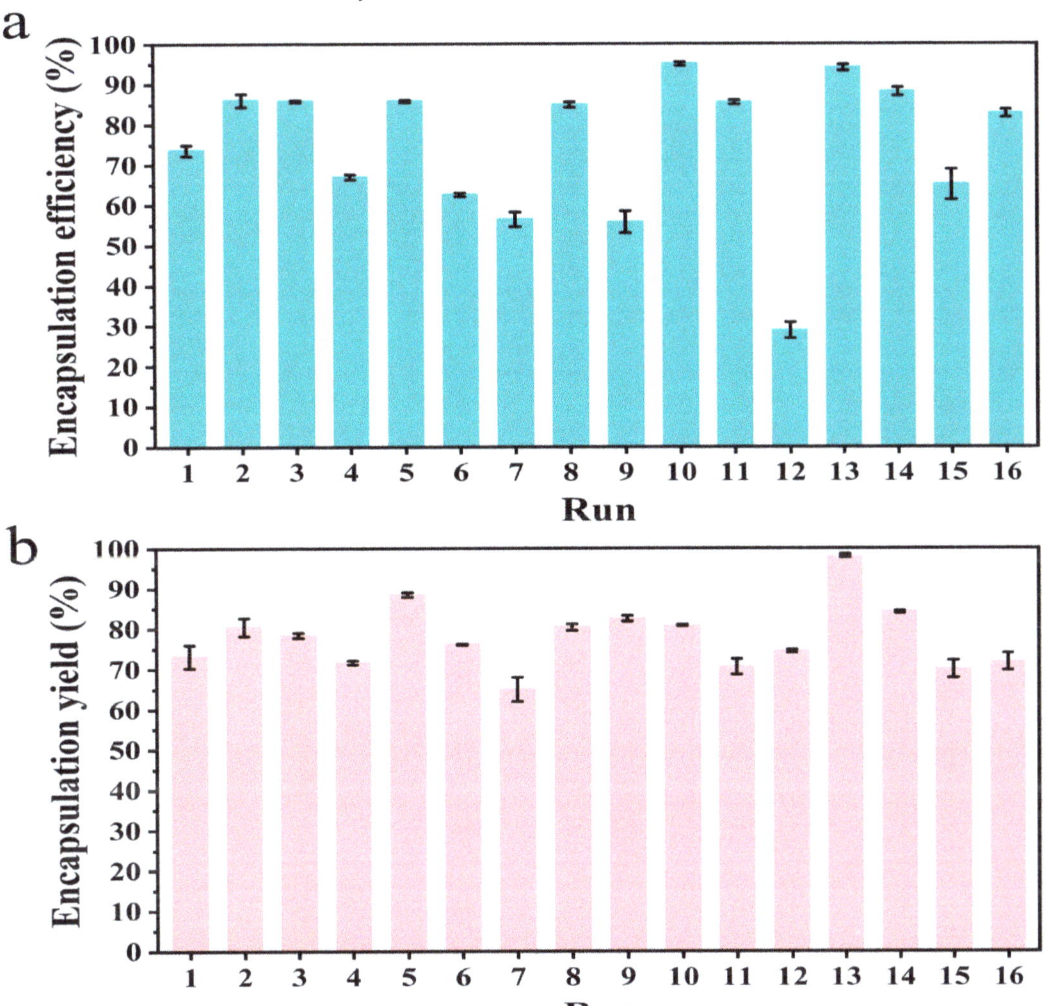

Figure 1. The encapsulation efficiencies (**a**) and encapsulation yields (**b**) of the flaxseed oil powders from the orthogonal *L16 (4)* [4] experiments.

3.2. Effect of Emulsification and Spray Drying Conditions on Spray-Dried Flaxseed Oil Powders

Based on the results of these experiments, the effects of emulsification and spray drying conditions on the characteristics of oil powders were then investigated. Homogenization pressure, homogenization cycles, and inlet air temperature were evaluated in this section, and the surface oil and total oil contents were selected as the key indicators. To optimize every single factor, a single variable was determined in the given scope: the homogenization pressures were tested at 20, 40, 60, and 80 MPa, the homogenization cycles at one, two, three, and four times, the inlet air temperature at 150, 160, 170, and 180 °C, and the emulsion feeding rate at 4, 6, 8, and 10 L/h. The raw material compositions were the same as the optimal determined in the preceding text.

During the emulsion preparation process, when the homogenization pressures increased from 20 to 40 MPa, the surface oil contents of the oil powders decreased significantly from 3.7% to 2.94%. However, with an excessive increment of this condition from 40 to 80 MPa, the surface oil contents of the oil powders barely changed. Another parameter that can affect the surface contents during the emulsion process is the homogenization cycle, and the surface oil contents will gradually be decreased from 3.12% to 2.18% when this factor increases from one time to four times. The results showed that moderate homogenization conditions could modulate the surface oil contents of spray-dried oil powders, and further increments of both homogenization pressures and cycles may cause no changes to surface oil contents. These phenomena were also observed in studies conducted by Quoc Dat Lai and Rudra Pangeni, in which they reported that moderate pressures and cycles could decrease the droplet size of emulsion and then affect the encapsulation efficiency or emulsion stability, while higher pressures and more cycles may not change these characteristics and even decrease them [21,22].

When the inlet air temperature was increased from 150 to 180 °C, the surface oil content decreased significantly from 2.78 to 1.07, and the inlet air temperature at 170 °C resulted in the lowest surface oil content. This might be due to the accelerated drying rate under higher inlet air temperature, promoting the formation of powder shells which can limit the leaching of oil from the powders. Similar behavior was reported in a study conducted by Mortaza Aghbashlo [23]. In addition, if the emulsion feeding rate was too high (10 L/h), the surface oil content was higher than 1.02%, while with a lower feeding rate at 4, 6, and 8 L/h, the surface oil content was significantly lower (~0.2%). Similar discoveries were reported by Phu Thuong Nhan Nguyen in the encapsulation of essential oil, and the increased surface oil contents under a higher feeding rate could be explained by larger droplets during the encapsulation process [24].

Therefore, through sequential experiments to determine the raw material compositions, emulsification conditions, and spray drying conditions, the optimal conditions for the preparation of flaxseed oil powders were obtained (Table 3). The conditions were wall ratio (1:0.2), core/wall ratio (1:1), solid concentration (50%), emulsifier content (1%), homogenization pressure (40 Mpa), homogenization cycles (three times), and inlet air temperature (170 °C). The availability of encapsulating a different oil type (safflower seed oil) in optimal conditions was further investigated, and the physicochemical properties of these powders were studied and compared.

Table 3. Effects of homogenization pressure, homogenization times, inlet air temperature, and emulsion feeding rate on surface oil contents of flaxseed oil powders.

Factors	Level	Surface Oil (%)
	20 MPa	3.70 ± 0.12
Homogenization pressure	40 MPa	2.94 ± 0.14
	60 MPa	2.88 ± 0.09
	80 MPa	2.90 ± 0.11
	1 time	3.12 ± 0.13
Homogenization cycles	2 times	2.90 ± 0.09
	3 times	2.22 ± 0.10
	4 times	2.18 ± 0.04
	150 °C	2.78 ± 0.22
Inlet air temperature	160 °C	2.22 ± 0.15
	170 °C	1.02 ± 0.07
	180 °C	1.07 ± 0.13
	4 L/h	0.22 ± 0.03
Emulsion feeding rate	6 L/h	0.23 ± 0.01
	8 L/h	0.19 ± 0.06
	10 L/h	1.02 ± 0.02

Through a series of experiments to optimize the optimal conditions for oil powders, an encapsulation system that has high encapsulation efficiency and yield was obtained. The possibility of encapsulating different plant-based oil types (flaxseed oil and safflower seed oil) and the properties of the obtained oil powders were evaluated.

3.3. Physicochemical Properties of Oil Emulsions and Powders

3.3.1. Emulsion Droplet Size

Figure 2 shows the emulsion droplet sizes of two emulsions from different oil types: flaxseed oil emulsion and safflower seed oil emulsion. The particle size distributions of both emulsions were comparable and displayed largely unimodal distributions. The surface-weighted mean diameter (D [2,3]) and volume-weighted mean diameter (D [3,4]) of flaxseed oil and safflower seed oil emulsions are 0.130 and 0.211 μm, and 0.123 and 0.190 μm, respectively. The emulsion droplet sizes are significantly smaller than the oil emulsions fabricated from similar wall materials [25,26]. The emulsion droplet size results in Table 4 were almost independent of the oil types, suggesting that the optimal wall material combinations and emulsification conditions were available to prepare stable feed emulsions necessary for microencapsulation by spray drying that contains a high oil loading content of up to 50% *w/w* oil in the final powder products.

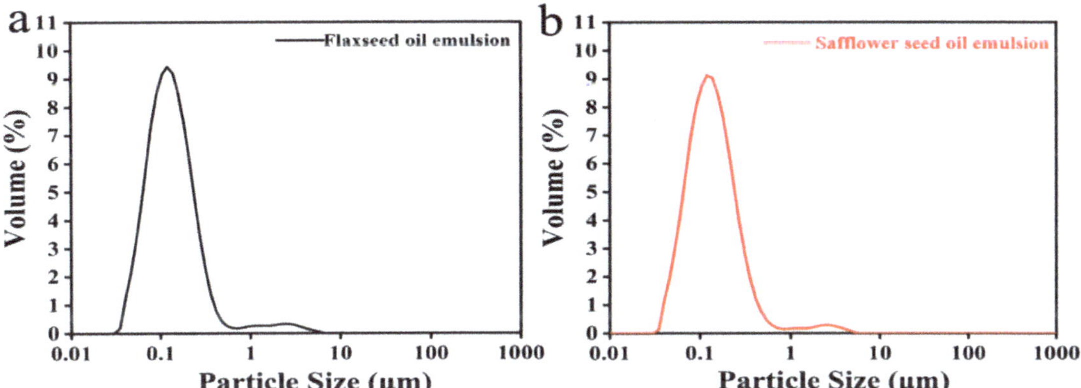

Figure 2. Particle size distributions of flaxseed oil emulsion (**a**) and safflower seed oil emulsion (**b**).

Table 4. Particle size results of flaxseed oil and safflower seed oil emulsions.

	Flaxseed Oil Emulsion	Safflower Seed Oil Emulsion
D [0.1]	0.067 μm	0.069 μm
D [0.5]	0.135 μm	0.140 μm
D [0.9]	0.301 μm	0.321 μm
D [3,2]	0.116 μm	0.120 μm
D [4,3]	0.234 μm	0.226 μm
Span	1.732	1.797

3.3.2. Powder Morphology

The microstructures of the oil powders containing flaxseed oil and safflower seed oil were observed by SEM. As shown in Figure 3a,b, both particles displayed spherical shapes with smooth surfaces, and little cracks or shrinkages were observed, suggesting that these wall material combinations are capable of retaining the core oil within the microcapsules for protection. The wrinkled surface in some powders can be attributed to the high inlet temperature that caused rapid shrinkage of sprayed emulsion droplets, which is inevitable during the spray drying process. Similar phenomena were also observed in previous research [27,28]. The mean diameter and size distribution of the particles were further determined by analyzing these images with ImageJ software (Figure 3c,d). The particle mean diameters of flaxseed oil and safflower seed oil powders were 30 ± 14 μm and 34 ± 15 μm, respectively, and these powders exhibited almost uniform size distributions ranging from 10 to 70 μm and 13 to 73 μm, respectively. These results demonstrate that encapsulating different oil samples using an encapsulation system has little influence on the microstructures and size distributions of the final products.

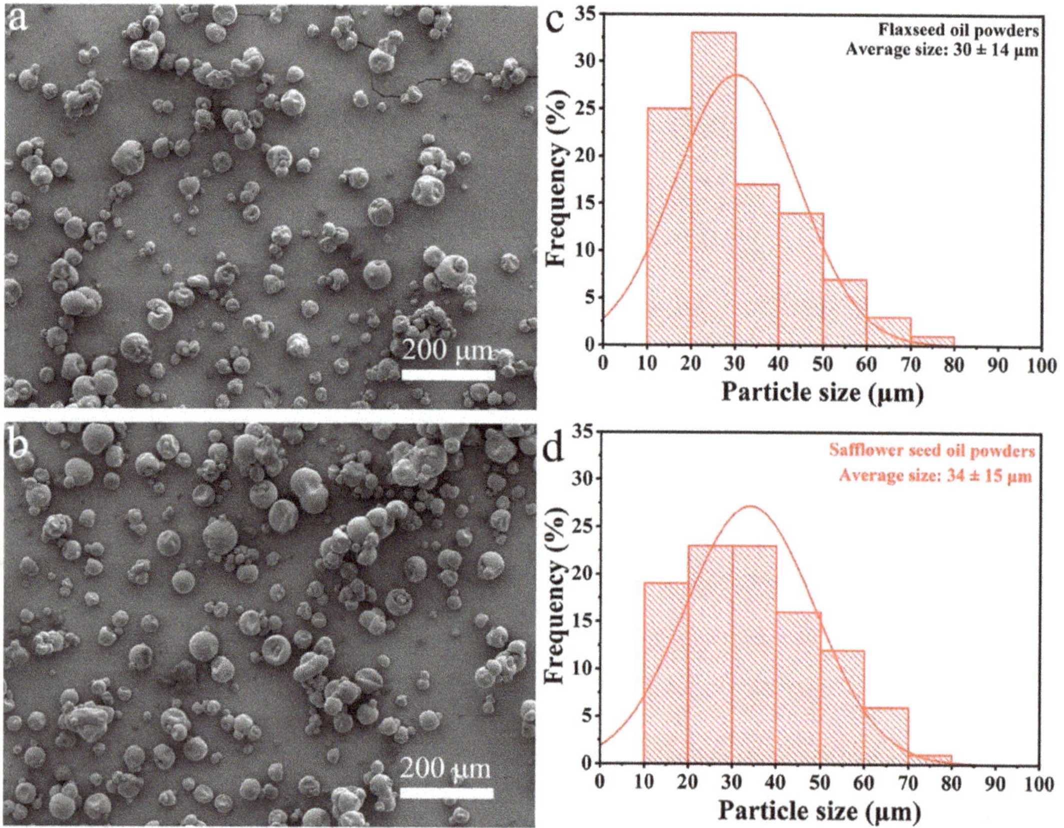

Figure 3. SEM images and corresponding particle size distributions of flaxseed oil powders (**a,c**) and safflower seed oil powders (**b,d**).

3.3.3. Physicochemical Properties of the Oil Powders

Table 5 shows some physicochemical characteristics of oil powders containing flaxseed oil and safflower seed oil prepared through optimal conditions. The moisture contents of flaxseed oil and safflower seed oil powders were 1.31% and 1.43%, respectively, lower than the recommended level of 4~5% for spray-dried powders used for food application [29,30]. Previous studies demonstrated that the high moisture content of powders can cause agglomeration, microbial growth, and accelerated deterioration, which are undesirable for the long-term storage of spray-dried powders [31]. A significantly low level of moisture content would minimize the risks and be beneficial to the storage of the powders.

Table 5. Physicochemical properties of the flaxseed oil and safflower seed oil powders.

Parameters	Flaxseed Oil Powders	Safflower Seed Oil Powders
Moisture content (%)	1.31 ± 0.04	1.43 ± 0.01
Total oil (%)	50.23 ± 0.22	50.20 ± 0.14
Surface oil (%)	0.19 ± 0.06	0.22 ± 0.02
Encapsulation efficiency (%)	99.6	99.6
Acidity value (mg/g)	0.60 ± 0.04	0.52 ± 0.03
Peroxide value (g/100 g)	0.09 ± 0.01	0.08 ± 0.02

The total oil contents were similar for these two kinds of oil powders and close to the theoretical value, with high total oil contents of ~50%. It is challenging to achieve such high oil loading as the increment of oil contents can result in increasing surface oil content, which is an important characteristic of oil powder. The presence of oil on the surface of the powders will make these powders susceptible to oxidation and cause deterioration of the quality of the powders. In this study, the surface oil contents of the flaxseed oil and safflower seed oil powders were only ~0.20%, which is significantly lower than some other spray-dried oil powders with the same level of oil loading. Through these two statistics, the encapsulation efficiencies can be obtained, which represent the proportion of oil that is surrounded by the wall materials and less exposed to the outer environment. Owing to the significantly low surface oil contents, the encapsulation efficiencies were obviously higher (~99.6%) than the other spray-dried oil powders at the same level of oil loading [32].

As the preparation of oil powders must go through a series of processing processes, the acidity values and peroxide values of the final powder products can be affected during this period. Compared with the initial oil, the encapsulated oil powders were more oxidized. The initial acidity values of flaxseed oil and safflower seed oil were 0.56 ± 0.02 and 0.48 ± 0.03 mg/g, and their initial peroxide values were 0.05 ± 0.03 and 0.06 ± 0.02 g/100 g, respectively. After emulsion preparation and spray drying processes, the acidity values of both powders increased slightly to 0.60 ± 0.04 and 0.52 ± 0.03 mg/g, and the peroxide values increased to 0.09 ± 0.01 and 0.08 ± 0.02 g/100 g, respectively. The peroxide values, which show the oxidation levels of oil before and after encapsulation, increased during this period, and this is due to the homogenization and high-temperature treatments during the encapsulation process. These results are in agreement with the observations of some similar studies, which obtained encapsulated oil samples with peroxide values ranging from 0.08 to 0.11 g/100 g (unit conversion from meq/kg oil) after spray drying treatments [5,33].

3.4. Storage Stability

The storage stability of the spray-dried powders was assessed to optimize the raw material compositions and operating conditions. Figure 4 shows some of the physicochemical properties of the flaxseed oil powders and safflower seed oil powders placed in 30 °C and 60% RH conditions over a storage period of 180 days. These powders had high stability in moisture, total oil, and surface oil contents, as the parameters of both oil powders were barely changed during this storage period. Though acidity values and peroxide values of both powders were gradually increased from 0 to 180 days, the acidity values were less than 1 mg/g, and peroxide values were less than 0.25 g/100 g. Compared with studies that had similar initial peroxide values after encapsulation, the initial oxidation levels did not guarantee oil powders with similar quality after long period storage. The peroxide values increased from 0.08 to 0.64 g/100 g after 45 days of storage at 25 °C and increased from 0.02 to 0.19 g/100 g after 30 days of storage at room temperature. Despite the lower storage temperature in these studies, the peroxide values were significantly higher than those observed during the same storage time in our work [33,34]. These results suggest that the proposed encapsulate system has high efficiency in maintaining the stability of flaxseed oil and safflower seed oil over a long period.

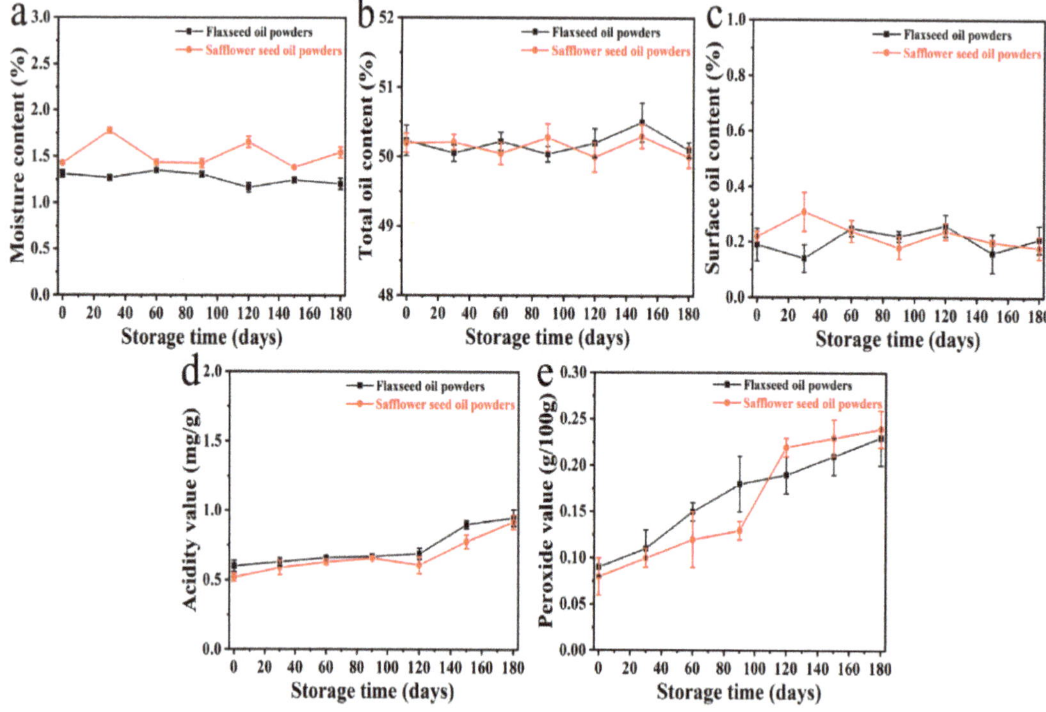

Figure 4. Moisture contents (**a**), total oil contents (**b**), surface oil contents (**c**), acidity values (**d**), and peroxide values (**e**) of the flaxseed oil powders and safflower seed oil powders over a storage period of 180 days.

3.5. Alleviation of DSS-Induced Colitis

An in vivo animal experiment focusing on the effect of oral treatment with the flaxseed oil powder on colitis induced by DSS was performed (Figure 5a). A progressive increase in body weight of the mice in the NC group during the experiment (4.1% at the endpoint) was observed, while DSS mice revealed significant body weight loss (−28.3%) at the end of the experiment. Mice supplied with flaxseed oil had less body weight, and mice administrated with flaxseed oil power had the least body weight loss when compared to DSS-treated colitis mice (Figure 5b). In addition, the colon length of DSS mice was shorter than that of NC mice, and both FS and FSP mice showed improved colon length compared to DSS mice (Figure 5b). Colon tissues in DSS mice exhibited a loss of goblet cells and epithelial cells, severe mucosal epithelial damage, and deformed crypt glands. After the flaxseed oil intervention, there was more complete mucosal structure than in the DSS mice group, particularly in the FSP group (Figure 5d). Several studies have shown that flaxseed oil rich in n-3 PUFA relieved colitis induced by DSS in rats through regulation of the oxidative stress, inflammatory response and intestinal microbiota [35], adjustment of the normal expression of the inflammatory eicosanoids and cytokines including interleukin-6 (IL-6) and NF-κB p65 subunit [36], and suppression of endotoxin-triggered inflammation by blocking the TLR4/MyD88/NF-κB pathway in the liver [37]. These results suggest that treatment with flaxseed oil alleviates DSS-induced colitis symptoms in line with previous reports. It is important to note that the effect of spray-dried flaxseed oil powder was superior to normal liquid flaxseed oil.

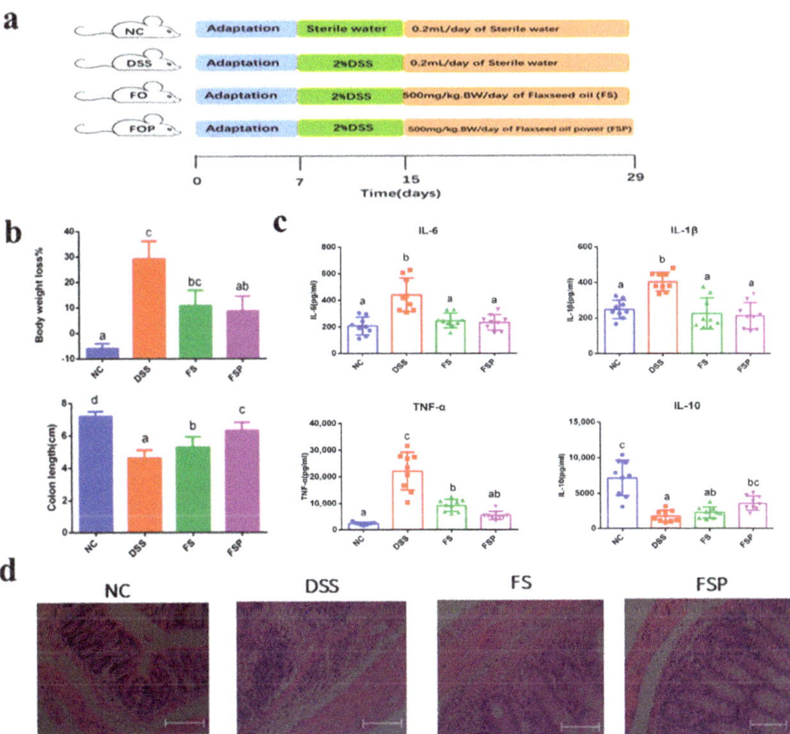

Figure 5. The schematic representation of the experiment (**a**), mice body weight change and colon length (**b**), inflammatory cytokine (IL-6, IL-1β, TNF-α and IL-10) levels in colonic tissue (**c**) and histopathological changes and assessment of colonic tissues from the experiment group (**d**). Groups are labeled from smallest to largest in the order of mean value. Mean values with different letters are significantly different ($p < 0.05$).

IBD is characterized by immune dysfunction, imbalanced cytokine network, and mucosal-associated inflammatory progression [38]. Patients with IBD showed elevated levels of pro-inflammatory cytokines in the intestinal mucosa, such as TNF-α, IL-1β, and IL-6 [39]. Therefore, the serum pro-inflammatory cytokines (IL-1β, TNF-α, IL-6) and anti-inflammatory cytokine (IL-10) levels in the tested mice were determined. As shown in Figure 5c, DSS treatment significantly increased serum TNF-α ($p < 0.01$), IL-1β ($p < 0.05$), and IL-6 ($p < 0.05$) levels compared to controls. The contents of these cytokines were significantly reversed in FS mice and FSP mice. When compared with the mice in the NC group, the cytokine IL-10 level in the colon tissue of the DSS group was significantly reduced ($p < 0.01$). However, compared with the DSS group, only the mice supplemented with flaxseed oil powder (FSP Group) showed an increase in IL-10 levels ($p < 0.05$). Researchers found that n-3 PUFAs could protect mice from intestinal inflammation via downregulating the TLR/NOD pathway [40], in which the latter leads to modulating pro-inflammatory cytokines, such as IL-1β and IL-18 [11]. Our results indicate that flaxseed oil could inhibit the release of cytokines, possibly by suppressing inflammasome activation and maturation and production of IL-1β, just like other oils rich in n-3 PUFA. Moreover, spray drying did not affect the ability of flaxseed oil to regulate inflammatory cytokines and even seemed to increase it.

3.6. Regulation of Gut Microbiota

16S rDNA sequencing technology was used to analyze the microbiota community variations of experimental mice. In terms of α-diversity indexes, Chao1 represents the community richness of gut microbiota, and Simpson, Shannon, and Pielou_e indexes indicate the community diversity. Figure 6a shows that the Shannon and Pielou_e indexes of the DSS group were higher than those of the NC group ($p < 0.05$). After treatment with flaxseed oil, the Chao1 index was relieved. Additionally, no significant difference in the Simpson index of gut microbiota was observed among the experimental groups. As shown in Figure 6b, principal coordinate analysis showed a distinct cluster of mice from the NC and DSS groups. The administration of flaxseed oil and flaxseed oil powder can change the structure of the intestinal flora induced by DSS treatment. The composition of the bacterial community at the phylum and genus levels was identified (Figure 6c,d). At the phylum level, the dominant taxa of the gut microbiota in each group were *Firmicutes*, *Bacteroidetes*, and *Proteobacteria*, accounting for ~95.5–99.8%. As many reports concluded, *Firmicutes* and *Bacteroidetes* are the absolute dominant phylum in both human and animal intestinal communities [42,43], and these two main communities are associated with host energy metabolism homeostasis [44]. In this study, *Firmicutes* and *Bacteroidetes* occupied dominance in the gut microbiota was not changed in the DSS-induced mice, which led us to pay more attention to the changes at genus levels or at other phyla. This result was consistent with our previous studies [17,45] and other research work [46].

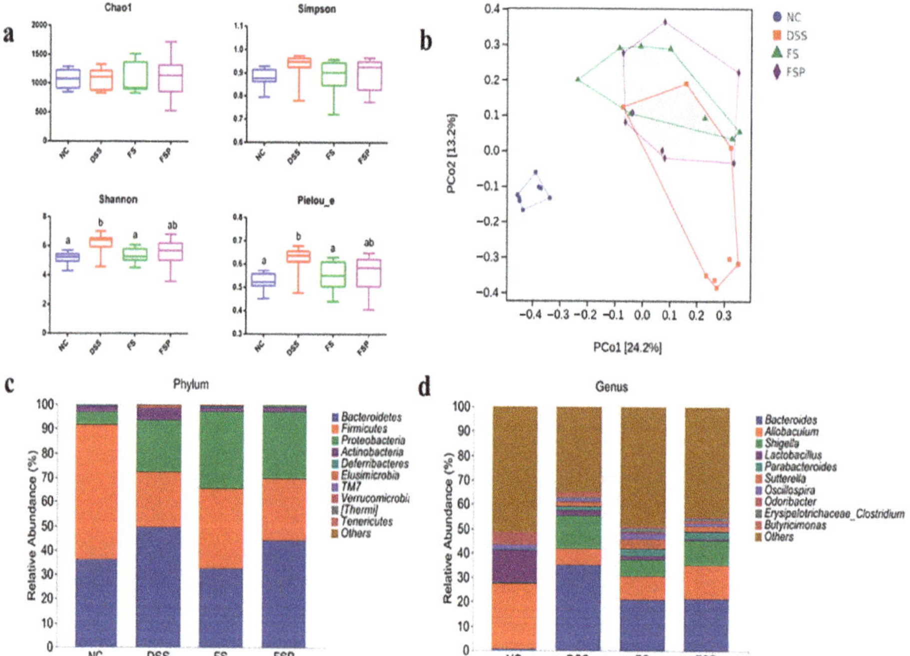

Figure 6. The alpha diversity (Chao1, Shannon, Simpson, and Pielou_e indexes) in different groups (**a**), PCoA plot of unweighted UniFrac distances of beta diversity (**b**), and community bar plots of phylum and genus level (**c,d**). Groups are labeled from smallest to largest in the order of mean value. Mean values with different letters are significantly different ($p < 0.05$).

At the genus level, the dominant bacterial genera detected in the tested group were different. *Allobaculum*, *Lactobacillus*, and *Odoribacter* were the top three taxa in the NC group, whereas *Bacteroides*, *Allobaculum*, and *Shigella* were the top three taxa in DSS, FS, and FSP groups. In addition, Figure 6d shows that the abundance of *Allobaculum* was significantly lower in the DSS group than in the NC group ($p < 0.05$), while supplementation with flaxseed oil powder (FSP group) but not flaxseed oil (FS group) increased the proportion of *Allobaculum* in DSS-induced mice. Additionally, in the FSP group, the relative abundance of *Bacteroides* significantly declined compared with the DSS group ($p < 0.05$). *Allobaculum* is considered a beneficial bacteria since it could inhibit mice weight gain by interfering with energy metabolism [47]. As a potential human intestinal mucin degrader, it is also reported to protect intestinal barrier function by producing short-chain fatty acids (SCFAs) [48]. On the other hand, pathogenic bacteria such as *Bacteroides* are reported to be negatively related to some special SCFAs [49]. Here, in the mice supplemented with flaxseed oil powder but not flaxseed oil, the drop of beneficial *Allobaculum* promoted the reparation of the broken intestinal barrier induced by DSS and inhibited the proliferation of harmful genera such as *Bacteroides* via the production of SCFAs.

LDA-LefSe was used to analyze the prominent taxa in each group and reveal changes in gut microbiota composition. According to Figure 7a,b, it is possible to observe that *Allobaculum*, *Lactobacillus*, and *Odoribacter* were enriched in the NC group, and *Bacteroides* were enriched in the DSS group. *Clostridium* and *Shigella* were enriched in the FS and FSP groups, respectively. Matastats were carried out to identify the significantly different taxa between each two group pairs (Figure 7c). Mice supplemented with flaxseed oil promoted more *Ruminococcus* and *Anaerotruncus* from *Firmicutes* phylum and less *Bacteroides* from *Bacteroidetes* phylum compared to untreated DSS-colitis mice. In addition to this effect, consumption of flaxseed oil powder induced more *Allobaculum* and *Odoribacter* after DSS treatment. These results matched the changed genus shown in Figure 6c,d, which suggests that FSP impacts the intestinal microbiome at different taxonomic levels by improving the proportion of SCFAs producers, *Allobaculum* and *Odoribacter*, and inhibiting the growth of bacterial species associated with diseases and inflammation such as *Bacteroides and Shigella*.

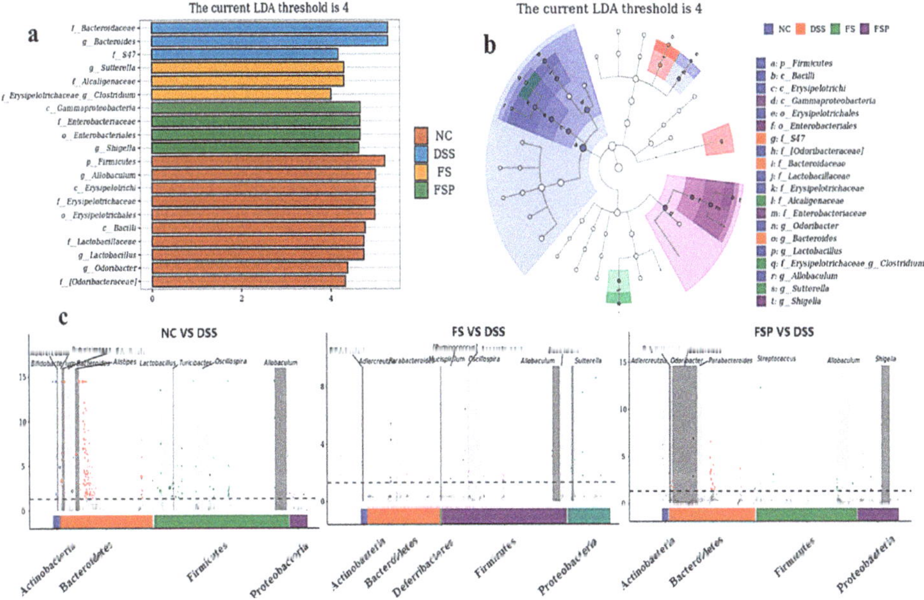

Figure 7. Taxonomic histogram (**a**) and cladogram (**b**) generated by LEfSe analysis and matastats analysis (**c**).

Intestinal dysbiosis induced by inflammatory bowel disease (IBD) was associated with intestinal barrier function via excessive inflammatory response [50]. Considering the fact that many bacterial species are closely associated with human diseases, the inner mucus layer was proved an important niche [51]. As discussed before, *Allobaculum* was known to be related to the host epithelium, which could maintain intestinal barrier function by secreting acetic acid [18] and 5-HT [52]. Here, both flaxseed oil and flaxseed oil powder displayed a positive regulatory effect on the bacterial community in colitis mice. Mice that accepted the flaxseed oil powder treatment had a remarkably increased abundance of *Allobaculum* and decreased proportion of infectious diseases associated the *Shigella* genus, which might improve the integrity and function of the intestinal barrier in colitis mice. However, the deeper mechanism between these regulated genera and the improvement of the intestinal barrier by the plant oil powder consumption needs to be further investigated.

4. Conclusions

This study systematically investigated the preparation of spray-dried plant-based oil powders that contain an oil loading of 50%. The optimal conditions, including raw material compositions, emulsification conditions, and spray drying conditions, were determined through designing a series of orthogonal experiments and single-factor experiments. The surface oil content was lower than 0.19%, and the encapsulation efficiency was significantly high (99%). The oil powders exhibited long-term stability under 30 °C and 60% relative humidity. The encapsulated flaxseed oil powders were shown to alleviate a series of DSS-caused inflammatory symptoms and modulate the intestinal microbiota in mouse models. Notably, flaxseed oil powder exhibited enhanced anti-colitis activity by improving the physiological indications and gut dysbiosis compared with lipid flaxseed oil.

Supplementary Materials: The following supporting information can be downloaded at: https://www.mdpi.com/article/10.3390/foods11192993/s1, Table S1: Fatty acid compositions of flaxseed oil and safflower seed oil; Table S2 and S3: Evaluation indices of the encapsulation efficiency and encapsulation yield of the orthogonal L_{16} $(4)^4$ experiments.

Author Contributions: Conceptualization, H.T. and C.Y.; methodology, H.Z.; software, Z.Q.; investigation and resources, X.W.; data curation, Z.X. and C.Y.; writing—original draft preparation, H.Z. and C.Y.; writing—review and editing, Z.X.; project administration and funding acquisition, F.H. All authors have read and agreed to the published version of the manuscript.

Funding: This work was supported by the Shandong Provincial Key Research and Development Program (2019GHZ031), the Central Public-interest Scientific Institution Basal Research Fund (No. Y2002LM04), the Earmarked Fund for China Agriculture Research System (CARS-14), the Taishan Scholar Project (to Fenghong Huang), and the Knowledge Innovation Program of Wuhan -Basic Research (3562).

Institutional Review Board Statement: This animal experiment was approved by the Animal Ethics Committee of Huazhong Agricultural University (Permission HZAUMO-2021-0170).

Informed Consent Statement: Not applicable.

Data Availability Statement: The data supporting the findings of this study are available within the article.

Conflicts of Interest: The authors declare no conflict of interest.

References

1. Simopoulos, A. An Increase in the Omega-6/Omega-3 Fatty Acid Ratio Increases the Risk for Obesity. *Nutrients* **2016**, *8*, 128. [CrossRef] [PubMed]
2. Bakry, A.M.; Fang, Z.; Ni, Y.Z.; Cheng, H.; Chen, Y.Q.; Liang, L. Stability of tuna oil and tuna oil/peppermint oil blend microencapsulated using whey protein isolate in combination with carboxymethyl cellulose or pullulan. *Food Hydrocolloid* **2016**, *60*, 559–571. [CrossRef]
3. Fioramonti, S.A.; Stepanic, E.M.; Tibaldo, A.M.; Pavon, Y.L.; Santiago, L.G. Spray dried flaxseed oil powdered microcapsules obtained using milk whey proteins-alginate double layer emulsions. *Food Res. Int.* **2019**, *119*, 931–940. [CrossRef] [PubMed]
4. Polavarapu, S.; Oliver, C.M.; Ajlouni, S.; Augustin, M.A. Physicochemical characterisation and oxidative stability of fish oil and fish oil–extra virgin olive oil microencapsulated by sugar beet pectin. *Food Chem.* **2011**, *127*, 1694–1705. [CrossRef]

5.   Carneiro, H.C.F.; Tonon, R.V.; Grosso, C.R.F.; Hubinger, M.D. Encapsulation efficiency and oxidative stability of flaxseed oil microencapsulated by spray drying using different combinations of wall materials. *J. Food Eng.* **2013**, *115*, 443–451. [CrossRef]

6.   Ahn, J.H.; Kim, Y.P.; Lee, Y.M.; Seo, E.M.; Lee, K.W.; Kim, H.S. Optimization of microencapsulation of seed oil by response surface methodology. *Food Chem.* **2008**, *107*, 98–105. [CrossRef]

7.   Bastioglu, A.Z.; Koc, M.; Yalcin, B.; Ertekin, F.K.; Otles, S. Storage characteristics of microencapsulated extra virgin olive oil powder: Physical and chemical properties. *J. Food Meas. Charact.* **2017**, *11*, 1210–1226. [CrossRef]

8.   Venugopalan, V.K.; Gopakumar, L.R.; Kumaran, A.K.; Chatterjee, N.S.; Soman, V.; Peeralil, S.; Mathew, S.; McClements, D.J.; Nagarajarao, R.C. Encapsulation and Protection of Omega-3-Rich Fish Oils Using Food-Grade Delivery Systems. *Foods* **2021**, *10*, 1566. [CrossRef]

9.   Kouame, K.J.E.P.; Bora, A.F.M.; Li, X.D.; Sun, Y.; Liu, L. Novel trends and opportunities for microencapsulation of flaxseed oil in foods: A review. *J. Funct. Foods* **2021**, *87*, 104812. [CrossRef]

10.   Perez, E.M.D.; De Alencar, N.M.N.; De Figueiredo, I.S.T.; Aragao, K.S.; Gaban, S.V.F. Effect of safflower oil (*Carthamus tinctorius* L.) supplementation in the abdominal adipose tissues and body weight of male Wistar rats undergoing exercise training. *Food Chem. Mol. Sci.* **2022**, *4*, 100083. [CrossRef]

11.   Stefani, F.D.; de Campo, C.; Paese, K.; Guterres, S.S.; Costa, T.M.H.; Flores, S.H. Nanoencapsulation of linseed oil with chia mucilage as structuring material: Characterization, stability and enrichment of orange juice. *Food Res. Int.* **2019**, *120*, 872–879. [CrossRef]

12.   *National Standard GB/T 5009.6–2016*; Determination of Fat in Food. China Standard Press: Beijing, China, 2016.

13.   *Chinese Aquaculture Industry Standard SC/T 3505-2006*; Microcapsules of Fish Oil. Ministry of Agriculture of PRC: Beijing, China, 2006.

14.   *National Standard GB/T 5009.3–2016*; Determination of Moisture Content in Food. China Standard Press: Beijing, China, 2016.

15.   *National Standard GB/T 5009.229–2016*; Determination of Acid Value in Food. China Standard Press: Beijing, China, 2016.

16.   *National Standard GB/T 5009.227–2016*; Determination of Peroxide Value in Food. China Standard Press: Beijing, China, 2016.

17.   Xu, Z.X.; Tang, H.; Huang, F.H.; Qiao, Z.X.; Wang, X.; Yang, C.; Deng, Q.C. Algal Oil Rich in n-3 PUFA Alleviates DSS-Induced Colitis via Regulation of Gut Microbiota and Restoration of Intestinal Barrier. *Front. Microbiol.* **2020**, *11*, 615404. [CrossRef] [PubMed]

18.   Yang, C.; Qiao, Z.X.; Xu, Z.X.; Wang, X.; Deng, Q.C.; Chen, W.C.; Huang, F.H. Algal Oil Rich in Docosahexaenoic Acid Alleviates Intestinal Inflammation Induced by Antibiotics Associated with the Modulation of the Gut Microbiome and Metabolome. *J. Agr. Food Chem.* **2021**, *69*, 9124–9136. [CrossRef]

19.   Mohseni, F.; Goli, S. Encapsulation of flaxseed oil in the tertiary conjugate of oxidized tannic acid-gelatin and flaxseed (*Linum usitatissimum*) mucilage. *Int. J. Biol. Macromol.* **2019**, *140*, 959–964. [CrossRef] [PubMed]

20.   Kaushik, P.; Dowling, K.; Mcknight, S.; Barrow, C.J.; Adhikari, B. Microencapsulation of flaxseed oil in flaxseed protein and flaxseed gum complex coacervates. *Food Res. Int.* **2016**, *86*, 1–8. [CrossRef]

21.   Lai, Q.D.; Doan, N.T.T.; Nguyen, T.T.T. Influence of Wall Materials and Homogenization Pressure on Microencapsulation of Rice Bran Oil. *Food Bioprocess. Tech.* **2021**, *14*, 1885–1896. [CrossRef]

22.   Pangeni, R.; Sharma, S.; Mustafa, G.; Ali, J.; Baboota, S. Vitamin E loaded resveratrol nanoemulsion for brain targeting for the treatment of Parkinson's disease by reducing oxidative stress. *Nanotechnology* **2014**, *25*, 485102. [CrossRef]

23.   Aghbashlo, M.; Mobli, H.; Madadlou, A.; Rafiee, S. Influence of Wall Material and Inlet Drying Air Temperature on the Microencapsulation of Fish Oil by Spray Drying. *Food Bioprocess. Tech.* **2013**, *6*, 1561–1569. [CrossRef]

24.   Nguyen, P.T.N.; Nguyen, H.T.A.; Hoang, Q.B.; Nguyen, T.D.P.; Nguyen, T.V.; Mai, H.C. Influence of spray drying parameters on the physicochemical characteristics of microencapsulated orange (*Citrus sinensis* L.) essential oil. *Mater. Today Proc.* **2022**, *60*, 2026–2033. [CrossRef]

25.   Delfanian, M.; Razavi, S.; Haddad Khodaparast, M.; Esmaeilzadeh Kenari, R.; Golmohammadzadeh, S. Influence of main emulsion components on the physicochemical and functional properties of W/O/W nano-emulsion: Effect of polyphenols, Hi-Cap, basil seed gum, soy and whey protein isolates. *Food Res. Int.* **2018**, *108*, 136–143. [CrossRef]

26.   Carvalho, A.G.S.; Silva, V.M.; Hubinger, M.D. Microencapsulation by spray drying of emulsified green coffee oil with two-layered membranes. *Food Res. Int.* **2014**, *61*, 236–245. [CrossRef]

27.   Edris, A.; Kalemba, D.; Adamiec, J.; Piątkowski, M. Microencapsulation of Nigella sativa oleoresin by spray drying for food and nutraceutical applications. *Food Chem.* **2016**, *204*, 326–333. [CrossRef] [PubMed]

28.   Li, X.; Wu, M.; Xiao, M.; Lu, S.; Wang, Z.; Yao, J.; Yang, L. Microencapsulated β-carotene preparation using different drying treatments. *J. Zhejiang Univ Sc. B* **2019**, *20*, 901–909. [CrossRef] [PubMed]

29.   Chew, S.C.; Tan, C.P.; Nyam, K.L. Microencapsulation of refined kenaf (*Hibiscus cannabinus* L.) seed oil by spray drying using β-cyclodextrin/gum arabic/sodium caseinate. *J. Food Eng.* **2018**, *237*, 78–85. [CrossRef]

30.   Patel, S.S.; Pushpadass, H.A.; Franklin, M.E.E.; Battula, S.N.; Vellingiri, P. Microencapsulation of curcumin by spray drying: Characterization and fortification of milk. *J. Food Sci. Tech. Mys.* **2022**, *59*, 1326–1340. [CrossRef]

31.   Polekkad, A.; Franklin, M.E.E.; Pushpadass, H.A.; Battula, S.N.; Rao, S.B.N.; Pal, D.T. Microencapsulation of zinc by spray-drying: Characterisation and fortification. *Powder Technol.* **2021**, *381*, 1–16. [CrossRef]

32.   Wang, B.; Adhikari, B.; Barrow, C.J. Optimisation of the microencapsulation of tuna oil in gelatin-sodium hexametaphosphate using complex coacervation. *Food Chem.* **2014**, *158*, 358–365. [CrossRef]

33. Elik, A.; Yanik, D.K.; Gogus, F. A comparative study of encapsulation of carotenoid enriched-flaxseed oil and flaxseed oil by spray freeze-drying and spray drying techniques. *Lwt Food Sci. Technol.* **2021**, *143*, 111153. [CrossRef]
34. Avramenko, N.A.; Chang, C.; Low, N.H.; Nickerson, M.T. Encapsulation of flaxseed oil within native and modified lentil protein-based microcapsules. *Food Res. Int.* **2016**, *81*, 17–24. [CrossRef]
35. Zhou, Q.; Ma, L.; Zhao, W.Y.; Zhao, W.; Han, X.; Niu, J.H.; Li, R.B.; Zhao, C.H. Flaxseed oil alleviates dextran sulphate sodium-induced ulcerative colitis in rats. *J. Funct. Foods* **2020**, *64*, 103602. [CrossRef]
36. Jangale, N.M.; Devarshi, P.P.; Bansode, S.B.; Kulkarni, M.J.; Harsulkar, A.M. Dietary flaxseed oil and fish oil ameliorates renal oxidative stress, protein glycation, and inflammation in streptozotocin-nicotinamide-induced diabetic rats. *J. Physiol. Biochem.* **2016**, *72*, 327–336. [CrossRef] [PubMed]
37. Wang, M.; Zhang, X.J.; Yan, C.; He, C.; Li, P.; Chen, M.; Su, H.; Wan, J.B. Preventive effect of α-linolenic acid-rich flaxseed oil against ethanol-induced liver injury is associated with ameliorating gut-derived endotoxin-mediated inflammation in mice. *J. Funct. Foods* **2016**, *23*, 532–541. [CrossRef]
38. Petrey, A.C.; de la Motte, C.A. The extracellular matrix in IBD: A dynamic mediator of inflammation. *Curr. Opin. Gastroen.* **2017**, *33*, 234–238. [CrossRef] [PubMed]
39. Loddo, I.; Romano, C. Inflammatory bowel disease: Genetics, epigenetics, and pathogenesis. *Front. Immunol.* **2015**, *6*, 551. [CrossRef]
40. Calder, P.C.; Ahluwalia, N.; Albers, R.; Bosco, N.; Bourdet-Sicard, R.; Haller, D.; Holgate, S.T.; Jonsson, L.S.; Latulippe, M.E.; Marcos, A.; et al. A Consideration of Biomarkers to Be Used for Evaluation of Inflammation in Human Nutritional Studies. *Ann. Nutr. Metab.* **2013**, *63*, 15. [CrossRef]
41. Nordlander, S.; Pott, J.; Maloy, K.J. NLRC4 expression in intestinal epithelial cells mediates protection against an enteric pathogen. *Mucosal. Immunol.* **2014**, *7*, 775–785. [CrossRef]
42. Weingarden, A.R.; Vaughn, B.P. Intestinal microbiota, fecal microbiota transplantation, and inflammatory bowel disease. *Gut Microbes* **2017**, *8*, 238–252. [CrossRef]
43. Rashidi, A.; Ebadi, M.; Rehman, T.U.; Elhusseini, H.; Nalluri, H.; Kaiser, T.; Holtan, S.G.; Khoruts, A.; Weisdorf, D.J.; Staley, C. Gut microbiota response to antibiotics is personalized and depends on baseline microbiota. *Microbiome* **2021**, *9*, 211. [CrossRef]
44. Musso, G.; Gambino, R.; Cassader, M. Gut microbiota as a regulator of energy homeostasis and ectopic fat deposition: Mechanisms and implications for metabolic disorders. *Curr. Opin. Lipidol.* **2010**, *21*, 76–83. [CrossRef]
45. Xu, Z.X.; Chen, W.C.; Deng, Q.C.; Huang, Q.D.; Wang, X.; Yang, C.; Huang, F.H. Flaxseed oligosaccharides alleviate DSS-induced colitis through modulation of gut microbiota and repair of the intestinal barrier in mice. *Food Funct.* **2020**, *11*, 8077–8088. [CrossRef]
46. Zhang, Y.; Tan, L.X.; Li, C.; Wu, H.; Ran, D.; Zhang, Z.Y. Sulforaphane alter the microbiota and mitigate colitis severity on mice ulcerative colitis induced by DSS. *Amb. Express* **2020**, *10*, 119. [CrossRef] [PubMed]
47. Cox, L.M.; Yamanishi, S.; Sohn, J.; Alekseyenko, A.V.; Leung, J.M.; Cho, I.; Kim, S.G.; Li, H.; Gao, Z.; Mahana, D.; et al. Altering the Intestinal Microbiota during a Critical Developmental Window Has Lasting Metabolic Consequences. *Cell* **2014**, *158*, 705–721. [CrossRef] [PubMed]
48. Ma, Q.T.; Li, Y.Q.; Wang, J.K.; Li, P.F.; Duan, Y.H.; Dai, H.Y.; An, Y.C.; Cheng, L.; Wang, T.S.; Wang, C.G.; et al. Investigation of gut microbiome changes in type 1 diabetic mellitus rats based on high-throughput sequencing. *Biomed. Pharm.* **2020**, *124*, 109873. [CrossRef] [PubMed]
49. Feng, G.L.; Mikkelsen, D.; Hoedt, E.C.; Williams, B.A.; Flanagan, B.M.; Morrison, M.; Gidley, M.J. In vitro fermentation outcomes of arabinoxylan and galactoxyloglucan depend on fecal inoculum more than substrate chemistry. *Food Funct.* **2020**, *11*, 7892–7904. [CrossRef]
50. Fry, L. Microbiome of chronic plaque psoriasis. In *The Human Microbiota and Chronic Disease Dysbiosis as a Cause of Human Pathology*, 1st ed.; Nibali, L., Henderson, B., Eds.; Wiley: New York, NY, USA, 2016; pp. 327–337.
51. van Muijlwijk, G.H.; van Mierlo, G.; Jansen, P.W.T.C.; Vermeulen, M.; Bleumink-Pluym, N.M.C.; Palm, N.W.; van Putten, J.P.M.; de Zoete, M.R. Identification of Allobaculum mucolyticum as a novel human intestinal mucin degrader. *Gut Microbes* **2021**, *13*, 1966278. [CrossRef] [PubMed]
52. Wu, M.; Tian, T.; Mao, Q.; Zou, T.; Zhou, C.J.; Xie, J.; Chen, J.J. Associations between disordered gut microbiota and changes of neurotransmitters and short-chain fatty acids in depressed mice. *Transl. Psychiat.* **2020**, *10*, 350. [CrossRef] [PubMed]

Article

Comparative Composition Structure and Selected Techno-Functional Elucidation of Flaxseed Protein Fractions

Xiaopeng Qin [1], Linbo Li [1], Xiao Yu [1,*], Qianchun Deng [2,*], Qisen Xiang [1] and Yingying Zhu [1]

[1] College of Food and Bioengineering, Zhengzhou University of Light Industry, Zhengzhou 450002, China; 15083059355@163.com (X.Q.); 15517587171@163.com (L.L.); xiangqisen2006@163.com (Q.X.); zhuying881020@163.com (Y.Z.)

[2] Oil Crops Research Institute, Chinese Academy of Agricultural Sciences, Wuhan 430062, China

* Correspondence: yuxiao@zzuli.edu.cn (X.Y.); dengqianchun@caas.cn (Q.D.)

Abstract: This study aimed to comparatively elucidate the composition structure and techno-functionality of flaxseed protein isolate (FPI), globulin (FG), and albumin (FA) fractions. The results showed that FA possessed smaller particle dimensions and superior protein solubility compared to that of FG ($p < 0.05$) due to the lower molecular weight and hydrophobicity. FA and FG manifested lamellar structure and nearly spherical morphology, respectively, whereas FPI exhibited small lamellar strip structure packed by the blurring spheres. The Far-UV CD, FTIR spectrum, and intrinsic fluorescence confirmed more flexible conformation of FA than that of FG, followed by FPI. The preferential retention of free phenolic acids was observed for FA, leading to excellent antioxidant activities compared with that of FG in FPI ($p < 0.05$). FA contributed to the foaming properties of FPI, relying on the earlier interfacial adsorption and higher viscoelastic properties. FA displayed favorable emulsifying capacity but inferior stability due to the limited interfacial adsorption and deformation, as well as loose/porous interface. By comparison, an interlayer anchoring but no direct interface coating was observed for lipid droplets constructed by FG, thereby leading to preferable emulsion stability. However, FPI produced lipid droplets with dense interface owing to the effective migration of FA and FG from bulk phase, concomitant with the easy flocculation and coalescence. Thus, the techno-functionality of flaxseed protein could be tailed by modulating the retention of albumin fraction and specific phenolic acids.

Keywords: flaxseed protein fractions; component structure; techno-functionality; phenolic acids; interfacial behavior

Citation: Qin, X.; Li, L.; Yu, X.; Deng, Q.; Xiang, Q.; Zhu, Y. Comparative Composition Structure and Selected Techno-Functional Elucidation of Flaxseed Protein Fractions. *Foods* **2022**, *11*, 1820. https://doi.org/10.3390/foods11131820

Academic Editor: Marie Alminger

Received: 8 May 2022
Accepted: 16 June 2022
Published: 21 June 2022

Publisher's Note: MDPI stays neutral with regard to jurisdictional claims in published maps and institutional affiliations.

1. Introduction

Considering milk allergies, lactose intolerance, vegetarians, and vegans, the transition from diets based on animal proteins towards diets primarily based on plant proteins has been occurring. Thus, the plant-based proteins with high functionality have been screened with respect to stabilize the multiphase food systems. Besides the bioactive α-linolenic acid, gum polysaccharides, lignans, and phenolic acids, flaxseed also contains appreciable amounts of protein with high nutritional quality [1]. Importantly, flaxseed protein and its hydrolysates possess desirable anti-hypertensive, anti-bacterial, and anti-diabetic activities, and also mitigated against ethanol or lead induced hepatotoxicity [2–4]. The techno-functionality of flaxseed protein, including water/oil-holding capacity, foaming, and emulsifying properties, had also been explored in previous studies [5,6]. Indeed, the naturally occurring gum polysaccharides could largely affect the hydrodynamic, foaming, and emulsifying properties of flaxseed protein obtained from whole or defatted flaxseed meal due to the noncovalent interaction between them [7,8]. Thus, the actual responsiveness between the molecular structure, spatial conformation, and techno-functionality of flaxseed protein and its fractions was still blurred [9]. This was particularly unfavorable for the optimal application of flaxseed-derived proteins in food-grade colloidal systems with desirable appearance or texture.

Whole flaxseed has an average protein content of 23%, containing approximately 70~85% of salt-soluble globulin and 15~30% of water-soluble albumin fractions, relying on the extraction substrates and methods [10–12]. The multiple polypeptides with higher molecular weight (Mw), hydrophobic and branched-chain amino acids were observed for globulin, whereas a single polypeptide with abundant negatively charged and sulfur-containing amino acids was prominent in albumin fraction [2]. As previously reported, flaxseed globulin and albumin fractions exhibited discrepant foam/lipid droplet formation capacity and stability [6,13]. Theoretically, protein fractions could spontaneously form the air–water or oil–water interfaces following the interfacial adsorption, structural re-organization, and sequential membrane expansion behavior against foam/lipid droplet instability [14]. Unfortunately, it was still undefined as to how the heterogeneous composition and structure between globulin and albumin fractions in flaxseed with stripping of gum polysaccharides affected their stabilizing and destabilizing properties of foam and lipid droplets based on the interfacial behaviors.

Increasing evidence had revealed that the phenolic compounds could naturally exist in protein fractions extracted from flaxseed due to the ionic bond, hydrogen bond, or hydrophobic interactions with phenolic hydroxyl and carboxyl groups [15]. The retention for phenolic compounds, including lignans and phenolic acids, could further affect the water-holding capacity or gel properties of flaxseed protein prepared by micellization and/or isoelectric precipitation [16]. Thus, the exogenous addition flaxseed polyphenols via simple complex or covalent binding had been conducted to positively regulate the antioxidant and emulsifying properties of flaxseed protein [17,18]. In particular, the free phenolic acids might be coexisted with globulin and albumin in protein bodies or embedded into membrane proteins of oil bodies against the in situ lipid oxidation of α-linolenic acid in flaxseed [19]. Actually, due to the distinct composition structure and spatial conformation between globulin and albumin, how the specific accumulation of phenolic compounds and ascending antioxidant potential occur was still undefined following the sequential extraction of flaxseed protein fractions [20]. Based on the above, the aim of this work was to conduct a comparative techno-functional elucidation, including the antioxidant, foaming, and emulsifying properties of flaxseed protein and its fraction globulin and albumin, focusing on their component structure, retention of phenolic acids, interfacial, and rheological behavior. This discrepancy in specific techno-functionality for flaxseed protein fractions could achieve the tailed application in foam- and emulsion-type foods, such as whipped cream, ice cream, mayonnaise, dressing, margarine, etc.

2. Materials and Methods

2.1. Chemicals and Materials

Dehulled flaxseed (variety: longya 13 #) was provided by Gansu academy of agricultural sciences (Lanzhou, China). Flaxseed oil was purchased from Hongjingyuan oil Co., Ltd. (Xilingol, China). Folin–ciocalteu reagent, rutin, and amino acid standards were purchased from were purchased from Beijing Solarbio Sciences and Technology Co., Ltd. (Beijing, China). 2,4,6-Tris(2-pyridyl)-striazine (TPTZ, 99%) and 2,2′-diphenyl-1-picrylhydrazyl radical (DPPH, 95%) were obtained from Shanghai Beyotime Biotechnology Co., Ltd. (Shanghai, China). Vanillin, coumarin, p-coumaric, gallic, protocatechuic, vanillic, caffeic, syringic, sinapic, ferulic, sallcylic, and cinnamic acids were provided by Shanghai Yuanye Biological Technology Co., Ltd. (Shanghai, China). Micro BCA Protein Assay Kit was obtained from Beyotime Biotechnology Co., Ltd. (Shanghai, China). Other analytical grade reagents were purchased from Sinopharm Chemical Reagent Co., Ltd. (Beijing, China).

2.2. Extraction and Isolation of Protein Fractions from Dehulled Flaxseed

The dehulled flaxseed was ground into fine powder using a coffee grinder, defatted twice using hexane at a ratio of 1:4 (*w*/*v*) for 2 h, and filtered using a filter paper disk. After being air-dried in a fume hood overnight, part of the defatted powder was dispersed in deionized water at a ratio of 1:15 (w/v), adjusted to pH 8.5 using 0.5 M NaOH solution, and stirred at 500 rpm for 2 h at 25 °C. Then, the supernatant was obtained following centrifugation at 8000× g for 20 min, adjusted to pH value of 3.8 with 0.5 M HCl to precipitate protein [21]. After centrifugation at 8000× g for 20 min again, the protein precipitation was dispersed in deionized water and adjusted to pH value of 6.8. In order to remove the non-bound salt ions, the above protein dispersion was subjected to dialysis using a dialysis bag with molecular cut off weight of 8–10 kDa against deionized water for 24 h at 4 °C, and then freeze-dried to obtain flaxseed protein isolate (FPI). The remaining part of defatted powder was dispersed in deionized water at 1:15 ratio (w/v) and stirred at 500 rpm for 2 h at 25 °C. After centrifugation at 8000× g for 20 min, collect the supernatant and repeat this step until the supernatant becomes transparent. The FA fraction was obtained after freeze-drying. The resulting precipitation was mixed, with 0.5 M NaCl at 1:10 ratio, stirred for 2 h at 25 °C and centrifuged at 8000× g for 20 min. The supernatant was subjected to dialysis using a dialysis bag with molecular cut off weight of 8–10 kDa, and then freeze-dried to obtain flaxseed globulin (FG) fraction [6].

2.3. Analysis of Proximate Composition of Flaxseed Protein Fractions

The contents of crude protein in flaxseed protein fractions were determined by automatic kjeldahl analyzer and calculated with a conversion factor of 6.25 [5]. For the amino acid composition, the protein samples were hydrolyzed with 6 M HCl for 24 h at 110 °C in a sealed tube and analyzed using a Biochrom 30 automatic amino acid analyzer. The individual amino acid was identified, and the results were expressed as the percentage of moisture content. The levels of total lipids, ash, and moisture were determined according to the AOAC official methods 920.85, 923.03, and 930.15, respectively. The contents of total sugars were obtained by the subtraction of crude protein, total lipids, ash, and moisture in individual protein samples.

2.4. Analysis of Physicochemical Properties of Flaxseed Protein Fractions

The mean particle size and zeta potential of flaxseed protein fractions in 50 mM phosphate buffer solution (PBS, pH 6.8) were evaluated using a ZetaSizer Nano-ZS 90 (Malvern Instruments Ltd., Worcestershire, UK). For analysis of protein solubility, the protein samples were dispersed in 50 mM PBS (pH 6.8) at a concentration of 0.1% (w/v), continuously stirred for 2 h at 25 °C, and centrifugated at 6000× g for 20 min. The protein content in supernatant was determined by the MicroBCA assay using bovine serum albumin as the standard and dividing the total protein content in the supernatant by the protein content in the initial sample, which was used to calculate the percentage of the soluble protein relative to the total protein content [22]. For analysis of surface hydrophobicity, a series of dilutions was achieved by mixing flaxseed protein dispersion (0.1%, w/v) into 50 mM PBS (pH 6.8) ranging from 0.002% to 0.01% (w/v). Then, 20 μL of the 8-anilino-1-naphthalene sulphonic acid (ANS) solution (8.0 mM) was added into 4 mL of each dilution and mixed thoroughly using a vortex mixer. After reacting for 10 min in the dark at 25 °C, the fluorescence intensity (FI) was determined using a F-7000 fluorescence spectrophotometer (Hitachi, Tokyo, Japan) with excitation and emission wavelength at 390 nm and 480 nm, respectively. The surface hydrophobicity (H_0) value was expressed by the initial slope of FI against sample concentration plot [23].

2.5. The Structural Properties of Flaxseed Protein Fractions

2.5.1. Scanning Electron Microscope

The surface morphology of flaxseed protein fractions was observed using high-resolution field emission scanning electron microscopy (FE-SEM) Regulus 8100 (Hitachi, Tokyo, Japan). Briefly, the power samples were attached to a sample stub with double-sided sticky tape, sputter coated with gold using a polaron sputter coater, and visualized at an accelerated voltage of 3.0 kV with magnification 10,000×.

2.5.2. Sodium Dodecyl Sulphate-Polyacrylamide Gel Electrophoresis (SDS-PAGE)

The aqueous dispersions of flaxseed protein fractions with equal protein amounts (1.0 mg/mL) were mixed with a reducing ($\times 4$) loading buffer, heated at 95 °C for 5 min, and separated by 10% sodium dodecyl sulphate polyacrylamide gel electrophoresis (SDS-PAGE). The gels were stained with 0.1% of coomassie brilliant blue solution under gentle shaking, washed with destaining solution, and imaged on a ChemiDoc XRS+ System (Bio-Rad, Hercules, CA, USA).

2.5.3. Intrinsic Fluorescence Spectra

The flaxseed protein fractions were diluted in deionized water (0.1 mg/mL) and analyzed for conformational characteristics by intrinsic fluorescence spectrum using a RF-7000 fluorescence spectrophotometer (Shimadzu, Kyoto, Japan) with the excitation and scanning wavelength of 280 nm and 300–450 nm, respectively.

2.5.4. Fourier Transform Infrared (FT-IR) and Far-UV Circular Dichroism (CD) Spectra

The infrared spectra of flaxseed protein fractions were assessed using a Fourier transform-infrared spectrum (FT-IR) (Vertex 70, Bruker, Germany). In brief, the freeze-dried power samples were fully mixed into a KBr pellet (1%, w/w), and recorded in the wavelength range 4000–400 cm^{-1}. The far-UV CD spectra were obtained using a Chirascan-plus spectropolarimeter (Applied Photophysics Ltd., Surrey, UK). The far-UV CD spectra were performed in a quartz cuvette of 2 mm with a protein concentration of 0.1 mg/mL in 50 mM PBS (pH 6.8). The samples were scanned from 190 to 260 nm at 25 °C. The contents of α-helix, β-strand, β-turns, and random coil were estimated from the far-UV CD spectrum using the deconvolution program (CDNN, version 2.1. Applied Photophysics Ltd., Surrey, UK).

2.6. In Vitro Antioxidant Activities and Free Phenolic Profiles of Flaxseed Protein Fractions

Flaxseed protein fractions were extracted with 5 mL of methanol aqueous solution (80%, v/v) by vortex mixing for 10 min and subsequent ultrasonic bath for 20 min. following centrifugation at $5000\times g$ for 20 min. The in vitro antioxidant activities of extracts from flaxseed protein fractions were performed by 2,2-Diphenyl-1-picrylhydrazyl (DPPH) and ferric reducing antioxidant power (FRAP) methods [24]. The values of DPPH and FRAP were expressed as mg ascorbic acid equivalents (AAE)/100 g protein samples, respectively. The total phenolic acids and flavonoids of extracts were determined using the Folin–Ciocalteu and aluminum nitrate assays. The results were expressed as mg gallic acid (GAE) and rutin equivalents (RE) per 100 g protein sample (dry basis), respectively [25]. The free phenolic acids of extracts were analyzed using an Agilent 1290 ultrahigh performance liquid chromatograph (Agilent, Santa Clara, CA, USA) coupled with a PDA detector and ACQUITY UPLC® BEH Shield RP18 column (2.1 × 100 mm, 1.7 μm). The results were quantified using the individual external standard and expressed as mg/100 g protein sample [26].

2.7. Foaming Properties of Flaxseed Protein Fractions
2.7.1. Foaming Capacity and Stability

In brief, 15 mL of flaxseed protein dispersion (1.0%, w/v) was accurately added into a graduated container and homogenized at 10,000 rpm for 2 min. The foaming capacity (FC, %), and foaming stability were calculated as the following equations

$$FC\ (\%) = V_0/V_L \times 100 \tag{1}$$

$$FS\ (\%) = V_0/V_1 \times 100 \tag{2}$$

where, V_L was the volume of non-shearing protein dispersion, 15 mL; V_0 was the volume of foams immediately after shearing; V_1 was the volume of foams 30 min after shearing.

2.7.2. Morphological Structure of Foams

Approximately 20 µL of foams were deposited on a glass slide with a groove and coverslip. The foam shape and size were acquired using an optical microscope (Scope. AI, Carl Zeiss, Oberkochen, Germany) equipped with a CCD camera (Jenoptik C14. Jenoptik Laser GmbH, Jena, Germany).

2.7.3. Air–Water Interfacial Activity and Microrheological Behavior of Flaxseed Protein Fractions

The interfacial pressure (π) of flaxseed protein fractions at air–water interface was determined with a tensiometer (K100, Kruss, Hamburg, Germany) using the Wilhelmy plate technique. In brief, the platinum plate was immersed in 40 mL of flaxseed protein dispersion (1.0%, w/v) to a depth of 2 mm. The time-dependent π values were continuously recorded for 3600 s at 25 °C.

The microrheological behavior of flaxseed protein fractions was determined by a microrheometer Rheolaser Master^{TM} (Formulation, Toulouse, France). A cylindrical glass tube containing the freshly prepared sample (20 mL) was placed in the microrheometer chamber and analyzed for 3 h at 25 °C. The results were expressed as the elasticity index (EI), macroscopic viscosity index (MVI), and fluidity index (FI) using the software Rheosoft Master 1.4. Then, the storage modulus (G') and loss modulus (G'') of samples were obtained in the range of 1–100 Hz when the measurement time was up to 10 min [27].

2.8. Emulsifying Properties of Flaxseed Protein Fractions

2.8.1. Preparation of Emulsions

The flaxseed protein fractions were dissolved into 50 mM PBS (pH 6.8) at the concentration of 1.0% (w/v) by continuous stirring for 2 h at 25 °C and were then stored overnight at 4 °C for a complete hydration. In order to avoid the potential influences of high-pressure homogenization on the emulsifying properties of flaxseed protein fractions, the coarse emulsions were prepared by mixing 10% (w/w) flaxseed oil into protein dispersion using a high-speed homogenizer (IKA, T25, Staufen, Germany) at 10,000 rpm for 2 min.

2.8.2. Physicochemical Properties and Physical Stability of Emulsions

The mean particle size, volume fraction, and zeta potential of emulsions were evaluated using a ZetaSizer Nano-ZS 90 (Malvern Instruments Ltd., Worcestershire, UK). The morphology of emulsion droplets was observed using optical microscope. After homogenization for 0 min and 10 min, 50 µL of the emulsions were immediately taken from the bottom of the beaker and diluted as 1:100 with 0.1% SDS solution. The absorbance of diluted emulsions was recorded at 500 nm. The emulsifying activity index (EAI) and emulsifying stability index (ESI) were calculated as the following equations [28]:

$$EAI\left(\mathrm{m^2/g}\right) = \frac{2 \times 2.303 \times A_0 \times DF}{C \times \varphi \times \theta \times 10,000} \tag{3}$$

$$ESI(\mathrm{min}) = \frac{A_0}{A_0 - A_{10}} \times 10 \tag{4}$$

where DF was the dilution factor (100), C was the protein concentration (g/mL), φ was the optical path (1 cm) and θ was the oil volume fraction (0.25), A_0 and A_{10} were the absorbance of the emulsions at 0 min and 10 min, respectively.

Then, the emulsions were subjected to the multiple light scattering (MLS) measurement (Turbiscan LAB, Formulaction, Toulouse, France). In brief, emulsions were transferred in a cylindrical glass cell and the curves of transmitted and backscattered light intensity versus the scanning height over a whole length of 40 mm were scanned at 25 °C for 30 min. The delta backscattering (ΔBS) was developed to evaluate determining emulsion stability by evolution of backscattering with time. The turbiscan stability index (TSI) was calculated from the BS of near-infrared light as a function of height.

2.8.3. Microrheological Behavior of Emulsions and Oil–Water Interfacial Activity of Flaxseed Protein Fractions

The microrheological behavior of emulsions was determined by a microrheometer as described above. Then, the interfacial pressure (π) values of flaxseed protein fractions at oil–water interface were determined with a tensiometer using the Wilhelmy plate technique. First, the platinum plate was immersed in 14 mL of protein dispersion (1.0%, w/v) to a depth of 2 mm. Then, 40 g of flaxseed oil was added into the dispersion to create the oil–water interface. The time-dependent π values of flaxseed protein fractions at the oil–water interface were continuously recorded for 3600 s at 25 °C.

2.8.4. Determination of Percentage of Adsorbed Proteins (AP%)

Percentage of adsorbed proteins were determined using the method described by Liang and Tang with slight modifications [22]. First, the 1mL emulsion was centrifuged at $8000 \times g$ for 20 min. Next, subnatant was carefully collected using a syringe, and protein concentration of the subnatant (C_S) was determined with the BCA method using BSA as the standard. The adsorbed proteins (AP) were calculated as the following equation. Where C_0 was protein concentration in the initial protein solutions.

$$AP(\%) = \frac{C_0 - Cs}{C_0} \times 100 \tag{5}$$

2.8.5. Microstructure of Emulsions

In brief, 2.0 μL of emulsions were frozen in liquid nitrogen, transferred into a chamber, cut into the cross section, and then sublimated at -80 °C and 1.3×10^{-6} mbar for 8 min using the PP3010T Cyro-SEM Preparation System (Quorum, Ringmer, UK). The in situ microstructure of lipid droplets and bulk continuous phase in emulsions were observed at 3 kV with magnification $2000\times$.

2.9. Statistical Analysis

The data were presented as mean $\pm$ standard deviations ($n = 3$) and carried out with SPSS 24 for Windows (SPSS Inc., Chicago, IL, USA). One-Way ANOVA, followed by Duncan test, was performed to analyze the significant differences between data ($p < 0.05$).

3. Results and Discussion

3.1. The Proximate Composition and Physicochemical Properties of Flaxseed Protein Fractions

As shown in Table 1, the highest moisture content was observed for FA, whereas the lowest value was found in FG, followed by FPI when subjected to the same freeze-drying condition. High and comparable crude protein contents were obtained for FPI, FG, and FA (>90%), which largely exceeded the results of protein fractions extracted from whole flaxseed [13]. In the current study, the excellent purity of protein fractions could be explained by the absolute removal of gum polysaccharides when dehulled flaxseed was selected as an extraction substrate [29]. Moreover, higher branched-chain, hydrophobic, and sulfur-containing amino acids, but lower negatively charged amino acids, were observed for FG in comparison to that of FA, which was in line with the findings reported by Madhusudhan and Singh [30]. Notably, the extremely low levels of total sugars were examined for FPI, FG, and FA due to the absolute removal of gum polysaccharides. FG had a higher content of total lipids when compared with that of FA (+56.00%, $p < 0.05$), followed by FPI (+42.67%, $p < 0.05$). The stronger surface hydrophobicity of FG might lead to easier oil–absorption capacity when compared with that of FA. The relatively low ash contents were determined for FA, FG, and FPI due to the dialysis process.

Table 1. The proximate composition of FPI, FG, and FA.

Parameters	FPI	FG	FA
Moisture (%)	3.43 ± 0.02 [b]	2.09 ± 0.03 [c]	4.61 ± 0.09 [a]
Total sugars (%)	1.60 ± 0.54 [b]	0.64 ± 0.20 [c]	2.36 ± 0.33 [a]
Crude proteins (%)	91.41 ± 1.41 [a]	93.28 ± 1.09 [a]	90.49 ± 1.25 [a]
Total lipids (%)	2.14 ± 0.13 [a]	2.34 ± 0.10 [a]	1.50 ± 0.05 [b]
Ash (%)	1.43 ± 0.09 [b]	1.65 ± 0.17 [a]	1.04 ± 0.13 [c]
Amino acid composition (%)			
Aspartic acid	10.73	9.80	9.32
Threonine	3.68	3.40	3.13
Serine	5.07	4.71	4.59
Proline	4.46	2.49	4.27
Glutamic acid	17.80	19.57	24.14
Glycine	5.06	5.73	6.24
Alanine	5.30	5.56	4.66
Cysteine	0.25	1.52	0.33
Valine	6.73	8.11	5.65
Methionine	2.25	2.45	1.95
Isoleucine	5.56	6.41	4.66
Leucine	6.55	7.33	6.49
Tyrosine	3.80	2.95	3.51
Phenylalanine	6.29	4.43	4.89
Hlstidine	2.49	2.21	1.80
Lysine	3.64	3.74	4.78
Argnine	10.34	9.59	9.58
Tryptophan	-	-	-
BCAA	18.84	21.85	16.8
HAA	26.64	31.38	23.74
NCAA	28.53	29.37	33.46
PCAA	16.47	15.54	16.16
SCAA	2.5	3.97	2.28

Means with different letters (a, b, c) on the same line were significant differences at $p < 0.05$ level. FPI: flaxseed protein isolate; FA: flaxseed albumin; FG: flaxseed globulin; BCAA: branched-chain amino acids; HAA: hydrophobic amino acids; NCAA: negatively charged amino acids; PCAA: positively charged amino acids; SCAA: sulfur-containing amino acids; -: not detected.

As seen in Figure 1a, a significantly larger hydrodynamic size was found for FPI, followed by FG when compared with that of FA (+4.21-fold, +1.18-fold; $p < 0.05$), which might be owing to the formation of protein aggregates between FA and FG [10]. The highest zeta potential value was observed for FPI when compared to those of FG and FA (+22.05%, +68.66%; $p < 0.05$). This could be explained by the mild conformation unfolding and deformation based on the intermolecular interactions between FA and FG. As presented in Figure 1b, an extremely lower protein solubility was found for FPI (−2.23-fold; $p < 0.05$), followed by FG (−0.91-fold; $p < 0.05$) when compared with those of FA (1.0%, w/v). FG displayed a highly stronger surface hydrophobicity than that of FPI and FA (+24.63%, +154.28%; $p < 0.05$). The basic subunits of FG with higher hydrophobic acids were usually buried in the inside of protein molecules in bulk aqueous phase. Thus, the surface hydrophobicity of FG depended on the outer acidic subunits oriented towards the aqueous phase due to the multi subunit tangling structure. The superior solubility of FA could be attributed to the single protein subunit with low molecular weight and hydrophobic amino acids. Upon the alkaline dissolution and acid precipitation process, the internalization of hydrophobic group might occur based on the noncovalent interactions between FG and FA, thereby resulting in mild reduction of the hydrophobic ability for FPI [31]. Concurrently, the formation of aggregates between FA and FG fractions definitely weakened the solubility of FPI, as evidenced by the largest hydrodynamic dimensions.

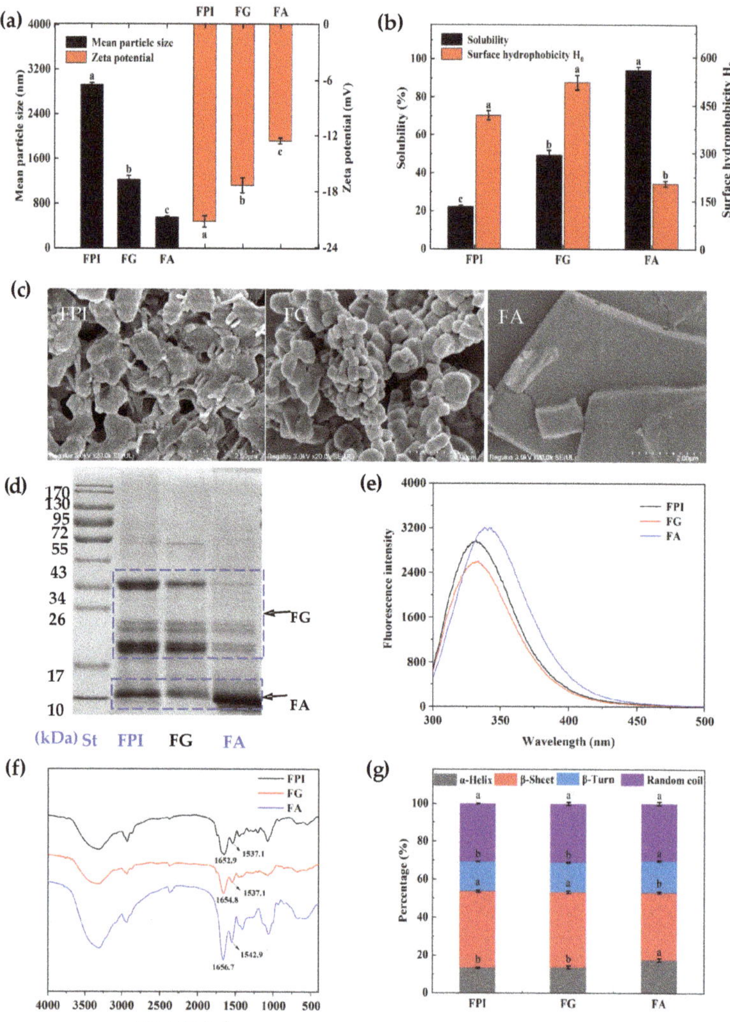

Figure 1. The physicochemical, morphological, and structural properties of FPI, FA, and FG. (**a**): the mean particle size (black bar) and zeta potential (red bar); (**b**): the solubility (black bar) and surface hydrophobicity (red bar); (**c**): the SEM imaging; (**d**): the SDS-PAGE analysis; (**e**,**f**): the intrinsic fluorescence and FTIR spectra; (**g**): the secondary structure obtained from CD spectra. Different alphabets in same index indicated significant differences at the $p < 0.05$ level. FPI: flaxseed protein isolate; FG: flaxseed globulin; FA: flaxseed albumin.

3.2. The Morphological and Structural Properties of Flaxseed Protein Fractions

As demonstrated in Figure 1c, FG displayed an approximate spherical morphology with relatively lager size distribution, further confirming the different degrees of aggregation between protein molecules. Comparatively, FA exhibited a large lamellar structure, which could be explained by the extended molecular conformation. Notably, FPI contained a relatively small lamellar strip structure (albumin), which was packed by the blurring spherical particles (globulin). Accompanied by the conformational extension and reconfiguration, the somewhat noncovalent interactions and complex aggregates between FA and FG might preferentially exist in FPI during the alkaline dissolution and acid precipitation. Undoubtedly, the surface morphology of FPI was partially inconsistent with the findings, as previously described in our study [23]. The gum polysaccharides with irregular block shape could naturally coexist in FA fraction during the extraction process, and thus affect the surface morphology of FPI obtained from whole flaxseed meal [7]. Moreover, the inconsistent isoelectric points between FA and FG also contributed to the different composition profiles in FPI obtained by the alkali dissolution and acid precipitation [32,33]. In addition, FPI could undergo varying structural alterations during different pretreatment and extraction process, thereby leading to different morphological characteristics compared with FA and FG prepared by ultracentrifugation and salting out, respectively [12,14].

As depicted in Figure 1d, FA had a major band at 10 kDa, followed by 18 and 20 kDa. In addition, an extremely weak intensity of band was also identified approximately at 35 kDa for FA, which might be explained by the coextraction of small amounts of FG due to the noncovalent interaction between them. By contrast, FG contained high intensity of bands at 35 and 18 kDa, followed by the light bands at about 55 and 20 kDa, respectively. However, FPI possessed similar subunit profiles but different subunit proportion when compared with that of FG, particularly for the evidently diminished intensity of band at 55 kDa and synchronously enhanced intensity of band at 35 kDa. Our findings were partly incompatible to the observation reported by Nwachukwu and Aluko, who prepared FA and FG from flaxseed protein meal, which could be further affected by the extraction matrix and methods [13]. Differential protein subunit profiles of albumin and globulin were obtained from other plant proteins, such as buckwheat and soapnut seeds [34,35].

The conformational properties of flaxseed protein fractions were sequentially analyzed by intrinsic fluorescence spectrum. As showed in Figure 1e, FPI presented the maximum fluorescence emission spectrum (λmax) at 331.6 nm. In comparison, FG exhibited slightly quenching maximum fluorescence intensity (FImax) at the similar λmax. However, the maximal value of FImax and 6.2 nm of red-shift of λmax were concurrently detected for FA in contrast to that of FG. Indeed, except for the high level of Trp, FA manifested flexible structural properties. Relatively, the closer chain packing of Trp residues within hydrophobic pocket and subsequent folded spatial conformation could be manifested for FG [36]. Although the gum polysaccharides were removed from protein fractions, the naturally coexisting phenolic compounds with varying abundances and profiles undoubtedly affected the spatial conformation of FA, FG, and FPI due to the noncovalent interaction between them [15].

The FT-IR spectrum of proteins possessed two primary features, the amide I (1620–1700 cm^{-1}) and amide II (1500–1560 cm^{-1}) bands raised from the specific stretching and bending vibrations of protein backbone C=O, C-N, and N-H. The amide I band of FPI had a blue shift of 1.9 cm^{-1} and 3.8 cm^{-1} relative to FG and FA. These indicated that the intermolecular hydrogen bonds were formed between the two components, and the electron cloud density of C=O decreased, which caused the absorption peak to shift to the lower wave number direction [37]. The detailed secondary structure content of different protein components was further obtained from the far-UV CD spectrum, as illustrated in Figure 1g. FPI contained 13.21% of α-helix, 40.44% of β-sheet, 15.58% of β-turns, and 30.75% of random coil, respectively, which was similar to that of FG. By contrast, obviously higher α-helix but lower β-sheet content was determined for FA when compared those of FG ($p < 0.05$), which could be explained by the more disulfide linkages and ordered structure as reported by Madhusudhan and Singh [30].

3.3. The Antioxidant Activities and Free Phenolic Acids of Flaxseed Protein Fractions

As depicted in Table 2, FA possessed the largest DPPH value, which was 5.49-fold higher than that of FG ($p < 0.05$), followed by FPI (-1.38-fold, $p < 0.05$). Similar changing trends were detected for the FRAP values of FA, FPI, and FG, revealing the specific accumulation of phenolic compounds during extraction process. As assessed by colorimetric methods, the contents of total phenolic acids and flavonoids in FPI reached 89.65 and 1.07 mg/100 g, respectively, which was obviously lower than the values of flaxseed protein prepared from whole flaxseed meals [23]. Indeed, flaxseed hulls also contained relatively abundant lignan oligomers in the secondary wall of sclerite cells, which consisted of secoisolariciresinol diglucoside (SDG), p-coumaric acid, ferulic acid, and herbacetin diglucosides [38]. The glycosidation and complex ester bonds in lignan oligomers could partially interfere with the contact between phenolic hydroxyl groups and free radicals. The steric hindrance effect of lignan could only produce relatively poor free radical scavenging activities when marginally migrated into flaxseed protein during extraction process [39]. Notably, the preferable accumulation of total phenolic acids and flavonoids was observed for FA in comparison to that of FG (+2.24-fold, +60.00%; $p < 0.05$). Nevertheless, equal contents of phenolic compounds were detected for both globulin and albumin prepared from flaxseed protein precipitated at the isoelectric point of 4.2 [31]. In fact, the different abundance and profiles of phenolic compounds in varying extraction substrates, such as whole/dehulled flaxseed, defatted flaxseed meals, flaxseed protein concentrates, etc., greatly affected the subsequent release and retention into FG and FA fractions. Thus, no signs of SDG, p-CouAG, and FeAG in lignans were identified for FPI, FA, and FG when the dehulled flaxseed was selected for extraction substrate, thereby resulting in no contribution to the antioxidant activities of protein fractions.

Table 2. The phenolic acid profiles and antioxidant activities of FPI, FG, and FA.

		FPA	FG	FA
Total phenolic acids (mg/100 g)		89.65 ± 9.65 [b]	81.83 ± 1.99 [b]	264.85 ± 13.22 [a]
Total flavonoids (mg/100 g)		1.07 ± 0.02 [a]	0.30 ± 0.04 [b]	0.48 ± 0.01 [b]
DPPH (mg AAE/100 g)		51.52 ± 3.23 [b]	22.30 ± 0.44 [c]	122.45 ± 4.41 [a]
FRAP (mg AAE/100 g)		56.67 ± 3.55 [b]	17.16 ± 0.62 [c]	101.19 ± 4.59 [a]
	Gallic acid	19.13 ± 0.25 [b]	16.31 ± 0.31 [c]	59.96 ± 0.27 [a]
	Protocatechuic acid	2.70 ± 0.01 [a]	0.83 ± 0.01 [b]	2.26 ± 0.03 [a]
	Vanillic acid	0.11 ± 0.01 [c]	0.15 ± 0.01 [a]	0.13 ± 0.01 [b]
	Caffeic acid	0.26 ± 0.01 [b]	-	0.39 ± 0.01 [a]
Free phenolic acids (mg/100 g)	Syringic acid	0.05 ± 0.01 [b]	0.27 ± 0.05 [a]	0.25 ± 0.02 [a]
	Vanillin	0.52 ± 0.03 [b]	0.23 ± 0.02 [b]	1.98 ± 0.02 [a]
	Sinapic acid	0.29 ± 0.03 [c]	0.36 ± 0.01 [b]	0.92 ± 0.02 [a]
	Coumarin	0.45 ± 0.01 [b]	0.22 ± 0.01 [b]	2.43 ± 0.03 [a]
	Sallcylic acid	0.59 ± 0.02 [a]	0.31 ± 0.01 [c]	0.48 ± 0.01 [b]
	Cinnamic acid	0.20 ± 0.01 [a]	0.08 ± 0.01 [b]	0.26 ± 0.01 [a]

Means with different letters (a, b, c) on the same line were significant differences at $p < 0.05$ level. FPI: flaxseed protein isolate; FA: flaxseed albumin; FG: flaxseed globulin; FPP: flaxseed polyphenols; -: not detected.

Instead, several free phenolic acids, including gallic acid, coumarin, protocatechuic acid, vanillin, sinapic acid, sallcylic acid, caffeic acid, cinnamic acid, syringic acid, and vanillic acid were identified in FA, accounting for 86.82%, 3.52%, 3.27%, 2.87%, 1.33%, 0.70%, 0.56%, 0.38%, 0.36%, and 0.19% of total amounts (69.06 mg/100g), respectively. By comparison, largely lower contents of free phenolic acids were obtained for FG when compared with that of FA (-72.84%, $p < 0.05$). In particular, higher proportion of gallic acid, protocatechuic acid, sinapic acid, sallcylic acid, syringic acid, vanillic acid, and coumarin (86.94%, 4.42%, 1.92%, 1.65%, 1.44%, 1.23%, 1.17%), but lower proportion of vanillic and cinnamic (0.80%, 0.43%) were identified in FG. However, the free caffeic acid was not detected in FG, suggesting obviously preferential migration into FA during extraction process. As previously reported, the free phenolic acids mainly coexisted with the extract supernatant of flaxseed protein and possessed higher antioxidant activity than bound phenolic acids detained in residue remaining after extraction [40]. Indeed, there was still no definite information about the specific in situ location of free phenolic acids in dehulled flaxseed [41]. Moreover, the dislocation migration of free phenolic acids between oleosins and storage proteins might occur owing to the newly established contact and strong noncovalent interaction between them upon high-speed shearing of flaxseed [40]. The lower molecular weight, more stretching conformation and superior molecular polarity contribute to the favorable migration of free phenolic acids into FA, inevitably leading to a weaker retention in FG following sequential extraction [34,42]. Most importantly, the proportion and noncovalent interactions between FA and FG definitely complicated the retention properties of free phenolic acids into FPI during the extraction process.

3.4. The Foaming Properties of Flaxseed Protein Fractions

The foam height can visually reflect the volume of foam produced by mechanical shearing of a bulk liquids and was considered as an index of foaming capacity of proteins. As shown in Figure 2a,b, the maximum foam volume was identified for FA, whereas the minimum value of foam was observed for FG (1.0%, w/v). Thus, the presence of FA fraction largely determined the foaming potential of FPI, which was undoubtedly inferior to that of individual FA (-23.9%, $p < 0.05$). The flexible spatial conformation, superior solubility, and lower molecular weights could enhance the migration efficiency of FA at the air–water interface, leading to the preferable foaming ability [43]. In comparison, the poor foaming activity of FG might be explained by the slow interface migration from the bulk aqueous phase into air/water interface due to the higher hydrophobic amino acids and molecular weights, respectively. As for FPI, the desirable interface absorption behavior of FG was positively driven by the coexisting low amounts of FA, which might occur due to the noncovalent interaction between them. Then, the foaming stability was evaluated by time-dependent subduction of foam height and corresponding changes in foam morphology (Figure 2b,c). After standing for 30 min at 25 °C, the foam volume produced by FPI, FG, and FA decreased by 22.3%, 57.5%, and 42.2% ($p < 0.05$), respectively, suggesting a desirable foaming stability of FPI relative to FA and FG. As evidenced by the microscopic imaging (Figure 2e), the foam freshly prepared by FG was distributed over a wide range of size, which inevitably could lead to an immediate coalescence or collapse after formation. In contrast, the sizes of foams freshly prepared by FA were distributed in a relatively narrow range, followed by FPI. Then, the foams obtained by FA were mainly subjected to rapid drainage, but no coalescence or collapse within 30 min of standing. However, a relatively slower drainage, coalescence, and collapse was displayed for foams prepared from FPI. Undoubtedly, the foaming stability of FPI was largely attributed to the coexistence of FA and FG with appropriate proportion at air/water interface. In depth, the interface intermolecular interactions between FA and FG in FPI could partially narrow the interface "holes", and subsequently suppress the gas migration between foams, leading to relative stability of foams prepared by FPI.

The abilities of flaxseed protein fractions to be absorbed at the air/water interface were further evaluated through time-dependent alteration of surface tension (π) values. As presented in Figure 2f, the initial π value of FPI (1.0%, w/v) was 41.60 mN m^{-1}, which slightly decreased with the equilibrium time extended to 3600 s. By comparison, FG possessed comparable initial π value, but displayed relatively stronger decline with the increase in equilibrium time (-6.39%, $p < 0.05$). However, FA manifested the lowest initial π value, but was comparable to that of FG when the equilibrium time reached 3600 s. The excellent performance of FA for early migrating into the air/water interface undoubtedly explained its desirable foaming capacity. However, the slight reduction of interfacial tension over time might suggest the relatively limited conformation reorganization at air/water interface due to the stretched conformation and weak intermolecular interactions, thereby leading to fast drainage upon standing [44]. In comparison, FG was unable to achieve early adsorption at air/water interface due to the multi subunit crosslinking structure, high molecular weight, and hydrophobicity. Instead, effective conformation adaptability and strong intermolecular crosslinks might inevitably occur for FG, leading to the formation of densely packed viscoelastic films. The higher initial value of interfacial pressure and slight reduction over time for FPI suggested that the low amounts of FA and noncovalent interactions with FG sufficiently resulted in low interfacial diffusion, adsorption, and conformation reorganization [43].

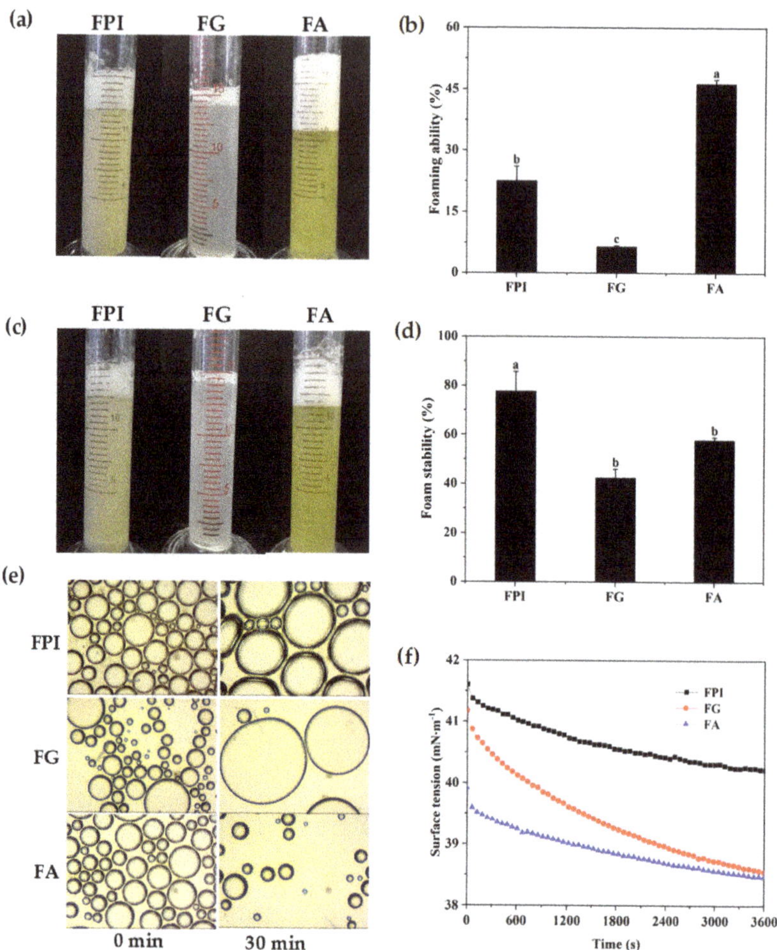

Figure 2. The foaming properties of FPI, FG, and FA. (**a**,**c**): photograph of foam formed by FPI, FG, and FA at 0 min and 30 min; (**b**): the foaming ability; (**d**): the foam stability; (**e**): the morphology of foams immediately after shearing and standing for 30 min; (**f**): the interfacial tension at air–water interface. Different alphabets in same index indicated significant differences at the $p < 0.05$ level. FPI: flaxseed protein isolate; FG: flaxseed globulin; FA: flaxseed albumin.

The EI calculated from the inverse of the mean square displacement (MSD) for a decorrelation period can quantify the elasticity of samples. The MVI, as the inverse slope of the MSD curves in double liner coordinates, can characterize the viscosity of samples. The fluidity index (FI) curve can reflect the fluidity properties of samples. The greater the fluidity of a sample, the weaker its viscosity. As seen in Figure 3a–c, FPI and FA possessed obviously higher viscoelastic properties, which might contribute to the delaying fluid drainage between bubbles with the interface shrunk, thereby accounting for their more desirable foaming ability when compared with that of FG. FPI displayed comparable fluidity to FA and FG, and then steeply dropped during the late test period, which might result from the intermolecular interactions between FA and FG in aqueous phase under no shearing condition, and partially explain the relatively ascendant foaming stability dependent on the densely packed monolayers. Actually, the naturally occurring phenolic compounds might also affect the structural adaptability and subsequent gas-holding capacity of protein fractions at the air–water interface. In particular, the coexistence of free phenolic acids in FA might compensate for its molecular flexibility and tend to form rigid and compact interface with desirable stability [45]. It can be seen that for all protein solution (1%), storage modulus (G′) was beyond the loss modulus (G″) in the range of 1–10 Hz, then G″ became greater than G′ with a continuous increase in frequency (Figure 3d). The bulk phase response was dominated by viscous effects, indicting that the protein in the bulk phase was adsorbed or aggregated to the interface so that loss modulus increased faster than the storage modulus [46,47].

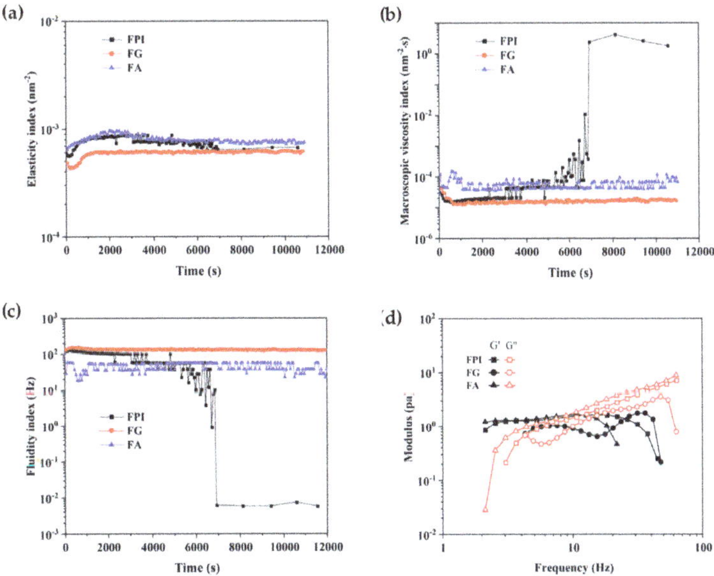

Figure 3. The microrheological properties of FPI, FG, and FA. (**a**): the elasticity index; (**b**): the macroscopic viscosity index; (**c**): the fluidity index; (**d**): storage modulus (G′) and loss modulus (G″). FPI: flaxseed protein isolate; FG: flaxseed globulin; FA: flaxseed albumin.

3.5. The Emulsifying Properties of Flaxseed Protein Fractions

3.5.1. The Physicochemical Properties and Physical Stability of Emulsions

As illustrated in Figure 4a,b, the mean lipid droplet sizes and zeta potential values of emulsions constructed by FPI were 1445.33 nm and −11.60 mV, respectively. By comparison, the mean particle size of emulsions fabricated by FG increased by 20.92% ($p < 0.05$), which was accompanied by the decrease in zeta potential value (−19.40%, $p < 0.05$). Relatively, FA behaved with superior emulsifying capacity as evidenced by the minimum particle dimensions, but comparable charge density to that of FPI. These indicated that the molecular conformation of FG in FPI might be interfered due to the noncovalent interactions with FA, which was favorable for the emulsion stability via the electrostatic repulsion between adjacent lipid droplets. According to the results from microscopic examination, FA produced the emulsion with relatively small and evenly distributed lipid droplets, whereas the lipid droplets produced by FG manifested larger mean particle sizes. As expected, the lipid droplets with more heterogeneous sizes had been observed for emulsion droplets stabilized by FPI. In particular, the beaded small lipid droplets and subsequent close contacts with large lipid droplets might weaken the emulsifying stability of FPI [29]. The different molecular composition and conformation of FA and FG definitely complicated the interfacial behavior and rheological properties of emulsions, which largely determined the physical stability of emulsions formulated by flaxseed protein fractions.

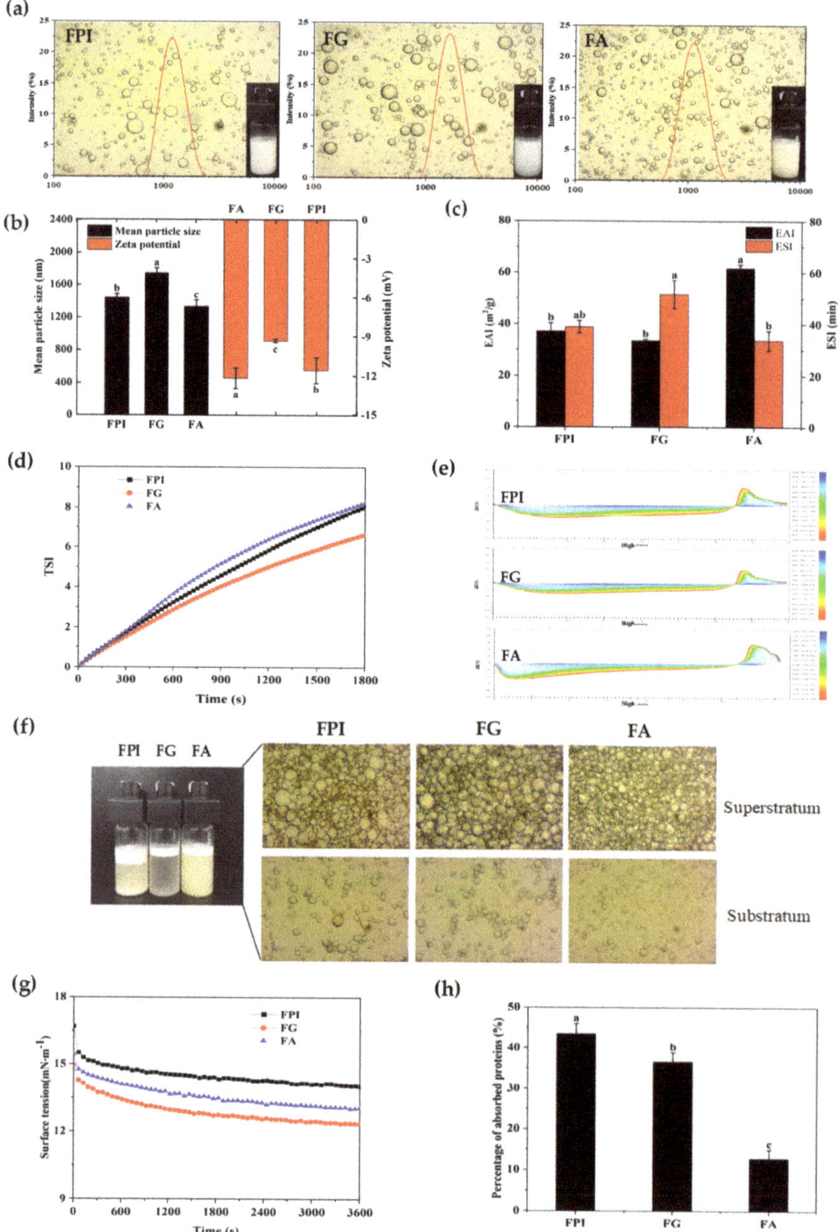

Figure 4. The physicochemical stability of emulsions constructed by FPI, FG, and FA. (**a**): the visual observation, optical micrographs, and particle size distribution; (**b**): the mean particle size (black bar) and zeta potential (red bar); (**c**): the EAI (black bar) and ESI (red bar); (**d**): the Turbiscan stability index (TSI); (**e**): the delta backscattering (ΔBS) curves; (**f**): the optical micrographs (stored at 4 °C for 48 h); (**g**): the surface tension at the oil–water interface; (**h**): the percentage of adsorbed protein fractions. Different alphabets in the same index indicated significant differences at the $p < 0.05$ level. FPI: flaxseed protein isolate; FG: flaxseed globulin; FA: flaxseed albumin.

As depicted in Figure 4c, FG and FPI had comparable but significantly lower EAI values when compared with that of FA (-45.61%, -39.63%, $p < 0.05$). However, the maximum ESI value was observed for FG, followed by FPI, whereas FA displayed the lowest levels of ESI. These indicated that the emulsifying activity and stability of FPI were contributed by the FA and FG, respectively. Consistent with the above results, the emulsion stability results obtained by the TSI values were as follows: FG > FPI $\approx$ FA (Figure 4d). As shown in Figure 4e, the ΔBS lines of emulsions formulated by FPI dramatically decreased with extending time in the measurement cell height range of 1.5 mm to 32 mm. Simultaneously, different increments in ΔBS curves of emulsions were observed in the measurement cell range of 32~38.5 mm, which was accompanied by a slight decline in ΔBS curves in the topmost part of the measurement cell (38.5~40 mm). These further confirmed that the instability of emulsions prepared by FPI was primarily due to the serious flocculation and coalescence of lipid droplets, followed by the lipid droplet flotation. As to the emulsion prepared by FG, the relatively smaller decrease in the middle curve of ΔBS revealed a weaker lipid droplet flocculation, coalescence, and floatation. In comparison, the large-scale reduction of ΔBS lines in the bottom part of measurement cell (0~10 mm) and synchronous promotion in the top part of measurement cell (32~37 mm) were found for the emulsions constructed by FA, manifesting the maximum lipid droplet flotation and oil perspiring, followed by flocculation and coalescence [48]. Similarly, Nwachukwu and Aluko found that the FG had better emulsifying stability than FA at neutral environment and specific concentration [13]. As depicted in Figure 4f, the relatively fast phase separation of emulsions prepared by FPI, FA, and FG indicated that the introduction of emulsifier, such as the gum polysaccharides and phospholipids, was still necessary to produce the gel network structure in bulk phase, applied for direct interface coating or narrowing the pore sizes of lipid droplets in emulsions obtained from FPI, FA, and FG [49].

3.5.2. The Interface Activities of Flaxseed Protein Fractions

The time-dependent tracking of interfacial tension for protein fractions mainly reflected their differences in instantaneous migration and adsorption capacity towards the oil–water interfaces. As presented in Figure 4f, the initial surface tension (π) value of FPI dispersion (1.0%, w/v) reached 16.69 mN m^{-1}, which gradually descended and achieved an ultimate reduction by 15.94% ($p < 0.05$) with the equilibrium time extended to 3600 s. By contrast, a lower initial π value (15.45 mN m^{-1}) with an almost similar ability to reduce the interfacial tension over time was observed for the FA dispersion (-15.48%, $p < 0.05$). However, FG dispersion exhibited the lowest initial π value (14.95 mN m^{-1}), and the maximum interfacial dropping trend (-17.73%, $p < 0.05$). It was considered that the primary driving force for the migration and absorption at the oil–water interfaces was the solubilization of nonpolar regions for protein molecules in oil phase. The higher affinity of FG to the oil–water interface could be demonstrated by the stronger hydrophobicity than that of FA fraction. As shown in Figure 4g, the percentages of adsorbed protein (AP%) at the interface of lipid droplets were approximately 43.44%, 36.58%, and 12.83% of the total protein for FPI, FG, and FA dispersion (1.0%, w/v), respectively. Thus, the more substantial conformation reorganization and greater adsorption capacity could also be elucidated from the effective interface adsorption kinetics and molecule loads for FG when compared with that of FA [50]. Importantly, the highest interface adsorption but limited interface deformation for FPI might be attributed to the formation of specific interface membrane due to the intermolecular interaction between FG with FA upon high-speed shearing.

3.5.3. The Microrheological Properties of Emulsions

According to Figure 5a,b, the emulsions stabilized by FPI and FG (1.0%, w/v) displayed the relatively low and almost equivalent viscoelasticity during the test period (0~3 h). By contrast, the emulsion produced by FA displayed the relatively higher viscoelasticity, consistent with the trend of EI and MVI curves of the emulsions produced by FPI and FG. As presented in Figure 5c, the results of FI were consistent with the MVI for emulsion prepared by FA, FG, and FPI. As illustrated in Figure 5d, the values of loss modulus (G″) of emulsion constructed by FPI apparently exceeded the storage modules (G′) over the frequency from 8.6 to 100 Hz, exhibiting the relatively weak gelling properties. Undoubtedly, the viscous properties of emulsion stabilized by FPI were largely contributed by the FA, but not FG fraction according to the G″ and G′ values of emulsion produced by FG and FA. Thus, it could be considered that the higher proportion of FA in bulk aqueous phase primarily determined the rheological behavior of emulsions.

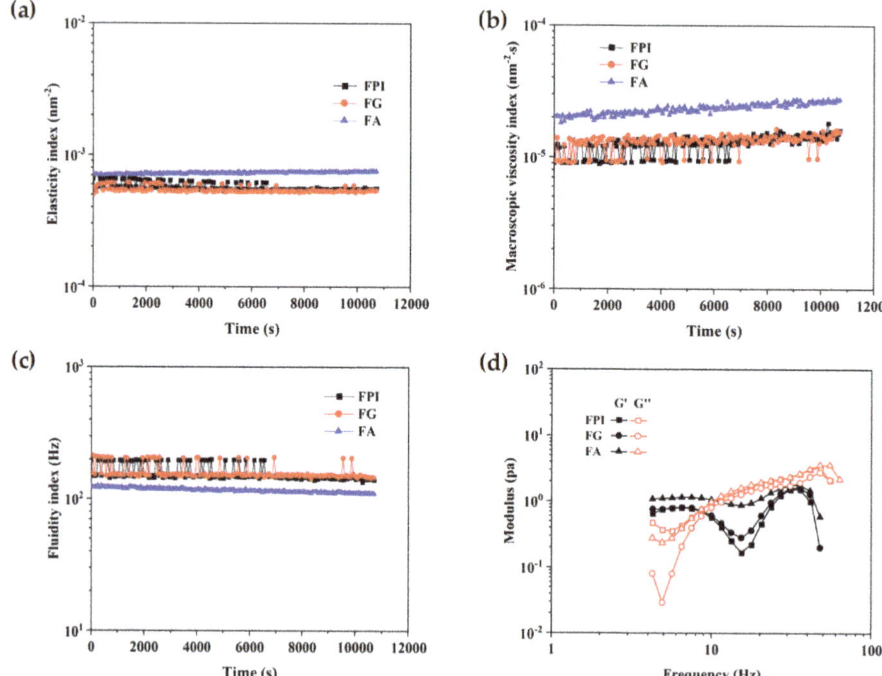

Figure 5. The microrheological properties of emulsions constructed by FPI, FG, and FA. (**a**): the elasticity index; (**b**): the macroscopic viscosity index; (**c**): the fluidity index; (**d**): the storage modulus G′ and loss modulus G″. FPI: flaxseed protein isolate; FG: flaxseed globulin; FA: flaxseed albumin.

3.5.4. The Microstructure of Emulsions

We further explored the microstructure of lipid droplets and bulk continuous phase in emulsions prepared by FA, FG, and FPI. As observed in Figure 6, a flimsy but relatively compact interfacial film was observed for the lipid droplets stabilized by FPI ranging from 4 μm to 20 μm based on the cryo-SEM imaging. Notably, FPI produced no spatial network structure in bulk continuous phase, which might be due to its desirable emulsifying capacity. Undoubtedly, the multiple close contacts between lipid droplets produced by FPI definitely contribute to the emulsion instability via the flocculation and coalescence, which further supported the results from the Turbiscan analysis. The small content of albumin in rapeseed protein isolate was demonstrated to play a major role in the formation of interface structure [50]. On the contrary, FG produced the lipid droplets with heterogeneous dimensions (2~20 μm), which was attached on the compact single, double, or multilayer lamellar gel network structure in bulk aqueous phase dependent on the particle sizes. These indicated that FG possessed the excellent potential of restricting the movement of lipid droplets via the interlayer anchoring but not direct interface coating. Notably, both the interface film of lipid droplets and lamellar gel structure of bulk aqueous phase were emerged for emulsion prepared by FA. In particular, the quite loose and porous interface microstructure of lipid droplets could be explained by the low interfacial adsorption and weak intermolecular interactions, as evidenced by the information from the tensiometer analysis. The relatively dispersed lipid droplets in bulk aqueous phase might explain the fast oil perspiring but relatively slow flocculation and coalescence rates when subjected to standing. The in situ interfacial microstructure of lipid droplets and continuous phase intuitively explained the different emulsifying properties of FG and FA in FPI, contributing to the specific stabilizing and destabilization behavior of emulsions.

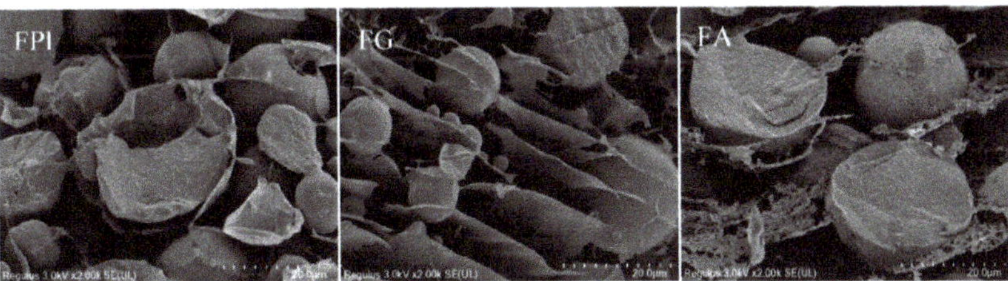

Figure 6. The cryo-SEM imaging of emulsions constructed by FPI, FG, and FA. FPI: flaxseed protein isolate; FG: flaxseed globulin; FA: flaxseed albumin.

4. Conclusions

In conclusion, FA possessed smaller particle size and desirable protein solubility due to lower molecular weight and surface hydrophobicity when compared to FG. Moreover, FA and FG manifested large lamellar and nearly spherical structure, respectively, which led to partial structure alteration for FPI due to the noncovalent interactions between them. The relatively flexible spatial conformation of FA resulted in favorable retention of free phenolic acids and increase in antioxidant activities compared to those of FG. Moreover, the foaming properties of FA were obviously superior to those of FG due to the effective adsorption behavior at air–water interface and higher viscoelastic properties of bulk phase. Notably, FA produced the emulsion droplets via direct interface coating, which was easily destabilized due to a quite loose and porous interface. FG exerted the interlayer anchoring and restricted the movement of lipid droplets by the spatial network structure. Notably, the emulsifying capacity and stability of FPI were determined by the synergistic interface and rheological behavior of FA and FG. Thus, the tailed retention of albumin could serve as an effective strategy to improve the selected techno-functionality of flaxseed protein based on the specific interaction with globulin fraction and naturally occurring phenolic acids.

Author Contributions: Conceptualization, X.Y. and Q.D.; methodology, X.Q., X.Y. and L.L.; software, X.Q. and L.L.; investigation, X.Y. and Q.D.; data curation, X.Y. and X.Q.; writing—original draft preparation, X.Y. and X.Q.; writing—review and editing, X.Y., Q.X. and Y.Z.; supervision, X.Y. and Q.D.; project administration, X.Y. and Q.D. All authors have read and agreed to the published version of the manuscript.

Funding: This research was financially supported by National Natural Science Foundation of China (32072267).

Institutional Review Board Statement: Not applicable.

Informed Consent Statement: Not applicable.

Data Availability Statement: All data are provided in the manuscript.

Conflicts of Interest: The authors declare no conflict of interest.

References

1. Bekhit, A.E.A.; Shavandi, A.; Jodjaja, T.; Birch, E.J.; Teh, S.; Mohamed Ahmed, I.A.; Juhaimi, F.Y.; Saeedi, P.; Bekhit, A.A. Flaxseed: Composition, detoxification, utilization, and opportunities. *Biocatal. Agr. Biotech.* **2018**, *13*, 129–152. [CrossRef]
2. Wu, S.; Wang, X.; Qi, W.; Guo, Q. Bioactive protein/peptides of flaxseed: A review. *Trends Food Sci. Tech.* **2019**, *92*, 184–193. [CrossRef]

3. Salma, N.U.; Peddha, M.S.; Setty, J. Ameliorative effect of flaxseed (*Linum usitatissimum*) and its protein on ethanol-induced hepatotoxicity in wistar rats. *J. Food. Biochem.* **2019**, *43*, e13047. [CrossRef] [PubMed]
4. Mohamed, R.S.; Fouda, K.; Akl, E.M. Hepatorenal protective effect of flaxseed protein isolate incorporated in lemon juice against lead toxicity in rats. *Toxicol. Rep.* **2020**, *7*, 30–35. [CrossRef]
5. Lan, Y.; Ohm, J.B.; Chen, B.; Rao, J. Physicochemical properties and aroma profiles of flaxseed proteins extracted from whole flaxseed and flaxseed meal. *Food Hydrocoll.* **2020**, *104*, 105731. [CrossRef]
6. Juodeikiene, G.; Zadeike, D.; Trakselyte-Rupsiene, K.; Gasauskaite, K.; Bartkiene, E.; Lele, V.; Viskelis, P.; Bernatoniene, J.; Ivanauskas, L.; Jakstas, V. Functionalisation of flaxseed proteins assisted by ultrasonication to produce coatings enriched with raspberries phytochemicals. *LWT Food Sci. Technol.* **2020**, *124*, 109180. [CrossRef]
7. Liu, J.; Shim, Y.Y.; Poth, A.G.; Reaney, M.J.T. Conlinin in flaxseed (*Linum usitatissimum* L.) gum and its contribution to emulsification properties. *Food Hydrocoll.* **2016**, *52*, 963–971. [CrossRef]
8. Kaushik, P.; Dowling, K.; Mcknight, S.; Barrow, C.J.; Wang, B.; Adhikari, B. Preparation, characterization and functional properties of flaxseed protein isolate. *Food Chem.* **2016**, *197*, 212–220. [CrossRef]
9. Nikbakht, M.N.; Goli, S.A.H.; Doost, A.S.; Dewettinck, K.; Van der Meeren, P. Bioparticles of flaxseed protein and mucilage enhance the physical and oxidative stability of flaxseed oil emulsions as a potential natural alternative for synthetic surfactants. *Colloid. Surf. B* **2019**, *184*, 110489. [CrossRef]
10. Rabetafika, H.N.; Remoortel, V.V.; Danthine, S.; Paquot, M.; Blecker, C. Flaxseed proteins: Food uses and health benefits. *Int. J. Food Sci. Technol.* **2011**, *46*, 221–228. [CrossRef]
11. Krause, J.; Schultz, M.; Dudek, S. Effect of extraction conditions on composition, surface activity and rheological properties of protein isolates from flaxseed (*Linum usitativissimum* L). *J. Sci. Food Agr.* **2002**, *82*, 970–976. [CrossRef]
12. Tirgar, M.; Silcock, P.; Carne, A.; Birch, E.J. Effect of extraction method on functional properties of flaxseed protein concentrates. *Food Chem.* **2017**, *215*, 417–424. [CrossRef] [PubMed]
13. Nwachukwu, I.D.; Aluko, R.E. Physicochemical and emulsification properties of flaxseed (*Linum usitatissimum* L.) albumin and globulin fractions. *Food Chem.* **2018**, *255*, 216–255. [CrossRef] [PubMed]
14. Yang, J.; Sagis, L.M.C. Interfacial behavior of plant proteins-novel sources and extraction methods. *Curr. Opin. Colloid Interface Sci.* **2021**, *56*, 101499. [CrossRef]
15. Alu'datt, M.H.; Rababah, T.; Kubow, S.; Alli, I. Molecular changes of phenolic-protein interactions in isolated proteins from flaxseed and soybean using Native-PAGE, SDS-PAGE, RP-HPLC, and ESI-MS analysis. *J. Food Biochem.* **2019**, *43*, e12849. [CrossRef]
16. Alu'datt, M.H.; Rababah, T.; Alli, I. Effect of phenolic compound removal on rheological, thermal and physico-chemical properties of soybean and flaxseed proteins. *Food Chem.* **2014**, *146*, 608–613. [CrossRef]
17. Silva, F.G.D.E.; Blanca Hernández-Ledesma, B.; Amigo, L.; Netto, F.M.; Miralles, B. Identification of peptides released from flaxseed (*Linum usitatissimum* L.) protein by alcalase (r) hydrolysis: Antioxidant activity. *LWT Food Sci. Technol.* **2017**, *76*, 140–146. [CrossRef]
18. Pham, L.B.; Wang, B.; Zisu, B.; Truong, T.; Adhikari, B. Microencapsulation of flaxseed oil using polyphenol-adducted flaxseed protein isolate-flaxseed gum complex coacervates. *Food Hydrocoll.* **2020**, *107*, 105944. [CrossRef]
19. Yu, X.; Nie, C.; Zhao, P.; Zhang, H.; Qin, X.; Deng, Q.; Huang, F.; Zhu, Y.; Geng, F. Influences of microwave exposure to flaxseed on the physicochemical stability of oil bodies: Implication of interface remodeling. *Food Chem.* **2022**, *368*, 130802. [CrossRef]
20. Ozdal, T.; Capanoglu, E.; Altay, F. A review on protein-phenolic interactions and associated changes. *Food Res. Int.* **2013**, *51*, 954–970. [CrossRef]
21. Marambe, H.K.; Shand, P.J.; Wanasundara, J. An in-vitro investigation of selected biological activities of hydrolysed flaxseed (*Linum usitatissimum* L.) proteins. *J. Am. Oil. Chem. Soc.* **2008**, *85*, 1155–1164. [CrossRef]
22. Liang, H.; Tang, C. pH-dependent emulsifying properties of pea [*Pisum sativum* (L.)] proteins. *Food Hydrocoll.* **2013**, *33*, 309–319. [CrossRef]
23. Yu, X.; Huang, S.; Nie, C.; Deng, Q.; Shen, R. Effects of atmospheric pressure plasma jet on the physicochemical, functional, and antioxidant properties of flaxseed protein. *J. Food Sci.* **2020**, *85*, 2010–2019. [CrossRef] [PubMed]
24. Gu, C.H.; Howell, K.; Dunshea, F.R.; Suleria, H. LC-ESI-QTOF/MS Characterisation of phenolic acids and flavonoids in polyphenol-rich fruits and vegetables and their potential antioxidant activities. *Antioxidants* **2019**, *8*, 405. [CrossRef] [PubMed]
25. Nie, C.; Qin, X.; Duan, Z.; Huang, S.; Yu, X.; Deng, Q.; Xiang, Q.; Geng, F. Comparative structural and techno-functional elucidation of full-fat and defatted flaxseed extracts: Implication of atmospheric pressure plasma jet. *J. Sci. Food Agr.* **2021**, *102*, 823–835. [CrossRef]
26. Deng, Q.; Yu, X.; Ma, F.; Xu, J.; Huang, F.; Huang, Q.; Sheng, F. Comparative analysis of the in-vitro antioxidant activity and bioactive compounds of flaxseed in china according to variety and geographical origin. *Int. J. Food Prop.* **2018**, *20*, S2708–S2722. [CrossRef]
27. Li, J.; Geng, S.; Zhen, S.; Lv, X.; Liu, B. Fabrication and characterization of oil-in-water emulsions stabilized by whey protein isolate/phloridzin/sodium alginate ternary complex. *Food Hydrocoll.* **2022**, *129*, 107625. [CrossRef]
28. Li, C.; Huang, X.; Peng, Q.; Shan, Y.; Xue, F. Physicochemical properties of peanut protein isolate-glucomannanconjugates prepared by ultrasonic treatment. *Ultrason. Sonochem.* **2014**, *21*, 1722–1727. [CrossRef]

29. Karaca, A.C.; Low, N.; Nickerson, M. Emulsifying properties of canola and flaxseed protein isolates produced by isoelectric precipitation and salt extraction. *Food Res. Int.* **2011**, *44*, 2991–2998. [CrossRef]

30. Madhusudhan, K.T.; Singh, N. Isolation and characterization of the major fraction (12S) of flaxseed proteins. *J. Agr. Food Chem.* **1985**, *33*, 673–677. [CrossRef]

31. Ayad, A.A. Characterization and Properties of Flaxseed Protein Fractions. Ph.D. Thesis, Mcgill University, Montreal, QC, Canada, 2010.

32. Nwachukwu, I.D.; Girgih, A.T.; Malomo, S.A.; Onuh, J.O.; Aluko, R.E. Thermoase-derived flaxseed protein hydrolysates and membrane ultrafiltration peptide fractions have systolic blood pressure-lowering effects in spontaneously hypertensive rats. *Int. J. Mol. Sci.* **2014**, *15*, 18131–18147. [CrossRef] [PubMed]

33. Marambe, H.K.; Shand, P.J.; Wanasundara, J. Release of angiotensin i-converting enzyme inhibitory peptides from flaxseed (*Linum usitatissimum* L.) protein under simulated gastrointestinal digestion. *J. Agr. Food Chem.* **2011**, *59*, 9596–9604. [CrossRef] [PubMed]

34. Tang, C.; Wang, X. Physicochemical and structural characterisation of globulin and albumin from common buckwheat (*Fagopyrum esculentum Moench* L.) seeds. *Food Chem.* **2010**, *121*, 119–126. [CrossRef]

35. Yin, S.; Chen, J.; Sun, S.; Tang, C.; Yang, X.; Wen, Q.; Qi, J. Physicochemical and structural characterisation of protein isolate, globulin and albumin from soapnut seeds (*Sapindus mukorossi Gaertn* L.). *Food Chem.* **2011**, *128*, 420–426. [CrossRef] [PubMed]

36. Chen, N.; Yang, B.; Wang, Y.; Zhang, N.; Li, Y.; Qiu, C.; Wang, Y. Improving the colloidal stability and emulsifying property of flaxseed 11S globulin by heat induced complexation with soy 7S globulin. *LWT-Food Sci. Technol.* **2022**, *161*, 113364. [CrossRef]

37. Jackson, M.; Mantsch, H. The use and misuse of FTIR spectroscopy in the determination of protein structure. *Crit. Rev. Biochem. Mol.* **1995**, *30*, 95–120. [CrossRef] [PubMed]

38. Attoumbré, J.; Bienaimé, C.; Dubois, F.; Fliniaux, M.A.; Chabbert, B.; Baltora-Rosset, S. Development of antibodies against secoisolariciresinol–application to the immunolocalization of lignans in *Linum usitatissimum* L. seeds. *Phytochemistry* **2010**, *71*, 1979–1987. [CrossRef]

39. Cheng, C.; Yu, X.; Huang, F.; Peng, D.; Chen, H.; Chen, Y.; Huang, Q.; Deng, Q. Effect of different structural flaxseed lignans on the stability of flaxseed oil-in-water emulsion: An interfacial perspective. *Food Chem.* **2021**, *357*, 129522. [CrossRef]

40. Alu'datt, M.H.; Rababah, T.; Alhamad, M.N.; Gammoh, S.; Ereifej, K.; Kubow, S.; Alli, I. Characterization and antioxidant activities of phenolic interactions identified in byproducts of soybean and flaxseed protein isolation. *Food Hydrocoll.* **2016**, *61*, 119–127. [CrossRef]

41. Matsumura, Y.; Sirison, J.; Ishi, T.; Matsumiya, K. Soybean lipophilic proteins-Origin and functional properties as affected by interaction with storage proteins. *Curr. Opin. Colloid Interface Sci.* **2017**, *28*, 120–128. [CrossRef]

42. Zhang, Q.; Cheng, Z.; Wang, Y.; Zheng, S.; Wang, Y.; Fu, L. Combining alcalase hydrolysis and transglutaminase-cross-linking improved bitterness and techno-functional properties of hypoallergenic soybean protein hydrolysates through structural modifications. *LWT-Food Sci. Technol.* **2021**, *151*, 112096. [CrossRef]

43. Drusch, S.; Klost, M.; Kieserling, H. Current knowledge on the interfacial behaviour limits our understanding of plant protein functionality in emulsions. *Curr. Opin. Colloid Interface Sci.* **2021**, *56*, 101503. [CrossRef]

44. Luo, Y.; Zheng, W.; Shen, Q.; Zhang, L.; Tang, C.; Song, R.; Liu, S.; Li, B.; Chen, Y. Adsorption kinetics and dilatational rheological properties of recombinant pea albumin-2 at the oil-water interface. *Food Hydrocoll.* **2021**, *120*, 106866. [CrossRef]

45. Rodriguez, S.D.; Staszewski, M.V.; Pilosof, A.M.R. Green tea polyphenols-whey proteins nanoparticles: Bulk, interfacial and foaming behavior. *Food Hydrocoll.* **2015**, *50*, 108–115. [CrossRef]

46. Derkach, S.R.; Levachov, S.M.; Kuhkushkina, A.N.; Novosyolova, N.V.; Kharlov, A.E.; Matveenko, V.N. Rheological properties of concentrated emulsions stabilized by globular protein in the presence of nonionic surfactant. *Colloid. Surf. A* **2007**, *298*, 225–234. [CrossRef]

47. Zhang, Z.; Liu, Y. Rheology of globular proteins: Apparent yield stress, high shear rate viscosity and interfacial viscoelasticity of bovine serum albumin solutions. *Curr. Opin. Chem. Eng.* **2017**, *16*, 48–55. [CrossRef]

48. Julio, L.M.; Ruiz-Ruiz, J.C.; Tomás, M.C.; Segura Campos, M.R. Chia (*Salvia hispanica* L.) protein fractions: Characterization and emulsifying properties. *J. Food Meas. Charact.* **2019**, *13*, 3318–3328. [CrossRef]

49. Sanchez, C.C.; Patino, J. Contribution of the engineering of tailored interfaces to the formulation of novel food colloids. *Food Hydrocoll.* **2021**, *119*, 106838. [CrossRef]

50. Krause, J.P.; Schwenke, K.D. Behaviour of a protein isolate from rapeseed (*Brassica napus* L.) and its main protein components-globulin and albumin-at air/solution and solid interfaces, and in emulsions. *Colloid. Surf. B* **2001**, *21*, 29–36. [CrossRef]

Article

Pickering Emulsion Stabilized by Tea Seed Cake Protein Nanoparticles as Lutein Carrier

Li Liang [1], Junlong Zhu [1], Zhiyi Zhang [1], Yu Liu [1], Chaoting Wen [1], Xiaofang Liu [1], Jixian Zhang [1], Youdong Li [1], Ruijie Liu [2], Jiaoyan Ren [3], Qianchun Deng [4], Guoyan Liu [1] and Xin Xu [1,*]

1   College of Food Science and Engineering, Yangzhou University, Yangzhou 225127, China; liangli0508@hotmail.com (L.L.); z17502110240@163.com (J.Z.); zhzhyi0928@163.com (Z.Z.); liuyu1998wang@163.com (Y.L.); chaoting@yzu.edu.cn (C.W.); liuxf@yzu.edu.cn (X.L.); zjx@yzu.edu.cn (J.Z.); liyoudonghh@163.com (Y.L.); yzufsff@163.com (G.L.)
2   National Engineering Research Center for Functional Food, Collaborative Innovation Center of Food Safety and Quality Control in Jiangsu Province, School of Food Science and Technology, Jiangnan University, Wuxi 214122, China; ruijieliu-2007@hotmail.com
3   School of Food Science and Engineering, South China University of Technology, Guangzhou 510640, China; jyren@scut.edu.cn
4   Hubei Key Laboratory of Lipid Chemistry and Nutrition, Key Laboratory of Oilseeds Processing, Ministry of Agriculture and Rural Affairs, Oil Crops Research Institute, Chinese Academy of Agricultural Sciences, Wuhan 430062, China; dengqianchun@caas.cn
*   Correspondence: xuxin@yzu.edu.cn; Tel.: +86-189-5273-1677

Abstract: To effectively deliver lutein, hydrothermally prepared tea seed cake protein nanoparticles (TSCPN) were used to fabricate Pickering emulsion, and the bioaccessibility of lutein encapsulated by Pickering emulsion and the conventional emulsion was evaluated in vitro. The results indicated that the average size and absolute value of zeta potential of TSCPN increased along with the increase in the protein concentration, and 2% protein concentration was adopted to prepare TSCPN. With the increase in the concentration of TSCPN, the size of Pickering emulsion decreased from 337.02 μm to 89.36 μm, and when the TSCPN concentration was greater than 0.6%, all emulsions exhibited good stability during the 14 days storage. Combined with the microstructure result, 1.2% TSCPN was used to stabilize Pickering emulsion. With the increase in ionic concentration (0–400 mM), the particle size of the emulsions increased while the absolute value of zeta potential decreased. TSCPN-based Pickering emulsion was superior to the conventional emulsion for both lutein encapsulation (96.6 ± 1.0% vs. 82.1 ± 1.4%) and bioaccessibility (56.0 ± 1.1% vs. 35.2 ± 1.2%). Thus, TSCPN-based Pickering emulsion in this study have the potential as an effective carrier for lutein.

Keywords: tea seed cake; protein; hydrothermal; nanoparticles; Pickering emulsion; lutein; delivery; in vitro digestion; bioaccessibility

Citation: Liang, L.; Zhu, J.; Zhang, Z.; Liu, Y.; Wen, C.; Liu, X.; Zhang, J.; Li, Y.; Liu, R.; Ren, J.; et al. Pickering Emulsion Stabilized by Tea Seed Cake Protein Nanoparticles as Lutein Carrier. *Foods* **2022**, *11*, 1712. https://doi.org/10.3390/foods11121712

Academic Editor: Harjinder Singh

Received: 30 April 2022
Accepted: 7 June 2022
Published: 10 June 2022

Publisher's Note: MDPI stays neutral with regard to jurisdictional claims in published maps and institutional affiliations.

1. Introduction

Lutein is a xanthophyll of the carotenoid family, and numerous research studies have revealed that it has antioxidative, anti-inflammatory, anti cancer, oculo-protective, cardioprotective and immunomodulatory effects [1]. However, lutein cannot be synthesized by humans, and it can only be ingested from one's diet [2], such as yellow corn, pepper, and egg yolks, etc. Lutein mainly relies on a passive diffusion pathway through the intestinal epithelial membrane to enter the human circulation, its hydrophobicity and environmental instability prevent it from being stably transported into the gut, resulting in limited bioaccessibility [3]. Therefore, it is necessary to develop lutein-encapsulated delivery systems to overcome these limitations.

The delivery systems that have been developed include emulsions, liposomes, nanoparticles and so on [4]. Due to the excellent stability, simple preparation process, and potential applications in the delivery of bioactive compounds, Pickering emulsions have attracted great attention in food, pharmaceuticals, and cosmetic fields. Pickering emulsions have an excellent stability to coalescence and even Ostwald ripening, mainly due to the formation of a flexible but robust physical colloidal barrier by solid particles at the liquid-liquid interface [5,6]. The low cost and environment-friendly processes, such as the hydrothermal method to make nanoparticles and one-step high-speed shearing at room temperature to make emulsion, make Pickering emulsion attractive for industry [7,8]. Pickering emulsions stabilized by starch, β-lactoglobulin, β-cyclodextrin nanoparticles, or complexed with polysaccharides, have been fabricated to prolong the storage time of lutein [9–15]; however, the bioaccessibility of lutein entrapped by Pickering emulsions has not been reported.

Increasing consumer demand for clean label, sustainable, and biocompatible products has made the search for plant-based particle stabilizers a hot topic in food research. The low solubility of plant proteins in both hydrophilic and hydrophobic media is consistent with the requirements of Pickering emulsion stabilizers [7,16]. Although soybean protein, peanut protein, etc., have been used as Pickering emulsion stabilizers [17,18], it is still necessary to develop more available proteins to find excellent stabilizers. Tea seed cake protein (TSCP) is extracted from the seed of tea tree (*Camellia sinensis* (L.) O, Kuntze), which is usually discarded or made into feed, causing a lot of waste. Our previous study revealed that TSCP had balanced amino acids composition and ordered structure [19], which are considered favorable factors for the formation of nanoparticles to stabilize Pickering emulsions [20]. Thus, it is necessary to further investigate whether nanoparticles fabricated from TSCP (TSCPN) can stabilize Pickering emulsions and deliver lutein effectively [21,22].

In this study, we first characterized the hydrothermally prepared TSCPN with different TSCP concentrations. Then, the TSCPN-stabilized Pickering emulsions were characterized, and the storage stability was continuously observed for 14 days. Finally, lutein was encapsulated by TSCPN-stabilized Pickering emulsion, and in vitro gastrointestinal digestion were performed to evaluate its bioaccessibility.

2. Materials and Methods

2.1. Materials

Tea seed was bought from Hubei, China, and the proximate composition was determined [19]. Nile red, Fluorescein isothiocyanate (FITC), and porcine pancreatic lipase (100–650 U/mg) were purchased from Sigma-Aldrich (St. Louis, MO, USA). Porcine pepsin (12 U/mg) and porcine pancreatin (40 U/mg) were obtained from Sunson (Beijing, China). Corn oil was obtained from a local grocery (Suguo, Yangzhou, China). All chemical reagents in this study were of analytical grade.

2.2. Preparation of TSCP

TSCP was prepared based on protein extraction methods previously optimized by our laboratory [19].

Tea saponins were removed firstly prior to protein extraction due to its bitter taste and haemolytic toxicity [23]. The TSC was crushed and dispersed in 90% ethanol, (1:10, w/v), stirred in a water bath at 30 °C for 2 h, and then centrifuged (L550 Cence, Changsha, Hunan, China) at 4000 rpm for 15 min. After removing the supernatant, the above operation was repeated until most saponins were removed. Vanillin-sulfuric acid method [24] was used to measure the content of tea saponin.

The processed TSC was dissolved in water (1:20, w/v), followed by adjusting to pH 10.5, incubated in water bath at 50 °C for 30 min, centrifuged at 4000 rpm for 30 min. The supernatant was collected and was adjusted to pH 3.6 using 1 M HCl, then the sample was centrifuged at 4000 rpm for 30 min. The precipitate was washed, dialyzed for 24 h, and freeze-dried (CTFD-18S, CREATRUST, Qingdao, China). The freeze-dried protein sample was kept at −20 °C before use.

2.3. Preparation and Characteristics of TSCPN

The TSCPN preparation method was slightly modified from Ren et al. [25]. In brief, TSCPN was prepared using the hydrothermal method. TSCP was dissolved in RO water and adjusted to pH 7, stirred in a water bath at 90 °C for 1.5 h, centrifuged with 4000 rpm to remove insoluble particles, and then the supernatant was stored in a refrigerator at 4 °C. In order to explore the effects of TSCP concentration on TSCPN features, different concentrations (0.5%, 1%, 2%, 3%, 4%) of TSCPN were prepared according to the above method.

The measurement methods for particle size and zeta potential were according to Liang et al. [26]. Specifically, after 100-fold dilution of TSCPN, the particle size and zeta potential of TSCPN were analyzed (Zetasizer Nano ZS, Malvern, UK), and the refractive indices of the dispersant and material were set to 1.330 and 1.470, respectively. All tests were repeated at least three times.

After lyophilization, the microstructure of TSCPN was characterized using field emission scanning electron microscopy (s-4800, HITACHI, Tokyo, Japan).

2.4. Preparation and Characteristics of TSCPN-Stabilized Pickering Emulsion

We aimed to investigate the effect of TSCPN concentration on emulsion. The Pickering emulsion was prepared using oil fractions ($\varphi = 0.4\ v/v$) and different concentrations of TSCPN [27]. In brief, 4 mL of corn oil was added to 6 mL of TSCPN solution with different concentrations (0.3%, 0.6%, 1.2%, 1.8%, and 2.4%, w/v), then, the corn oil phase and TSCPN were mixed by high-speed homogenizer (T18, IKA, Staufen, Germany) at 20,000 rpm at ice-water bath for 3 min. The prepared emulsion was stored at 4 °C in the dark.

This type of TSCPN-stabilized Pickering emulsion was determined according to the drop test method [28]. In brief, emulsion was added to RO water and corn oil, respectively. If the droplets of emulsions agglomerated in the water and dispersed in the corn oil, it was considered as water-in-oil emulsions, otherwise, it was regarded as oil-in-water emulsions.

The method for determining the average particle size of the emulsion is the same as that for the nanoparticles described above. All tests were repeated at least five times. The microstructures of the emulsions were observed by laser scanning confocal microscope (LSM 880NLO, Carl Zeiss AG, Oberkochen, Germany). Before observation, Lipid and TSCPN were stained by 2–5 drops 1 mg/mL Nile red and 1 mg/mL FITC, respectively. The excitation and emission spectrum were 543 and 605 nm for Nile red, whereas they were 488 and 515 nm for FITC. In this case, Nile red fluoresces red, whereas FITC fluoresces green.

In addition, in order to investigate the influence of ionic strength. NaCl (final concentration 0–0.4 M) was added to Pickering emulsion, the preparation and characterization of the emulsions were carried out according to the above procedure.

2.5. Storage Stability

In order to explore the storage stability of Pickering emulsion at 4 °C, creaming index (*CI*) was evaluated. The of *CI* value was calculated as follows

$$CI\% = \frac{H_0}{H_t} \times 100$$

where H_s and H_t represent the height of the serum and all phases, respectively.

2.6. In Vitro Digestion Model

The ability of TSCPN-stabilized Pickering emulsions to deliver lutein was evaluated by measuring the release rate of lutein at different stages and the bioaccessibility after digestion in an in vitro digestion model. In addition, particle size and microscopic changes during digestion were measured as described in Section 2.4.

The in vitro digestion model is modified from the method of Liang et al. [26]. First, gastric and intestinal fluid storage solutions were prepared. Gastric stock solution was made up of 35.16 mM NaCl and 227.58 mM HCl; intestinal fluid stock solution was made up of 3.75 M NaCl and 247.53 mM CaCl. Bile salts had to be prepared 12 h in advance, and 0.5352 g of pig bile salts were dissolved in 10 mL of neutral phosphate buffer.

Due to the short retention time of the emulsion in the oral cavity, the influence of oral phase was not considered. Before entering the gastric phase, the emulsion was diluted to a 1% fat concentration for optimal digestion [29].

Gastric phase. We took 20 mL of gastric juice stock solution and incubated it at 37 °C, then mixed 20 mL of the diluted emulsion with 20 mL of gastric working solution, adjusted to pH 2.5, added 1.33 g pepsin and adjusted to pH 2.5, kept the incubation in a constant temperature shaker at 37 °C, 100 rpm culture for 2 h.

Small Intestinal Phase. We took 30 mL of gastric chyme, incubated it at 37 °C to adjust pH 7.0, then added 1.5 mL of intestinal fluid storage solution and 3.5 mL of bile salts, and adjusted to pH 7.0 again. We added 0.6 g trypsin and 0.4 g trypsin. 37 °C, 100 rpm constant temperature shaker for 2 h.

Lutein was extracted from chyme with 3 volumes of ethanol and acetone = 1:1 (v/v), and the mixture was shaken for 1 min and centrifuged at 5000 rpm for 5 min. The supernatant was collected to measure the absorbance at 450 nm. The release rate of lutein was calculated by the following equation:

$$RR(\%) = \frac{C_F}{C_I} \times 100\%$$

where C_F is the concentration of free lutein, and C_I is the theoretical concentration of lutein at different stages.

Bioaccessibility was calculated according to the report of E Fernández-García et al. [30]. After gastrointestinal simulation, the upper micelle was collected after centrifugation (8000 rpm, 30 min), and lutein was measured as above. Bioaccessibility is calculated according to the following equation:

$$a(\%) = \frac{C_M}{C_R} \times 100\%$$

where C_M is the lutein concentration in the micelle phase and C_R is the concentration of lutein in the entire intestinal phase.

The degree of protein hydrolysis was measured by neutral formaldehyde titration method [31]. In short, 10 mL of chyme was added to 5 drops of 30% hydrogen peroxide, adjusted to pH 7.5, 15 mL of neutral methanol was added, and titrated with 0.15 M NaOH after 1 min of reaction, and then the consumption volume of NaOH was recorded. The undigested TSCPN was set as blank control. The degree of hydrolysis can be calculated by the following formula:

$$C(\%) = \frac{(V_1 - V_0) \times V_w \times C_{NaOH} \times 14.01 \times 6.25}{1000 \times M}$$

where V_1 is the volume of NaOH solution consumed by digesta(mL); V_0 is the volume of NaOH solution consumed by blank (mL); V_w is the volume of water for ingredients (mL); C_{NaOH} is the concentration of NaOH solution for titration (0.15 M); 14.01 is the molar mass of nitrogen; 6.25 is the coefficient of nitrogen conversion to protein; M is the mass of TSCP.

2.7. Statistical Analysis

Triplicate tests were performed for each sample. The data were handled as the mean values $\pm$ standard deviation. The differences among the tests were evaluated by one-way ANOVA using Origin 2021 software ($p < 0.05$).

3. Result and Discussion

3.1. Effect of TSCP Concentrations on the Properties of TSCPN

3.1.1. Effect of TSCP Concentrations on the Particle Size and Zeta Potential of TSCPN

The properties of nanoparticles determine their ability to stabilize Pickering emulsions. Among them, particle size and potential are two important indicators. Specifically, the smaller the particle size, the faster the nanoparticles adsorb to the oil-water interface [32]. In addition, the electrostatic repulsion plays a vital role in the stabilization of particles against aggregation [33]. It has been reported that protein concentration has a significant effect on the properties of hydrothermally prepared protein nanoparticles [34]. As shown in Figure 1 (histogram), the average particle size of the TSCPN increased significantly with the increase in TSCP concentration (except for 1.0–2.0%), which may be related to the increased chance of particle-to-particle interactions [35]. Similar results were found in the study of soy protein nanoparticles by Fu et al. [34]. It is worth noting that, the suitable particle size of protein nanoparticles developed to stabilize Pickering emulsions is mostly in the range of 100–350 μm [36]. In addition, the particle size of TSCPN prepared with 0.5–2.0% concentration of TSCP was within this range. Moreover, it can be seen from Figure 1 (line graph) that with the increase in TSCP concentration, the absolute value of zeta potential of TSCPN increased significantly. The higher zeta potential means that the nanoparticles have greater electrostatic repulsion, which is beneficial to the stabilization of the water–oil interface in the emulsions [37]. Under the premise of suitable particle size, TSCPN prepared with 2.0% TSCP had the high absolute value of zeta potential (-32.5 ± 1.6 mV). Therefore, the TSCPN prepared by TSCP with a concentration of 2.0% had the potential to stabilize the Pickering emulsion.

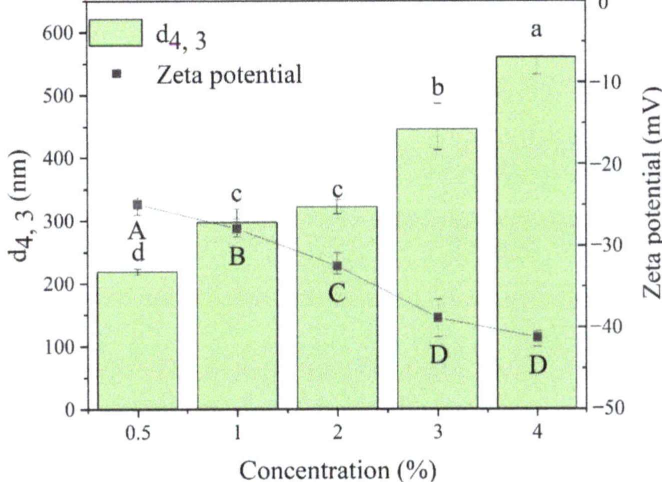

Figure 1. Effect of tee seed cake protein (TSCP) concentrations on the particle size (volume mean diameter $d_{4,3}$ histogram) and zeta potential (line) of nanoparticles fabricated from TSCP (TSCPN). Different lowercase letters indicated significant differences of particle size between treatments ($p < 0.05$), and different uppercase letters indicated significant differences of zeta potential between treatments ($p < 0.05$).

3.1.2. The Microstructure of TSCP and TSCPN

The morphology of solid nanoparticles can affect their ability to stabilize Pickering emulsions, specifically the particles' shape governs their behavior at the interface [38]. According to the above results, the microstructure of TSCPN prepared with a concentration of 2.0% TSCP was observed. After hydrothermal treatment, the porous spherical shape of TSCP has been changed significantly, and a thin plate structure composed of microspheres of TSCPN was formed (Figure 2). The non-deformable nanoparticle structure of TSCPN had the characteristics of a large specific surface area, thus they can adsorb to the water–oil interface quickly, which was necessary to stabilize Pickering emulsions [39].

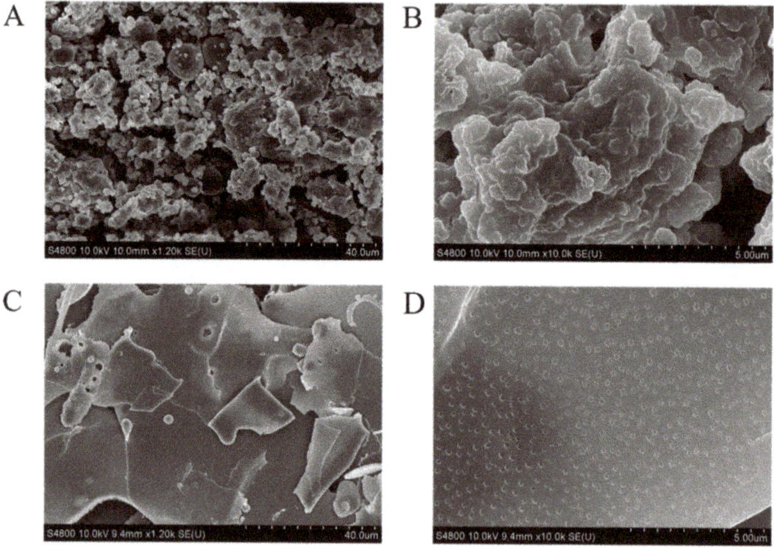

Figure 2. SEM image of TSCP (**A,B**) and TSCPN (**C,D**). The magnifications of (**A,C** and **B,D**) are 1200 and 10,000 times, respectively.

3.2. Characteristics of TSCPN-Stabilized PICKERING Emulsion

3.2.1. Type of TSCPN-Stabilized Pickering Emulsion

The TSCPN-stabilized Pickering emulsion was added dropwise to water and oil, respectively, and the type of emulsion was determined according to its dispersion and aggregation state. According to titration experiments, TSPN-stabilized Pickering emulsions aggregated in oil and diffused in water, which indicated that the TSPN-stabilized Pickering emulsion was an O/W emulsion.

3.2.2. Effect of TSCPN Concentrations on the Particle Size and Micromorphology of Pickering Emulsions

The stability of Pickering emulsions is significantly affected by the concentration of solid particles [40]. In addition, the stability of the emulsion can be reflected by particle size, because the smaller the particle size, the less aggregation. Figure 3 showed the average particle size of Pickering emulsions stabilized by different concentrations of TSCPN at a constant oil phase fraction of 0.4. The average particle size of Pickering emulsions decreased significantly with increasing TSCPN concentration. This may be due to the fact that more TSCPN adsorbed to the water–oil interface, resulting in a larger specific surface area of the emulsion [41]. This phenomenon was also observed in Pickering emulsions stabilized by soy protein nanoparticles [34] and rapeseed protein nanogels [42].

CLSM images can intuitively reflect the behavior of nanoparticles at the water–oil interface. The microstructure of Pickering emulsions stabilized by different concentrations of TSCPN (Figure 4) presented consistent results with their particle sizes (Figure 3). When the concentration of TSCPN was low (0.3–0.6%), there was not enough TSCPN to form a complete physical barrier at the water–oil interface, and lots of oil droplets agglomerated. When the concentration of TSCPN increased to 1.2–2.4%, the oil droplets size gradually decreased and the distribution became uniform, which was attributed to more TSCPN adsorbed to the water–oil interface, and it formed a dense physical barrier. In addition, higher concentrations of TSCPN increased the contact between the droplets, resulting in greater electrostatic repulsion, which also contributed to the stability of Pickering emulsions [43]. Combining the average emulsion size and the CLSM image, it can infer that 1.2–2.4% TSCPN can stabilize the Pickering emulsion with an oil phase of 0.4.

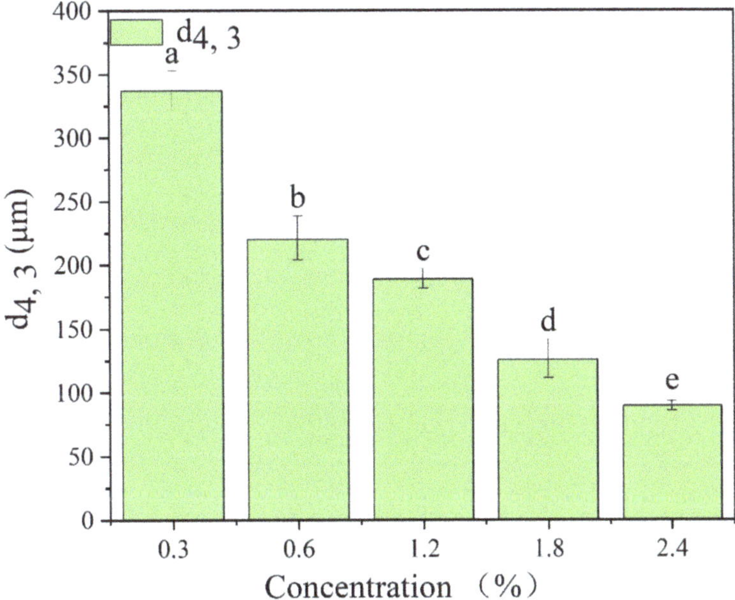

Figure 3. Volume mean diameter (d4, 3) of Pickering emulsions stabilized by different concentrations of TSCPN. Different lowercase letters indicated significant differences ($p < 0.05$).

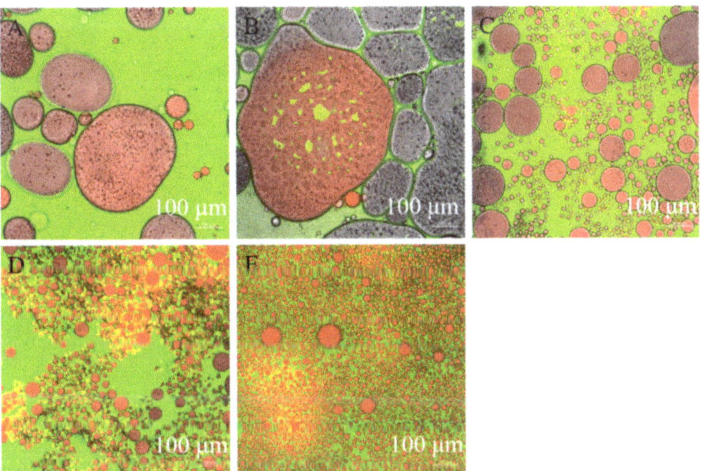

Figure 4. CLSM images of Pickering emulsions stabilized by different concentrations of TSCPN. (**A**), 0.3%; (**B**), 0.6%; (**C**), 1.2%; (**D**), 1.8%; (**E**), 2.4%. Oil was dyed red, and protein was dyed green.

3.2.3. Effect of TSCPN Concentrations on the Storage Stability of Pickering Emulsions

Good storage stability is a prerequisite for the use of emulsions in the food industry [44]. CI can reflect the aggregation and stratification of the emulsion, and then evaluate the storage stability of the emulsion, which has been used in the evaluation of various emulsions [42,45]. Within 8 h after fresh preparation, CI and its growth rate of the TSCPN-stabilized Pickering emulsion decreased significantly with increasing concentration of TSCPN (Figure 5B). This may be due to the smaller average size of Pickering emulsions stabilized by higher concentrations of TSCPN (Figure 3), which is helpful to slow down the aggregation of the emulsions. The same phenomenon was also observed in long-term storage over 14 days (Figure 5A), and CI of all Pickering emulsions entered a plateau at the third day of storage. This process has also been reported in other Pickering emulsions [42,46]. This suggests that the concentration of TSCPN is important not only to the formation, but also to the stabilization of Pickering emulsions on storage. A TSCPN concentration of 1.2% was chosen for the following experiments.

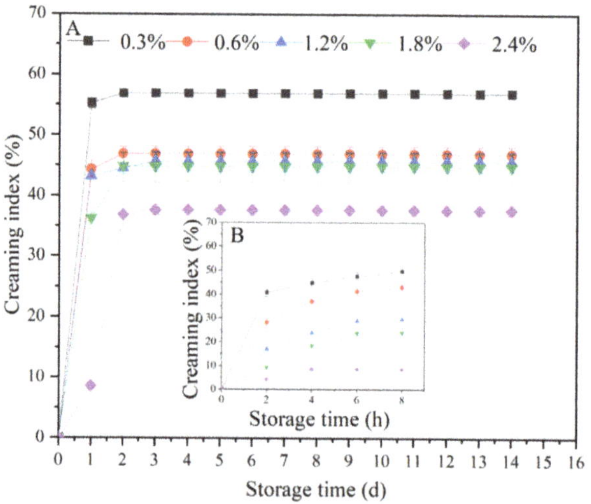

Figure 5. Changes in CI of Pickering emulsions stabilized by different concentrations of TSCPN over a storage period of 14 days (**A**) and within 8 h after preparation (**B**).

3.2.4. Effect of Ionic Strength on the Particle Size and Zeta Potential of Pickering Emulsions

Salt ions are ubiquitous in food matrices, and many studies have shown that ionic strength has a significant effect on Pickering emulsions [47,48]. Thus, it is necessary to investigate the effect of ionic strength on TSCPN-stabilized Pickering emulsions. Figure 6 (histogram in E) showed that the particle size of the emulsions increased along with the increase in ionic concentration, and the microstructural image also indicated that the droplets aggregation had occurred (Figure 6A–D). The result might be attributed to the electrostatic screening effect of ions, which can cause the reduction in electrostatic repulsion and aggregation of droplets. The significant decrease in the absolute value of the zeta potential of Pickering emulsions confirmed this hypothesis (Figure 6 line in E). It was worth noting that a slight decrease in droplet size occurred at 200 mM ionic strength. Although the increase in ionic strength negatively impacted the stability of Pickering emulsions overall, appropriate ionic strength will reduce the electrostatic repulsion of nanoparticles, and then free nanoparticles adsorb on the water–oil interface. The result could be evidenced by their microstructural images as well. The droplet exhibited irregular shape at low 100 mM of ionic concentration, and the droplets reverted to spherical shape at higher concentrations of ionic strength. Wang et al. also observed similar phenomena in the rapeseed-protein-nanogel-stabilized Pickering emulsion [42].

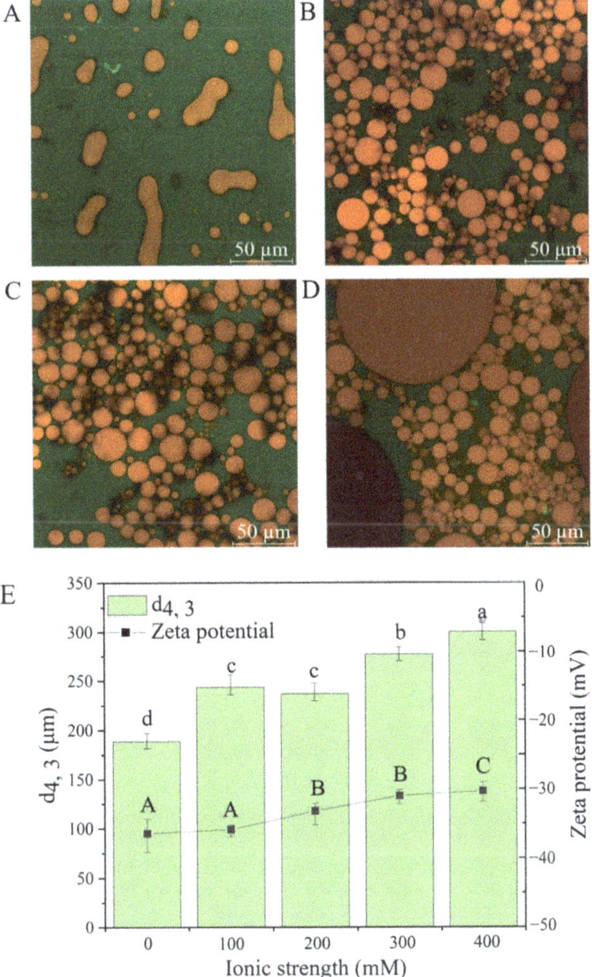

Figure 6. CLSM images (**A–D**) of the effect of different ion concentrations (100–400 mM) on the Pickering emulsion. Effect of ionic strength on particle size $d_{4,3}$ (histogram in (**E**)) and zeta potential (line in **E**) of Pickering emulsion. Different lowercase letters indicate significant difference of particle size between treatments ($p < 0.05$), and different uppercase letters indicated significant differences of zeta potential between treatments ($p < 0.05$).

3.3. Encapsulation and Delivery of Lutein: An In Vitro Digestion Model

The lutein delivery capacity of TSCPN-stabilized Pickering emulsion was evaluated in a simulated gastrointestinal digestion system, and TSCP-stabilized emulsion was prepared as a control. It can be seen from Figure 7 that there is 3.4 ± 1.0% and 17.9 ± 1.4% of free lutein initially in TSCPN-stabilized Pickering emulsion and TSCP-stabilized emulsion, respectively. The significantly higher encapsulation rate of Pickering emulsion may be due to the fact that TSCPN could quickly adsorb to the water–oil interface and form a dense physical barrier. Compared with the currently developed lutein delivery vehicles, such as nanocapsules with chitosan, nano lipid carriers with ω-3 fatty acid, etc., the TSCPN-stabilized Pickering emulsion has a higher encapsulation rate [1].

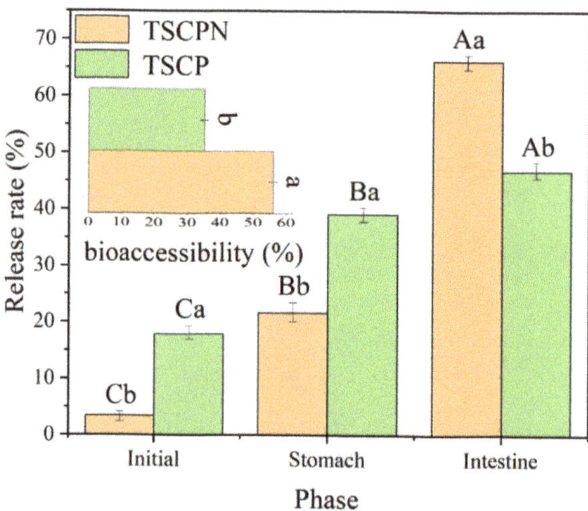

Figure 7. Lutein release rates (vertical bar graphs) of TSCPN-stabilized Pickering emulsions and TSCP-stabilized emulsions at initial stage, post-gastric, and post-intestinal; TSCPN-stabilized Pickerings and TSCP-stabilized post-intestinal Lutein bioaccessibility of emulsion delivery (horizontal bar graph). Different lowercase letters indicated significant differences between TSCPN-stabilized Pickering emulsion and TSCP-stabilized emulsion ($p < 0.05$), different capital letters indicate significant difference among different digestion phases ($p < 0.05$).

After gastric digestion, 21.6 ± 1.8% and 39.1 ± 1.2% of lutein was released from the TSCPN-stabilized Pickering emulsion and TSCP-stabilized emulsion, respectively. Compared with the initial emulsion, more lutein was released from TSCP-stabilized emulsion than from TSCPN-stabilized Pickering emulsion. The result may be attributed to the different aggregation state of the droplets after gastric digestion, which can be confirmed from their average droplet size (Figure 8B). Compared with the initial droplet size, the droplet size of the TSCPN-stabilized Pickering emulsion was significantly increased, indicating more aggregation, and thus possibly inhibiting the release of lutein. The same phenomenon was also observed in the zein/soluble polysaccharide composite nanoparticle system [49].

Compared with the stomach phase, the cumulative release rate of lutein in both emulsions increased significantly after intestinal digestion. Although the release rate of lutein from TSCP-stabilized emulsions in the gastric phase was higher than that of TSCPN-stabilized Pickering emulsions, both the release rate and the bioaccessibility of lutein after intestinal digestion were significantly higher in the latter than in the former emulsion, indicating that TSCPN-stabilized Pickering emulsion has better delivery efficiency than the TSCP-stabilized emulsion. The difference of release rate of lutein may be due to the difference of the hydrolysis degree of protein in the interface. We verified the assumption by measuring the hydrolysis degree of protein after intestinal digestion in both emulsions, and the hydrolysis degree of protein in both emulsion for TSCPN and TSCP was 56.80 ± 2.55%and 27.89 ± 2.59%, respectively. In addition, the particle size measurement of the micellar phase found that the size of the micelles obtained from the TSCPN-stabilized Pickering emulsion was 279.81 ± 21.08 nm, whereas the size of the micelles obtained from the TSCPN-stabilized emulsion was 306.41 ± 17.58 nm. Studies have demonstrated that a smaller micellar phase can increase the bioaccessibility of lutein [50]. There are no studies evaluating Pickering emulsion delivery of lutein in vitro, so comparisons can only be made with other types of delivery systems. Yu et al. developed the composite nanoparticles from gum Arabic and carboxymethylcellulose-modified stauntonia brachyyanthcra seed albumin for lutein delivery, and this system can increase the bioaccessibility of lutein to 46.8% [51]. Zhang et al. utilized glycosylated zein as a delivery vehicle for lutein and obtained a bioaccessibility of about 30% [52]. Compared with these developed carriers, TSCPN-stabilized Pickering emulsions exhibited better lutein delivery capacity during digestion in vitro.

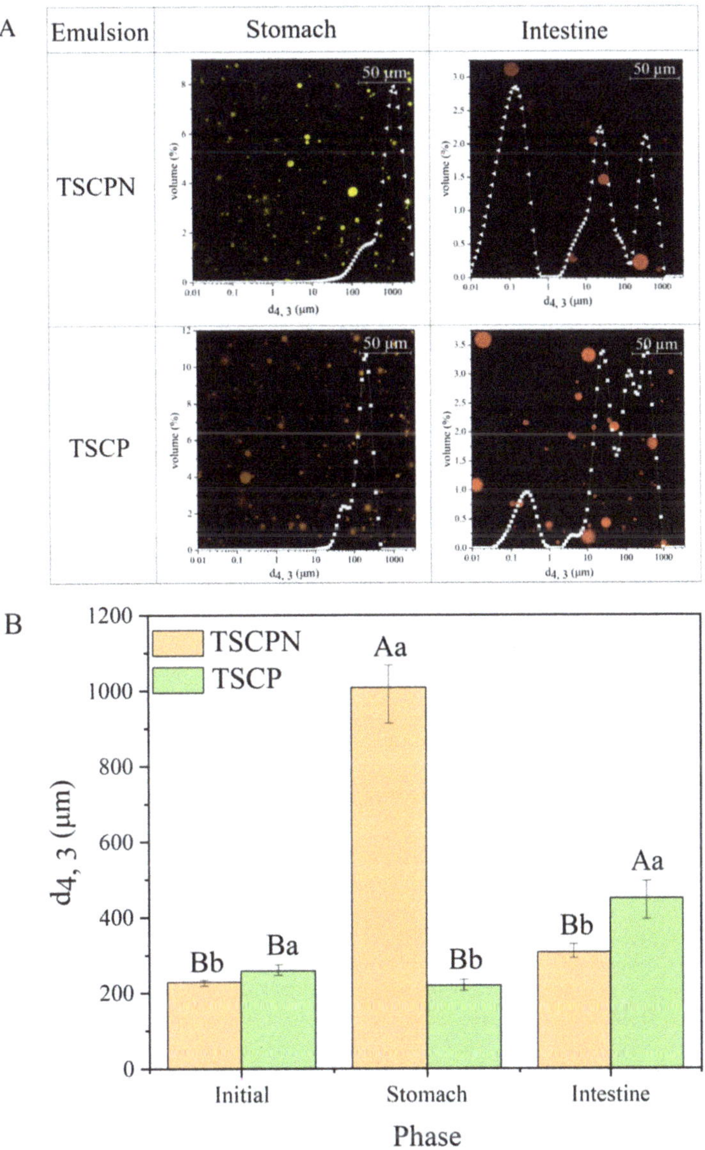

Figure 8. CLSM images, particle size distribution (**A**) and particle size (d4, 3) (**B**) of TSCPN-stabilized Pickering emulsion and TSCP-stabilized emulsion at different digestion periods. Different lowercase letters indicate significant differences between TSCPN-stabilized Pickering emulsions and TSCP-stabilized emulsions ($p < 0.05$), and different uppercase letters indicate significant differences at different digestion stages ($p < 0.05$).

4. Conclusions

A novel lutein delivery vehicle was prepared using TSCPN-stabilized Pickering emulsion. The particle size of TSCPN was affected by the concentration of TSCP, but the average particle size is between 321.55 ± 11.05–548.40 ± 8.7 nm. Pickering emulsion stabilized by 1.2% TSCPN showed good storage stability during the 14 days storage period. In addition, TSCPN-stabilized Pickering emulsion could significantly improve the bioaccessibility of lutein (56.0 ± 1.1%) in vitro, which was attributed to the high degree of hydrolysis of TSCPN. Therefore, the Pickering emulsion prepared in this study can be used as a good carrier to improve the bioaccessibility of lutein in beverages, salad dressing, etc.

Author Contributions: Conceptualization, L.L., C.W., G.L. and X.X.; data curation, L.L. and J.Z. (Junlong Zhu); formal analysis, L.L., J.Z. (Junlong Zhu) and G.L.; funding acquisition, L.L.; investigation, J.Z. (Junlong Zhu), Z.Z. and Y.L. (Yu Liu); methodology, L.L., J.Z. (Junlong Zhu), Z.Z., Y.L. (Yu Liu), C.W., X.L., J.Z. (Jixian Zhang), Y.L. (Youdong Li), R.L., J.R. and Q.D.; project administration, X.X.; writing—original draft, J.Z. (Junlong Zhu); writing—review and editing, L.L., C.W., X.L., J.Z. (Jixian Zhang), Y.L. (Youdong Li), R.L., J.R., Q.D. and G.L.; supervision, G.L. and X.X.; validation, L.L., G.L. and X.X. All authors have read and agreed to the published version of the manuscript.

Funding: This work was supported by the Program of National Natural Science Foundation of China (31901651).

Institutional Review Board Statement: Not applicable.

Informed Consent Statement: Not applicable.

Data Availability Statement: Data are contained within the article.

Acknowledgments: The author thanks all the members of the laboratory for helping the experiment.

Conflicts of Interest: The authors declare no conflict of interest.

References

1. Madaan, T.; Choudhary, A.N.; Gyenwalee, S.; Thomas, S.; Mishra, H.; Tariq, M.; Vohora, D.; Talegaonkar, S. Lutein, a versatile phyto-nutraceutical: An insight on pharmacology, therapeutic indications, challenges and recent advances in drug delivery. *PharmaNutrition* **2017**, *5*, 64–75. [CrossRef]
2. Sajilata, M.; Singhal, R.; Kamat, M. The carotenoid pigment zeaxanthin—A review. *Compr. Rev. Food Sci. Food Saf.* **2008**, *7*, 29–49. [CrossRef]
3. Kijlstra, A.; Tian, Y.; Kelly, E.R.; Berendschot, T.T. Lutein: More than just a filter for blue light. *Prog. Retin Eye Res.* **2012**, *31*, 303–315. [CrossRef] [PubMed]
4. Shegokar, R.; Mitri, K. Carotenoid lutein: A promising candidate for pharmaceutical and nutraceutical applications. *J. Diet. Suppl.* **2012**, *9*, 183–210. [CrossRef] [PubMed]
5. Yang, H.; Su, Z.; Meng, X.; Zhang, X.; Kennedy, J.F.; Liu, B. Fabrication and characterization of Pickering emulsion stabilized by soy protein isolate-chitosan nanoparticles. *Carbohyd. Polym.* **2020**, *247*, 116712. [CrossRef] [PubMed]
6. Marefati, A.; Rayner, M. Starch granule stabilized Pickering emulsions: An 8-year stability study. *J. Sci. Food Agric.* **2020**, *100*, 2807–2811. [CrossRef] [PubMed]
7. Sarkar, A.; Dickinson, E. Sustainable food-grade Pickering emulsions stabilized by plant-based particles. *Curr. Opin. Colloid Interface Sci.* **2020**, *49*, 69–81. [CrossRef]
8. Marto, J.; Ascenso, A.; Simoes, S.; Almeida, A.J.; Ribeiro, H.M. Pickering emulsions: Challenges and opportunities in topical delivery. *Expert Opin. Drug Del.* **2016**, *13*, 1093–1107. [CrossRef]
9. Geng, S.; Li, Y.; Lv, J.; Ma, H.; Liang, G.; Liu, B. Fabrication of food-grade Pickering high internal phase emulsions (HIPEs) stabilized by a dihydromyricetin and lysozyme mixture. *Food Chem.* **2022**, *373*, 131576. [CrossRef]
10. Li, S.; Zhang, B.; Li, C.; Fu, X.; Huang, Q. Pickering emulsion gel stabilized by octenylsuccinate quinoa starch granule as lutein carrier: Role of the gel network. *Food Chem.* **2020**, *305*, 125476. [CrossRef]
11. Liu, Z.; Geng, S.; Jiang, Z.; Liu, B. Fabrication and characterization of food-grade Pickering high internal emulsions stabilized with beta-cyclodextrin. *LWT* **2020**, *134*, 110134. [CrossRef]
12. Liu, Z.; Li, Y.; Geng, S.; Mo, H.; Liu, B. Fabrication of food-grade Pickering high internal phase emulsions stabilized by the mixture of beta-cyclodextrin and sugar beet pectin. *Int. J. Biol. Macromol.* **2021**, *182*, 252–263. [CrossRef]
13. Su, J.; Guo, Q.; Chen, Y.; Dong, W.; Mao, L.; Gao, Y.; Yuan, F. Characterization and formation mechanism of lutein pickering emulsion gels stabilized by beta-lactoglobulin-gum arabic composite colloidal nanoparticles. *Food Hydrocoll.* **2020**, *98*, 105276. [CrossRef]

14. Su, J.; Guo, Q.; Chen, Y.; Mao, L.; Gao, Y.; Yuan, F. Utilization of beta-lactoglobulin-(-)-Epigallocatechin-3-gallate(EGCG) composite colloidal nanoparticles as stabilizers for lutein pickering emulsion. *Food Hydrocoll.* **2020**, *98*, 105293. [CrossRef]

15. Wani, T.A.; Shah, A.G.; Wani, S.M.; Wani, I.A.; Masoodi, F.A.; Nissar, N.; Shagoo, M.A. Suitability of Different Food Grade Materials for the Encapsulation of Some Functional Foods Well Reported for Their Advantages and Susceptibility. *Crit. Rev. Food Sci. Nutr.* **2016**, *56*, 2431–2454. [CrossRef]

16. Ebert, S.; Grossmann, L.; Hinrichs, J.; Weiss, J. Emulsifying properties of water-soluble proteins extracted from the microalgae Chlorella sorokiniana and Phaeodactylum tricornutum. *Food Funct.* **2019**, *10*, 754–764. [CrossRef]

17. Yan, X.; Ma, C.; Cui, F.; McClements, D.J.; Liu, X.; Liu, F. Protein-stabilized Pickering emulsions: Formation, stability, properties, and applications in foods. *Trends Food Sci. Technol.* **2020**, *103*, 293–303. [CrossRef]

18. da Silva, A.M.M.; Almeida, F.S.; Sato, A.C.K. Functional characterization of commercial plant proteins and their application on stabilization of emulsions. *J. Food Eng.* **2021**, *292*, 110277. [CrossRef]

19. Xiao, Y. *Study on the Preparation of Tea Seed Cake Protein and Its Functional and Digesion and Absorption Properties*; Yangzhou University: Yanghzhou, China, 2021.

20. Yang, Y.; Jiao, Q.; Wang, L.; Zhang, Y.; Jiang, B.; Li, D.; Feng, Z.; Liu, C. Preparation and evaluation of a novel high internal phase Pickering emulsion based on whey protein isolate nanofibrils derived by hydrothermal method. *Food Hydrocoll.* **2022**, *123*, 107180. [CrossRef]

21. Aziz, A.; Khan, N.M.; Ali, F.; Khan, Z.U.; Ahmad, S.; Jan, A.K.; Rehman, N.; Muhammad, N. Effect of protein and oil volume concentrations on emulsifying properties of acorn protein isolate. *Food Chem.* **2020**, *324*, 126894. [CrossRef]

22. Cui, Z.; Chen, Y.; Kong, X.; Zhang, C.; Hua, Y. Emulsifying properties and oil/water (O/W) interface adsorption behavior of heated soy proteins: Effects of heating concentration, homogenizer rotating speed, and salt addition level. *J. Agric. Food Chem.* **2014**, *62*, 1634–1642. [CrossRef] [PubMed]

23. Zhu, Z.; Wen, Y.; Yi, J.; Cao, Y.; Liu, F.; McClements, D.J. Comparison of natural and synthetic surfactants at forming and stabilizing nanoemulsions: Tea saponin, Quillaja saponin, and Tween 80. *J. Colloid Interface Sci.* **2019**, *536*, 80–87. [CrossRef] [PubMed]

24. Zhang, X.-F.; Yang, S. L.; Han, Y.-Y; Zhao, L.; Lu, G.-L.; Xia, T.; Gao, L.-P. Qualitative and quantitative analysis of triterpene saponins from tea seed pomace (*Camellia oleifera* abel) and their activities against bacteria and fungi. *Molecules* **2011**, *19*, 7568–7580. [CrossRef] [PubMed]

25. Ren, Z.; Chen, Z.; Zhang, Y.; Lin, X.; Li, B. Novel food-grade Pickering emulsions stabilized by tea water-insoluble protein nanoparticles from tea residues. *Food Hydrocoll.* **2019**, *96*, 322–330. [CrossRef]

26. Liang, L.; Zhang, X.; Wang, X.; Jin, Q.; McClements, D.J. Influence of dairy emulsifier type and lipid droplet size on gastrointestinal fate of model emulsions: In vitro digestion study. *J. Agric. Food Chem.* **2018**, *66*, 9761–9769. [CrossRef]

27. Boostani, S.; Hosseini, S.M.H.; Golmakani, M.-T.; Marefati, A.; Hadi, N.B.A.; Rayner, M. The influence of emulsion parameters on physical stability and rheological properties of Pickering emulsions stabilized by hordein nanoparticles. *Food Hydrocoll.* **2020**, *101*, 105520. [CrossRef]

28. Feng, Y.; Lee, Y. Surface modification of zein colloidal particles with sodium caseinate to stabilize oil-in-water pickering emulsion. *Food Hydrocoll.* **2016**, *56*, 292–302. [CrossRef]

29. Zhang, R.; Zhang, Z.; Zhang, H.; Decker, E.A.; McClements, D.J. Influence of lipid type on gastrointestinal fate of oil-in-water emulsions: In vitro digestion study. *Food Res. Int.* **2015**, *75*, 71–78. [CrossRef]

30. Fernández-García, E.; Carvajal-Lérida, I.; Pérez-Gálvez, A. In vitro bioaccessibility assessment as a prediction tool of nutritional efficiency. *Nutr. Res.* **2009**, *29*, 751–760. [CrossRef]

31. Rutherfurd, S.M. Methodology for determining degree of hydrolysis of proteins in hydrolysates: A review. *J. AOAC Int.* **2010**, *93*, 1515–1522. [CrossRef]

32. Liang, H.-N.; Tang, C.-H. pH-dependent emulsifying properties of pea [*Pisum sativum* (L.)] proteins. *Food Hydrocoll.* **2013**, *33*, 309–319. [CrossRef]

33. Wu, J.; Ma, G.H. Recent studies of Pickering emulsions: Particles make the difference. *Small* **2016**, *12*, 4633–4648. [CrossRef] [PubMed]

34. Liu, F.; Tang, C.-H. Emulsifying properties of soy protein nanoparticles: Influence of the protein concentration and/or emulsification process. *J. Agric. Food Chem.* **2014**, *62*, 2644–2654. [CrossRef]

35. Liu, F.; Tang, C.-H. Soy protein nanoparticle aggregates as pickering stabilizers for oil-in-water emulsions. *J. Agric. Food Chem.* **2013**, *61*, 8888–8898. [CrossRef]

36. Albert, C.; Beladjine, M.; Tsapis, N.; Fattal, E.; Agnely, F.; Huang, N. Pickering emulsions: Preparation processes, key parameters governing their properties and potential for pharmaceutical applications. *J. Control. Release* **2019**, *309*, 302–332. [CrossRef]

37. Luo, M.; Olivier, G.K.; Frechette, J. Electrostatic interactions to modulate the reflective assembly of nanoparticles at the oil–water interface. *Soft Matter* **2012**, *8*, 11923–11932. [CrossRef]

38. Ortiz, D.G.; Pochat-Bohatier, C.; Cambedouzou, J.; Bechelany, M.; Miele, P. Current trends in Pickering emulsions: Particle morphology and applications. *Engineering* **2020**, *6*, 468–482. [CrossRef]

39. Zhang, T.; Xu, J.; Chen, J.; Wang, Z.; Wang, X.; Zhong, J. Protein nanoparticles for Pickering emulsions: A comprehensive review on their shapes, preparation methods, and modification methods. *Trends Food Sci. Technol.* **2021**, *113*, 26–41. [CrossRef]

40. Levine, S.; Bowen, B.D.; Partridge, S.J. Stabilization of emulsions by fine particles I. Partitioning of particles between continuous phase and oil/water interface. *Colloids Surf.* **1989**, *38*, 325–343. [CrossRef]
41. Frelichowska, J.; Bolzinger, M.-A.; Chevalier, Y. Effects of solid particle content on properties of o/w Pickering emulsions. *J. Colloid Interface Sci.* **2010**, *351*, 348–356. [CrossRef]
42. Wang, Z.; Zhang, N.; Chen, C.; He, R.; Ju, X. Rapeseed protein nanogels as novel pickering stabilizers for oil-in-water emulsions. *J. Agric. Food Chem.* **2020**, *68*, 3607–3614. [CrossRef] [PubMed]
43. Shi, A.; Feng, X.; Wang, Q.; Adhikari, B. Pickering and high internal phase Pickering emulsions stabilized by protein-based particles: A review of synthesis, application and prospective. *Food Hydrocoll.* **2020**, *109*, 106117. [CrossRef]
44. Chappat, M. Some applications of emulsions. *Colloids Surf. A* **1994**, *91*, 57–77. [CrossRef]
45. El Bouchikhi, S.; Ibrahimi, A.; Bensouda, Y. Creaming behavior prediction of argan oil in water emulsion stabilized by lacto-fermentation: Creaming index. *BMC Biotech.* **2021**, *21*, 53. [CrossRef]
46. Ding, M.; Zhang, T.; Zhang, H.; Tao, N.; Wang, X.; Zhong, J. Gelatin molecular structures affect behaviors of fish oil-loaded traditional and Pickering emulsions. *Food Chem.* **2020**, *309*, 125642. [CrossRef]
47. Araiza-Calahorra, A.; Sarkar, A. Pickering emulsion stabilized by protein nanogel particles for delivery of curcumin: Effects of pH and ionic strength on curcumin retention. *Food Struct.* **2019**, *21*, 100113. [CrossRef]
48. Zhu, Y.; McClements, D.J.; Zhou, W.; Peng, S.; Zhou, L.; Zou, L.; Liu, W. Influence of ionic strength and thermal pretreatment on the freeze-thaw stability of Pickering emulsion gels. *Food Chem.* **2020**, *303*, 125401. [CrossRef]
49. Li, H.; Yuan, Y.; Zhu, J.; Wang, T.; Wang, D.; Xu, Y. Zein/soluble soybean polysaccharide composite nanoparticles for encapsulation and oral delivery of lutein. *Food Hydrocoll.* **2020**, *103*, 105715. [CrossRef]
50. Mamatha, B.S.; Baskaran, V. Effect of micellar lipids, dietary fiber and β-carotene on lutein bioavailability in aged rats with lutein deficiency. *Nutrition* **2011**, *27*, 960–966. [CrossRef]
51. Yu, N.; Wang, J.; Jiang, C.; Nie, X.; Hu, Z.; Ye, Q.; Meng, X.; Xiong, H. Development of composite nanoparticles from gum Arabic and carboxymethylcellulose-modified Stauntonia brachyanthera seed albumin for lutein delivery. *Food Chem.* **2022**, *372*, 131269. [CrossRef]
52. Chang, Y.; Jiao, Y.; Li, D.-J.; Liu, X.-L.; Han, H. Glycosylated zein as a novel nanodelivery vehicle for lutein. *Food Chem.* **2022**, *376*, 131927. [CrossRef]

 *foods*

Article

Effect of Fibril Entanglement on Pickering Emulsions Stabilized by Whey Protein Fibrils for Nobiletin Delivery

Fangcheng Jiang [1,2], Chunling Chen [1], Xinlan Wang [1], Wenjing Huang [1], Weiping Jin [1,*] and Qingrong Huang [1,2,*]

[1] Key Laboratory for Deep Processing of Major Grain and Oil, School of Food Science and Engineering, Wuhan Polytechnic University, Wuhan 430023, China; fj109@scarletmail.rutgers.edu (F.J.); ccl3529151143@163.com (C.C.); wxl19980609@163.com (X.W.); huangwenjingbest@163.com (W.H.)

[2] Department of Food Science, Rutgers University, 65 Dudley Road, New Brunswick, NJ 08901, USA

* Correspondence: jwpacademic@outlook.com (W.J.); qhuang@sebs.rutgers.edu (Q.H.)

Abstract: The aim of the study was to investigate the effects of whey protein isolate (WPI) fibrils entanglement on the stability and loading capacity of WPI fibrils-stabilized Pickering emulsion. The results of rheology and small-angle X-ray scattering (SAXS) showed the overlap concentration (C*) of WPI fibrils was around 0.5 wt.%. When the concentration was higher than C*, the fibrils became compact and entangled in solution due to a small cross-sectional radius of gyration value (1.18 nm). The interfacial behavior was evaluated by interfacial adsorption and confocal laser scanning microscopy (CLSM). As the fibril concentration increased from 0.1 wt.% to 1.25 wt.%, faster adsorption kinetics (from 0.13 to 0.21) and lower interfacial tension (from 11.85 mN/m to 10.34 mN/m) were achieved. CLSM results showed that WPI fibrils can effectively absorb on the surface of oil droplets. Finally, the microstructure and in vitro lipolysis were used to evaluate the effect of fibrils entanglement on the stability of emulsion and bioaccessibility of nobiletin. At C* concentration, WPI fibrils-stabilized Pickering emulsions exhibited excellent long-term stability and were also stable at various pHs (2.0–7.0) and ionic strengths (0–200 mM). WPI fibrils-stabilized Pickering emulsions after loading nobiletin remained stable, and in vitro digestion showed that these Pickering emulsions could significantly improve the extent of lipolysis (from 36% to 49%) and nobiletin bioaccessibility (21.9% to 62.5%). This study could provide new insight into the fabrication of food-grade Pickering emulsion with good nutraceutical protection.

Keywords: whey protein isolate; Pickering emulsion; fibrils; interfacial structure; overlap concentration; bioaccessibility; nobiletin

Citation: Jiang, F.; Chen, C.; Wang, X.; Huang, W.; Jin, W.; Huang, Q. Effect of Fibril Entanglement on Pickering Emulsions Stabilized by Whey Protein Fibrils for Nobiletin Delivery. *Foods* 2022, 11, 1626. https://doi.org/10.3390/foods11111626

Academic Editors: Qianchun Deng, Ruijie Liu and Xin Xu

Received: 30 April 2022
Accepted: 24 May 2022
Published: 31 May 2022

Publisher's Note: MDPI stays neutral with regard to jurisdictional claims in published maps and institutional affiliations.

1. Introduction

Pickering emulsions have been widely applied in foods, pharmaceuticals, and cosmetics due to their outstanding characteristics, such as extraordinary stability, controlled release of components, and targeted nutraceutical delivery [1–3]. The formation and stability mechanisms of Pickering emulsions are affected by some factors, such as particle concentration, oil fraction, pH, ionic strength, and temperature [4–6].

The particle concentration is one of the most important and easily adjusted factors during Pickering emulsion formation [7]. In general, droplet size decreases with increasing particle concentration (at a fixed oil-to-water ratio) until it reaches a plateau level. This effect is attributed to two factors: (1) more particles are available to cover the oil droplet surfaces with concentration increasing; (2) sufficient particles cover all the oil droplet surfaces against re-coalescence [8]. The droplet sizes of emulsion also play an important role in emulsion stability due to the lower interfacial layer caused by large droplet sizes [9] Feng et al. reported that the apparent viscosity and modulus, and stability of food-grade gelatin nanoparticle–stabilized Pickering emulsions were improved by increasing the nanoparticle concentration [7]. Gao et al. found that Pickering emulsions stabilized by protein fibrils showed excellent physical stability at appropriate fibril concentrations (0.5–2.0%), while excessive fibril concentrations ($\geq$2.5%) resulted in larger emulsion droplet size [10]. This is attributed to the accumulation and aggregation of excess particles in the bulk phase. The aggregation not only increases the particle size but also forms a network on the surface of emulsion droplets or a gel-like structure in the continuous phase. When the particle concentration of pea protein isolate (PPI) nanoparticles is larger than 2 g/100 mL, the gel-like whole emulsion without creaming is obtained [11,12]. In addition, particle concentration plays a key role in the dynamic adsorption at the oil–water interface, which can further affect the properties of Pickering emulsions and their applications [13]. It was reported that the diffusion at the interface is a concentration-dependent process, and the emulsification performance of the nanoparticles and the stability of the corresponding emulsions were improved at high concentrations (>0.5%, w/v) [14]. Although the effect of particle concentration on Pickering emulsions has attracted much attention in recent years, few works have focused on fibril, especially if the fibril concentration is higher than overlap concentration (C*) in solutions.

Whey protein isolate (WPI), a byproduct of the cheese-making process, comprises proteins such as β-lactoglobulin, α-lactalbumin, and bovine serum albumin [15,16]. WPI could form fibrils through heat processing at acidic pH and low ionic strength [17,18]. These fibrils are usually a few nanometers in diameter but up to several micrometers in length, and the phase behavior in solution is similar to that of polymers. The critical overlap concentration (C*) is the intermediate state of the polymer between the dilute solution and the semi-concentrated solution. When the polymer concentration is lower than C*, the molecules move freely in the solution, while if the polymer concentration is higher than C*, molecules begin to interact and entangle [19]. In applications, WPI fibrils are potential emulsifiers and have been used to deliver lipophilic ingredients [20–22]. The effect of the C* of WPI fibrils on the oil-water interfacial behavior, stability of emulsions, and delivery efficiency has not been confirmed.

Nobiletin, a poly-methoxylated flavone (PMF) isolated from the peels of various citrus fruits, is a model for lipophilic ingredients [23,24]. It has lots of potential health benefits, but the poor water solubility (~0.9 μM in the aqueous phase) and low bioaccessibility limit nobiletin applications in the food industry [24,25]. This work aims to study the influence of concentration on the structural and interfacial properties of WPI fibrils. Then, Pickering emulsions stabilized by WPI fibrils were prepared to evaluate the in vitro bioaccessibility of nobiletin and to understand the relationship between interfacial structure and performance of WPI fibril-stabilized Pickering emulsions in delivering lipophilic nutraceuticals.

2. Materials and Methods

2.1. Materials

Whey protein isolate (WPI, Hilmar™ 9020) (protein content of 89.5%) was purchased from Hilmar Company (Hilmar Cheese, CA, USA). Soybean oil was supplied by Yihai Kerry Arawana Oils and Grains Industries Ltd. (Shanghai, China). Nobiletin (HPLC purity 98%) was obtained from Sichuan Weikeqi Biological Technology, Co., (Chengdu, China). Pancreatin from porcine pancreas and pepsin from porcine gastric mucosa were purchased from Sigma-Aldrich (St. Louis, MO, USA). All solutions were prepared with Millipore water (Millipore, MA, USA). All other chemical reagents, including thioflavin T (ThT), hydrogen chloride (HCl), and sodium hydroxide (NaOH), were analytical grade.

2.2. Preparation of WPI Fibrils

WPI fibrils were prepared according to our previous study [6]. Briefly, 2 wt.% WPI solution was adjusted to pH 2.0 and heated at 90 °C for 5 h. Non-fibril peptides were removed via dialysis with pH 2.0 water for three days. WPI fibrils were dispersed in water at various concentrations (from 0.1 wt.% to 1.25 wt.%).

2.3. Rheology Analysis

The viscosity of WPI fibrils was measured by a rheometer (Malvern Panalytical Limited, Malvern, England, UK) using a 60 mm cone and plate, with a shear rate ranging from 0.01 to 1000 s^{-1}. All samples were measured at 25 °C. The obtained viscosity, as a function of shear rate data, was fitted to Cross models (Equation (1)) to obtain the relevant parameters [26]. The shape and curvature of a flow curve can be described through this model and the behavior at unmeasured shear rates can be predicted as well.

$$\frac{\eta - \eta_\infty}{\eta_0 - \eta_\infty} = \frac{1}{1 + (K\dot{\gamma})^m} \tag{1}$$

where η_0 is the zero-shear viscosity; η_∞ is the infinite shear viscosity; K is the cross constant, which is indicative of the onset of shear-thinning; $\dot{\gamma}$ is the shear rate or strain rate; and m is the shear thinning index, which ranges from 0 (Newtonian) to 1 (infinitely shear thinning).

2.4. SAXS Analysis

The samples were prepared with a series of WPI fibril concentrations (0.1, 0.25, 0.5, 1.0, 1.25 wt.%). A detector located 2631 mm from the sample was set to collect the scattering intensity, and the wavelength of X-ray radiation was 1.033 Å. The final scattering data were obtained by averaging 20 curves, with the scattering vector Q ranging from 1.0×10^{-2} to 4.1×10^{-1} Å^{-1}.

2.5. Dilatational Interfacial Rheology

The change in the interface tension (mNm^{-1}) at the oil–water interface with adsorption time (t) was measured by droplet shape analysis (DSA30S, KRUSS, Hamburg, Germany). The WPI fibril solution was collected in a syringe with a diameter of 1.8 mm. The needle tip was immersed in a rectangular glass tank filled with soybean oil. The interfacial tension value was recorded at 1 frame/s for the first 1 h and 0.1 frame/min thereafter, and the value was monitored continuously for 10,800 s. All measurements were performed at 25 °C.

2.6. Preparation of WPI Fibril-Stabilized Pickering Emulsion with and without Nobiletin

2.6.1. Preparation of Pickering Emulsions Stabilized by WPI Fibrils

Samples were mixed with different WPI fibril concentrations (0.1–1.25 wt.%) and a constant oil phase volume (φ = 50%). The samples were emulsified using a high-speed homogenizer at 10,000 rpm for 3 min (Ultra Turrax, T18 digital, IKA, Staufen, Germany).

2.6.2. Preparation of Nobiletin-Loaded Pickering Emulsions Stabilized by WPI Fibrils

Nobiletin was dissolved in soybean oil by heating (100 °C for 1 h) in an oil bath to a final concentration of 0.5 wt.%. The nobiletin-loaded Pickering emulsion was prepared as described above.

2.7. Characterization of WPI Fibril-Stabilized Pickering Emulsion

Confocal Laser Scanning Microscopy

Confocal laser scanning microscopy (FV1200, Olympus, Tokyo, Japan) was performed to confirm the formation of Pickering emulsions. Prior to confocal laser scanning microscopy analysis, samples were dyed by ThT. Images were obtained using a 20× magnification lens at an excitation wavelength of 488 nm. ImageJ software was used to estimate the mean droplet diameters.

2.8. Optical Microscopy

The microstructure images of the Pickering emulsion stabilized by WPI fibrils were obtained using an optical microscope (CX40, Sunny Optical Technology Co., Ltd., Yuyao, China) equipped with a camera. Thirty microliters of Pickering emulsion were deposited on the glass slide and covered with a coverslip. All samples were measured with a magnification of $20\times$ at room temperature. Nobiletin-loaded Pickering emulsions were characterized using the same method described for the unloaded Pickering emulsions.

2.9. Physicochemical Stability of WPI Fibril- Stabilized Pickering Emulsions

2.9.1. Effect of pH

Freshly prepared samples of Pickering emulsions were diluted 10 times using deionized water, and the pH was adjusted to pH 2.0, 3.0, 4.0, 5.0, 6.0, and 7.0, respectively, with 0.1 M NaOH solutions.

2.9.2. Effect of Ionic Strength

Freshly prepared samples of Pickering emulsions were diluted 10 times using 0 mM to 200 mM NaCl solutions.

2.9.3. Effect of Storage Time

After the preparation of emulsions, the freshly prepared sample emulsions were stored at room temperature. The microstructure of the emulsions was determined at regular storage periods (1, 7, and 28 d).

2.10. In Vitro Digestion Analysis, Free Fatty Acid Release, and Bioaccessibility of Nobiletin

2.10.1. Digestion of Pickering Emulsions Stabilized by WPI Fibrils

This study used an in vitro model consisting of oral, gastric, and intestine phases slightly modified from Brodkorb et al. [27]. Salivary amylase was not added in the oral digestion phase in this study. Fresh emulsions were prepared with an oil fraction of 50% (w/w). The samples, including soybean oil and WPI fibril-stabilized Pickering emulsion containing 0.25 g of oil, were mixed with 4 mL of simulated salivary fluid (SSF), 0.025 mL of 0.3 M $CaCl_2$, and 0.225 mL of pure water for 2 min. Eight milliliters of simulated gastric fluid (SGF), 0.005 mL of 0.3 M $CaCl_2$, 0.4 mL of 5 M HCl, and 0.448 mL of pure water were added to oral digestive juice successively, and freshly dissolved pepsin was added to achieve enzyme activity of 2000 U/mL and to initiate the gastric digestion process. Subsequently, the reaction mixture was incubated under continuous stirring in a temperature-controlled oil bath ($37.0 \pm 0.1\ ^\circ$C) for 2 h. Eight milliliters of simulated intestinal fluid (SIF), 0.04 mL of 0.3 M $CaCl_2$, 10 mM bile salt, and 3.16 mL of pure water were added to the gastric digesta, and the pH was adjusted to 7.0 with the addition of NaOH. Pancreatin was added to gastric digesta to obtain an enzyme activity of 100 U/mL to start intestinal digestion. The mixture was incubated under continuous agitation in an oil bath ($37.0 \pm 0.1\ ^\circ$C) for 2 h, and 0.1 M NaOH was added manually to maintain the pH at 7.0 during lipolysis. The volume of added NaOH solution was recorded over time during the intestinal digestion. Lipolysis in samples was characterized by the release of free fatty acids (**FFA**). The fraction of **FFA** released was calculated as follows:

$$\%\mathbf{FFA} = 100 \times \frac{M_{\text{lipid}} \times V_{\text{NaOH}} \times m_{\text{NaOH}}}{W_{\text{lipid}} \times 2} \tag{2}$$

where M_{lipid} is the molecular mass of the triacylglycerol oil (in g/mol); V_{NaOH} is the volume of NaOH solution used to neutralize the released **FFA** (in L); m_{NaOH} is the molarity of NaOH solution (in mol/L); and W_{lipid} is the total mass of the initial triacylglycerol oil. The molecular mass of soybean oil was 876.56 g/mol.

2.10.2. High-Performance Liquid Chromatography (HPLC) Analysis of Nobiletin

The digests were collected after gastrointestinal digestion and centrifuged at $10{,}000\times$ g for 40 min. The clear micelle phase was collected, and the nobiletin content in the micelle phase was determined by an UltiMate 3000 HPLC system (Dionex, Sunnyvale, CA, USA), with a SunFire C18 column (150 mm $\times$ 4.6 mm, 5 μm). The HPLC mobile phase consisted of (A) acetonitrile and (B) water. The elution conditions were as follows: 45% (A) and

55% (B), and the running time was 15 min. The detection wavelength was 333 nm, and the concentration of nobiletin was determined using a standard curve of nobiletin.

2.10.3. Determination of Nobiletin Bioaccessibility

After HPLC quantification of nobiletin in the micelle phase, nobiletin bioaccessibility was determined by the following equation [28]:

$$\text{bioaccessibility} = \frac{\text{nobiletin content in the micelle phase}}{\text{total nobiletin content in the formulations}} \times 100\% \tag{3}$$

2.11. Statistical Analysis

Each experiment was conducted in triplicate. All statistical analyses were performed using OriginPro 9.0. Duncan's test of SPSS Statistics 26 was used, and significance was set at $p < 0.05$.

3. Results and Discussion

3.1. Rheology

In polymer solution, there are three regimes: dilute, semi-dilute, and concentrated. The transition from the dilute to the semi-dilute regime depends on the concentration of the polymer solution, and entanglement occurs at concentrations above the overlap threshold (C*) [29]. The C* can be determined by measuring the concentration at which the viscosity of the polymer solution increases suddenly. Nanofibrils are, to some extent, similar to polymers, which have a few nanometers in diameter but up to several micrometers in length. As shown in Figure 1A, the viscosity of WPI fibril solutions at the same concentration decreased with increasing shear rates, indicating that all samples showed shear-thinning behavior, a type of non-Newtonian behavior. At the same shear rates, the viscosity of WPI fibrils increased with increasing WPI fibril concentration, which was attributed to molecule chain entanglement [26]. There was more entanglement among protein molecules as the WPI fibril concentration increased. In addition, at low enough shear rates, shear-thinning samples showed a constant viscosity, which was defined as the zero-shear viscosity (η_0). Therefore, to further understand the steady flow behavior of WPI fibrils, the Cross model was applied in Figure 1B. At low concentrations (0.1–0.25 wt.%), the η_0 of the WPI fibrils did not change significantly, indicating that the WPI fibrils were flexible and moved freely in the solution. As the concentration increased to 0.5 wt.%, the η_0 increased abruptly, which indicated that it was a critical overlap concentration (C*) [26]. At this concentration, WPI fibrils were closer to each other, and entanglement occurred between the WPI fibrils. As the concentration increased further, the density of entanglement between the WPI fibrils increased, which led to an increase in the viscosity of the WPI fibrils.

3.2. SAXS

Small-angle X-ray scattering (SAXS) is a powerful method that probes the structure of biomolecules at the nanoscale [30]. Figure 2A shows the scattering intensity profiles of WPI fibrils at various concentrations. In a previous study, β-Lg fibrils exhibited a rigid rod-like structure at low concentrations (0.2 and 0.3 wt.%) since the slope of the low Q region of the scattering profiles was −1 [6]. Guinier analysis is a straightforward approach to determining the cross-sectional radius of gyration (R_c), and the shape of the primary nanostructure is calculated via Guinier fitting in the Q range of 0.03–0.05 Å [31]. Guinier plots of WPI fibrils are presented in Figure 2B. The R_c of WPI fibrils from 0.1 wt.% to 1.25 wt.% was 11.45 Å, 15.55 Å, 13.75 Å, 12.57 Å, and 11.84 Å, respectively. The value of R_c increased when WPI fibril concentration increased, and it decreased after the concentration exceeded 0.5 wt.%, which suggested that there were two distinct, concentration dependent scaling regions in the R_c of WPI fibril solutions. These regions are analogous to the threshold between dilute and semi-dilute regions in WPI fibrils dissolved in isolated form. Moreover, there was hardly any intermolecular interaction when the concentration of WPI fibrils was lower than 0.5 wt.%. Above 0.5 wt.%, the fibril size decreased upon the increase in fibril concentration, which may be due to intermolecular interpenetration [32]. The threshold concentration between the two distinct scaling regions was very close to the C* of the WPI fibril solution.

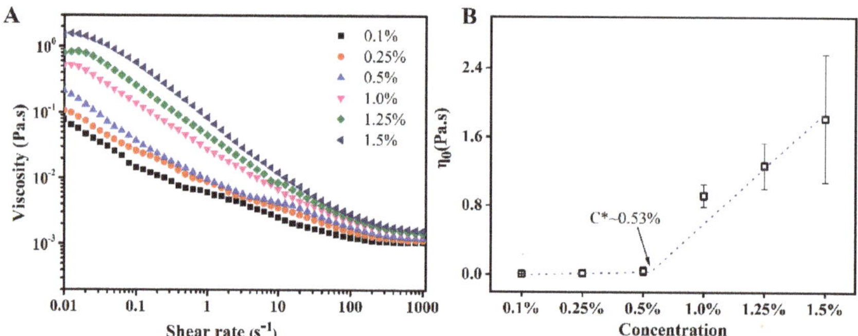

Figure 1. Rheology diagram of WPI fibrils solutions with the concentrations of 0.1–1.25 wt.% (**A**) Steady flow curves and (**B**) the relationship between η_0 and of WPI fibrils concentrations (pH 2.0).

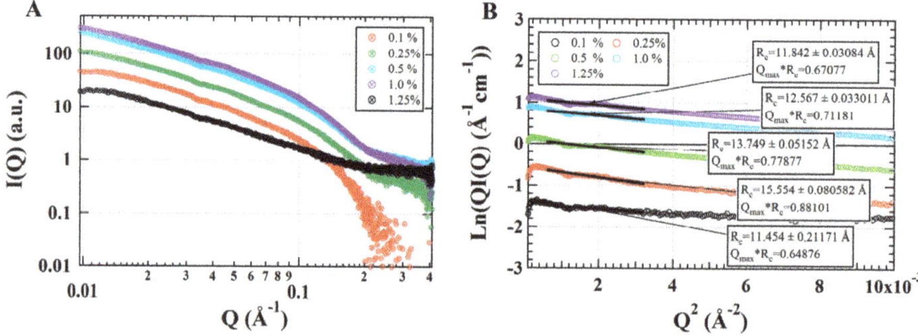

Figure 2. Small-angle X-ray scattering (SAXS) at different WPI fibrils concentration. (**A**) Profiles of SAXS intensity I(q) and (**B**) Guinner plots of WPI fibrils with different concentrations (0.1–1.25%) at pH 2.0.

3.3. Interfacial Adsorption Behavior

To evaluate the role of the WPI fibril concentration in the formation dynamics of the films at the oil–water interface, the interfacial adsorption behavior of the WPI fibrils in relation to concentration was investigated. Figure 3A shows the time evolution of interfacial tension for the WPI fibrils (0.1–1.25 wt.%) at the oil–water interface. Initially, the interfacial tension values of WPI fibrils with concentrations of 0.1–1.25 wt.% were 15.96 mN/m, 15.36 mN/m, 15.08 mN/m, 13.98 mN/m, and 13.47 mN/m, respectively. This suggested that a significant reduction in interfacial tension occurred when the concentration was higher than C*. The interfacial tension values of WPI fibrils rapidly decreased with adsorption time and exhibited a notable dependence on fibril concentration, which suggests a high concentration could positively promote the adsorption process [33]. The diffusion rate was used to explain the migration of proteins from the bulk phase to the oil–water interface [34]. K_{diff} is an estimation of the rate of initial diffusion-controlled migration, and it was dependent on the concentration in the bulk phase [35]. The diffusion of fibrils from the bulk phase to the interface occurred at relatively short adsorption times (up to about 47.8 s) [35], and gradually reached equilibrium [36]. Figure 3B shows that the K_{diff} values increased with increasing WPI fibril concentration. Thus, the rate of WPI adsorption was faster at higher concentrations. This suggests the concentration gradient is the driving force for the diffusion of WPI fibrils, which was supported by the previous results for whey protein isolate at the oil–water interfaces [35]. Liu et al. studied the diffusion of soy glycinin when the particle concentration was low (0.01 wt.%). While diffusion-controlled adsorption occurred at a high concentration (>0.5 wt.%) during the initial periods of adsorption, it can be inferred that diffusion is related to the C* [14].

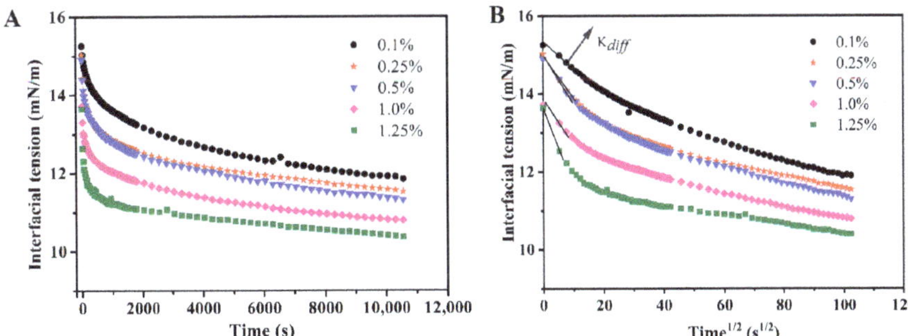

Figure 3. Interfacial behavior of WPI fibrils at the oil/water interface. (**A**) Time evolution of the interfacial tension (γ) for WPI fibrils (0.1–1.25 wt.%) at pH 2.0 and (**B**) dynamic interfacial tension vs. square root of time.

3.4. Characterization of WPI Fibril-Stabilized Emulsions

The visual observation of freshly prepared Pickering emulsions at different concentrations is shown in Figure 4A. The WPI fibrils with various concentrations stabilized the Pickering emulsions and exhibited obvious creaming. The droplet sizes of the emulsions stabilized by WPI fibrils, with a fixed oil-phase fraction of 0.5, are presented in Figure 4B. The droplet size decreased with increasing WPI fibril concentration (from 0.1 wt.% to 0.5 wt.%), and the average droplet size remained relatively constant when the WPI fibril concentration increased from 0.5 wt.% to 1.25 wt.%. This suggested that there was a minimal concentration (0.5 wt.%) for the formation of homogenous WPI fibril-stabilized Pickering emulsions. CLSM was used to observe the morphology and interface properties of Pickering emulsions. The CLSM images of WPI-stabilized Pickering emulsions, with different concentrations and a fixed oil-phase fraction of 0.5, are shown in Figure 4C. WPI fibrils were labeled with ThT, which can only bind with fibrillar structures, and bright green fluorescence was observed at various concentrations, indicating all WPI fibrils were able to absorb at the oil–water interface [35]. In addition, when the concentration was low (<0.5 wt.%), the image showed relatively large emulsion droplets, which may be due to insufficient coverage of the surface of the emulsion droplets by WPI fibrils [10]. When the WPI fibril concentration increased to 0.5 wt.%, the droplet size became smaller and homogenous. This suggests the surface of droplets was sufficiently covered by fibrils at high concentrations (from 0.5 wt.% to 1.25 wt.%), and the emulsion became stable due to the electrostatic or/and steric repulsions [37]. Thus, when the WPI concentration was lower than C*, single-molecule chains existed in isolation with a low amount of WPI fibrils. When the WPI fibril concentration was greater than C*, the viscosity of the system increased due to a high degree of polymer entanglement [26], which could hinder the movement and collision of droplets.

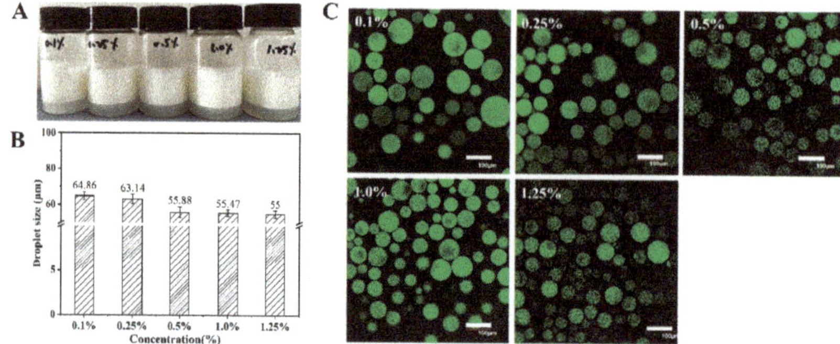

Figure 4. Microstructure of WPI fibrils stabilized Pickering emulsions. (**A**) Appearance, (**B**) average droplet size, and (**C**) confocal microscopic images of Pickering emulsion with different fibrils concentrations (0.1–1.25 wt.%) at pH 2.0. The scale bar was 100 μm.

3.5. Physicochemical Stability of WPI Fibril-Stabilized Pickering Emulsions

Figure 5 illustrates the appearance and microstructure of the Pickering emulsion stabilized by WPI fibrils (0.1–1.25 wt.%, *v/v*) at pH 2 during storage for up to 28 d. The appearance of the samples did not change significantly when the emulsions were stored for 28 d. Additionally, the droplet was still stable without aggregation when the WPI fibril concentration was low (0.1 wt.%), suggesting that WPI fibrils are outstanding emulsifiers. Figure 6 shows the emulsion microstructures at various pH values and ionic strengths. No strong aggregation or droplet collapse was observed, which indicated that highly stable emulsion droplets were formed [38]. A slight increase in emulsion droplet size was observed when the pH was closer to the isoelectric point (pI), which was possibly due to a decrease in the electrostatic repulsion force at pH closer to the pI [39,40].

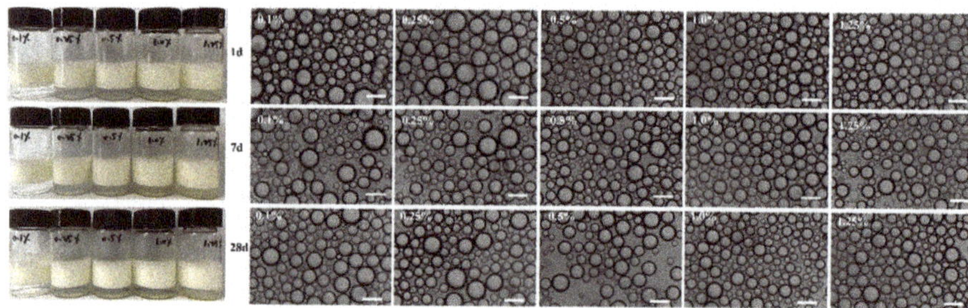

Figure 5. Appearance and corresponding microstructure of WPI fibrils stabilized Pickering emulsions when stored at 1 day, 7 days and 28 days. The scale bar was 100 μm.

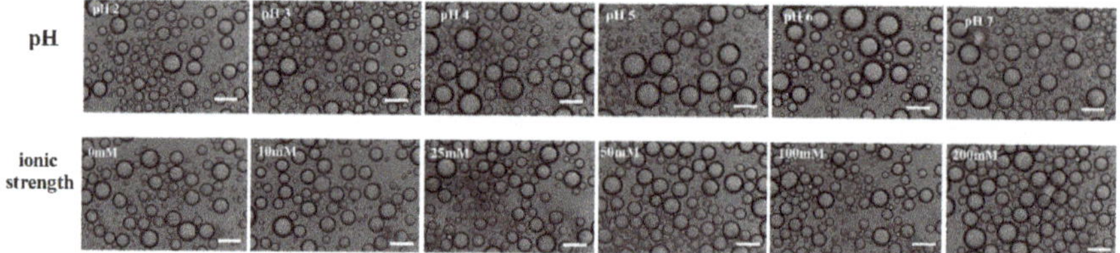

Figure 6. Microstructure of WPI fibrils-stabilized emulsions under different pH values (pH 2.0–7.0) and different ionic strength (0–200 mM NaCl). The scale bar was 100 μm.

3.6. Characterization of Nobiletin-Loaded WPI Fibrils Stabilized Pickering Emulsion

Figure 7 illustrates the visual appearance, microstructure, and average droplet size of the Pickering emulsion stabilized by WPI fibrils (0.1–1.25 wt.%) at pH 2 after loading 0.5 wt.% nobiletin. The emulsion droplets presented structural integrity without droplet collapse, and there was no obvious droplet coalescence. The results suggest WPI fibrils functioned well as Pickering emulsifiers to deliver nobiletin. The average droplet size of the Pickering emulsion decreased from 44.4 μm to 38.5 μm when the concentration of WPI fibril increased from 0.1 wt.% to 0.5 wt.%, and the droplet size became constant at around 39 μm, with increasing WPI fibril concentration. When the concentration of WPI fibrils was low (0.1 wt.% and 0.25 wt.%), the amount of WPI fibrils was insufficient to ensure particle surface coverage, and a smaller interfacial area was needed to prevent the coalescence of droplets. To decrease the interfacial area, the droplet size must increase [41]. When increasing the concentration of WPI fibrils to 0.5 wt.%, the sufficient coverage of WPI fibrils can facilitate the occurrence of smaller emulsion droplets with a larger surface area.

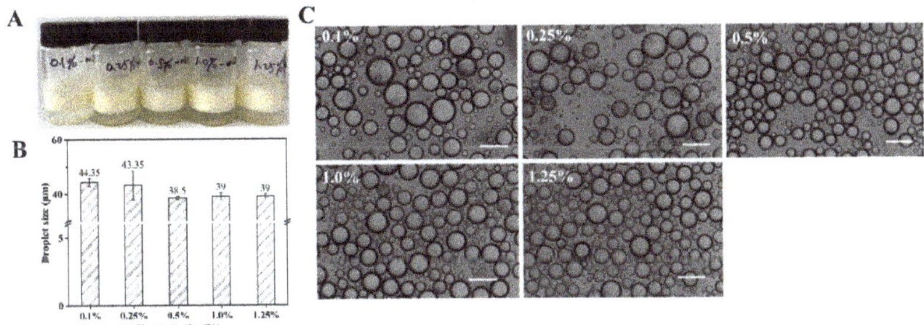

Figure 7. WPI fibrils-stabilized Pickering emulsion loaded with nobiletin at different fibrils concentrations (0.1–1.25%) (**A**) Appearance, (**B**) average droplet size, and (**C**) microstructure. The scale bar was 100 μm.

3.7. Lipolysis and Bioaccessibility of Nobiletin in WPI Fibril-Stabilized Pickering Emulsions

Hydrophobic compounds have higher solubility in lipids and can be incorporated into the micelle core and then absorbed through the intestinal lining when lipids are hydrolyzed by lipase and micellized with bile salts, which can make the hydrophobic component become bioaccessible [28]. The in vitro lipolysis model is a useful tool to evaluate the lipid digestion kinetics and the bioaccessibility of target compounds in the delivery system [28]. During lipid digestion, pH will decrease due to the continuous release of fatty acids. To maintain the optimum pH for enzymatic digestion, sodium hydroxide was added to the digestion buffer [42]. In this study, the rate of lipolysis was determined by monitoring the volume of 0.1 M sodium hydroxide solution. Figure 8A shows the total amount of **FFAs** released from WPI fibril Pickering emulsion and soybean oil. During lipid digestion, all samples were digested rapidly in the initial stage (10–30 min), which indicated that lipase could access the emulsified lipids and catalyze the conversion of triacylglycerols into **FFA** and monoacylglycerols [43]. However, there were some differences between the extent of digestion and the digestion rates of Pickering emulsion and soybean oil. The **FFA** released from Pickering emulsions stabilized by all concentrations of WPI fibrils was significantly higher than that in soybean oil, except for a concentration of 0.1 wt.% WPI fibrils. The more **FFA** released, the higher the degree of lipolysis. Since lipolysis is an interfacial process and WPI fibril-stabilized Pickering emulsion droplets showed a larger interfacial area with a small droplet size, the Pickering emulsion had more opportunity to come in contact with the digestive enzymes [44]. In addition, the interface composition of the Pickering emulsion plays an important role in lipid digestion [45]. As shown in Figure 8B, the bioaccessibility of nobiletin in the Pickering emulsion with 0.5 wt.% WPI fibril was 62.53% ± 0.19%, which was much higher than that in soybean oil (21.91% ± 0.10%). It suggested that the higher the degree of lipid digestion, the more micelles are formed, and nobiletin will enter the micelles and be absorbed effectively.

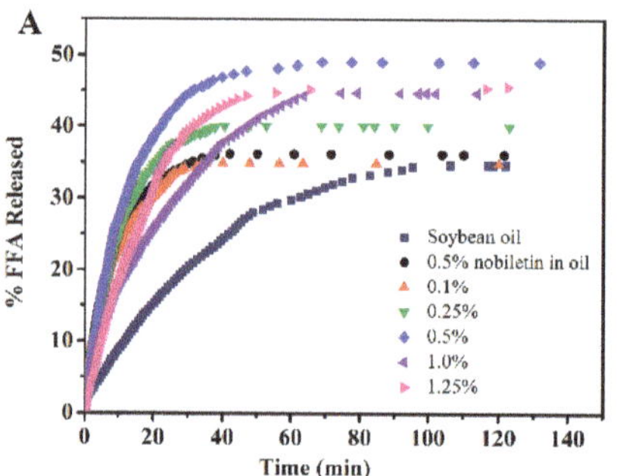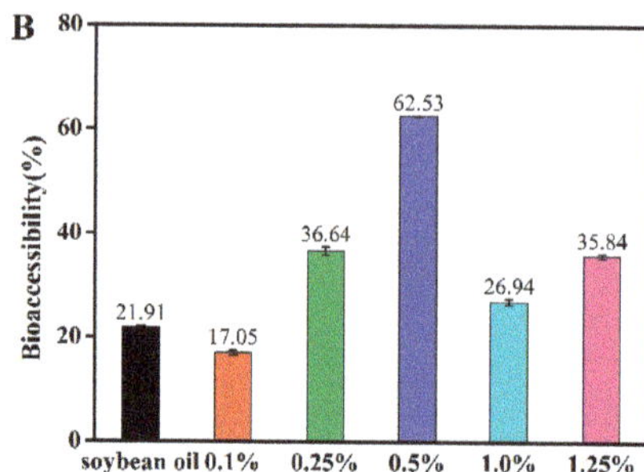

Figure 8. In vitro digestion of Pickering emulsion (**A**) Release profile of free fatty acids (**FFA**) and (**B**) bioaccessibility of nobiletin in soybean oil and Pickering emulsion stabilized by WPI fibrils after in vitro digestion.

The effect of WPI fibrils concentration on the interfacial structure of emulsions and the bioaccessibility of nobiletin is depicted in a schematic diagram (Figure 9). When the concentration is less than C*, WPI fibrils move freely in the aqueous solution and have a high steric probability of adsorption to the oil-water interface. However, the amount of WPI fibrils is not enough to cover the full interface, resulting in large droplet size and uneven distribution of Pickering emulsions. When the concentration is close to C*, the oil droplets can be effectively stabilized, and the emulsion exhibited homogenous at an appropriate concentration necessary for full surface coverage. If continued to increase WPI fibrils, the fibrils penetrated each other and became aggregated and entangled on the surface of oil droplets. The flocculation or aggregates in emulsions may slow their rate of lipid digestion because the floc or aggregates would prevent lipase molecules from reaching the lipid droplets [46]. The in vitro bioaccessibility of nobiletin was mainly affected by the degree of lipolysis. After the lipid is digested, nobiletin enters the micelle's structure along with it and is further absorbed by the small intestinal epithelial cells. When the interfacial structure is dense (c~C*) or there are still numerous fibrils existing in the aqueous phase (c > C*), un-adsorbed fibrils might exert a depletion effect and make Pickering emulsions have less opportunity to come in contact with the digestive enzymes.

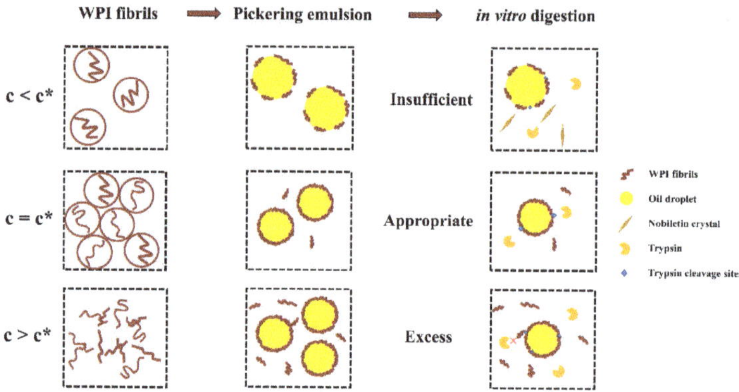

Figure 9. Schematic diagram of influence of WPI fibrils concentration on the emulsion structure and the bioaccessibility of nobiletin.

4. Conclusions

In conclusion, the overlap concentration (C*) of WPI fibrils was around 0.5 wt.%. The interfacial absorption ability was improved with the concentration increasing. WPI fibrils at various concentrations can effectively stabilize Pickering emulsion and showed a long-term stability at room temperature. At C* of WPI fibrils, the droplet size of the Pickering emulsion became homogenous, and the emulsion was stable at various pHs and ionic strengths. WPI fibrils stabilized Pickering emulsion could significantly improve the degree of lipolysis and the bioaccessibility of nobiletin, especially at C* of WPI fibrils, indicating that the entanglement of WPI fibrils was an important parameter for designing food-grade Pickering emulsions. It also provides an important insight for broadening the applications of hydrophobic nutraceuticals.

Author Contributions: Conceptualization, F.J. and C.C.; methodology, X.W. and W.H.; software, X.W.; formal analysis, W.H.; resources, Q.H.; writing—original draft preparation, F.J.; writing—review and editing, W.J.; supervision, Q.H.; project administration, W.J. and Q.H.; funding acquisition, W.J. All authors have read and agreed to the published version of the manuscript.

Funding: This research was funded by National Natural Science Foundation of China, grant number No. 32072152. The APC was funded by Wuhan Polytechnic University ESI Team Building.

Acknowledgments: We thank Yunqi Li and Ce Shi for their assistance during the SAXS experiments, and gratefully acknowledge Qingyun Lv for assistance with HPLC experiments.

Conflicts of Interest: The authors declare no conflict of interest.

References

1. Wei, Z.; Cheng, Y.; Zhu, J.; Huang, Q. Genipin-crosslinked ovotransferrin particle-stabilized Pickering emulsions as delivery vehicles for hesperidin. *Food Hydrocoll.* **2019**, *94*, 561–573. [CrossRef]
2. Su, J.; Guo, Q.; Chen, Y.; Dong, W.; Mao, L.; Gao, Y.; Yuan, F. Characterization, and formation mechanism of lutein Pickering emulsion gels stabilized by β-lactoglobulin-gum arabic composite colloidal nanoparticles. *Food Hydrocoll.* **2020**, *98*, 105276. [CrossRef]
3. Yan, X.; Ma, C.; Cui, F.; McClements, D.J.; Liu, X.; Liu, F. Protein-stabilized Pickering emulsions: Formation, stability, properties, and applications in foods. *Trends Food Sci. Technol.* **2020**, *103*, 293–303. [CrossRef]
4. Yang, Y.; Jiao, Q.; Wang, L. Preparation and evaluation of a novel high internal phase Pickering emulsion based on whey protein isolate nanofibrils derived by hydrothermal method. *Food Hydrocoll.* **2021**, *123*, 107180. [CrossRef]
5. Dai, L.; Zhan, X.; Wei, Y.; Sun, C.; Mao, L.; Mcclements, D.J.; Gao, Y. Composite zein-propylene glycol alginate particles prepared using solvent evaporation: Characterization and application as Pickering emulsion stabilizers. *Food Hydrocoll.* **2018**, *85*, 281–290. [CrossRef]
6. Jiang, F.; Pan, Y.; Peng, D.; Huang, W.; Shen, W.; Jin, W.; Huang, Q. Tunable self-assemblies of whey protein isolate fibrils for Pickering emulsions structure regulation. *Food Hydrocoll.* **2022**, *124*, 107264. [CrossRef]
7. Feng, X.; Dai, H.; Ma, L.; Fu, Y. Properties of Pickering emulsion stabilized by food-grade gelatin nanoparticles: Influence of the nanoparticles concentration. *Colloids Surf. B Biointerfaces* **2020**, *196*, 111294. [CrossRef]
8. Bai, L.; Huan, S.; Gu, J.; McClements, D.J. Fabrication of oil-in-water nanoemulsions by dual-channel microfluidization using natural emulsifiers: Saponins, phospholipids, proteins, and polysaccharides. *Food Hydrocoll.* **2016**, *61*, 703–711. [CrossRef]
9. Aziz, A.; Khan, N.M.; Ali, F.; Khan, Z.U.; Muhammad, N. Effect of protein and oil volume concentrations on emulsifying properties of acorn protein isolate. *Food Chem.* **2020**, *324*, 126894. [CrossRef]
10. Gao, Z.; Zhao, J.; Huang, Y.; Yao, X.; Zhang, K.; Fang, Y.; Nishinari, K.; Phillips, G.O., Jiang, F.; Yang, H. Edible Pickering emulsion stabilized by protein fibrils. Part 1: Effects of pH and fibrils concentration. *LWT-Food Sci. Technol.* **2017**, *76*, 1–8. [CrossRef]
11. Liang, H.-N.; Tang, C.-H. Pea protein exhibits a novel Pickering stabilization for oil-in-water emulsions at pH 3.0. *LWT-Food Sci. Technol.* **2014**, *58*, 463–469. [CrossRef]
12. Shao, Y.; Tang, C.-H. Gel like pea protein Pickering emulsions at pH 3.0 as a potential intestine-targeted and sustained-release delivery system for beta-carotene. *Food Res. Int.* **2016**, *79*, 64–72. [CrossRef]
13. Bİtcher, S.; Keppler, J.K.; Drusch, S.J.C.; Physicochemical, S.A. Mixtures of Quillaja saponin and beta-lactoglobulin at the oil/water interface: Adsorption, interfacial rheology and emulsion properties. *Colloids Surf. A Physicochem. Eng. Asp.* **2016**, *518*, 46–56. [CrossRef]
14. Liu, F.; Tang, C. Soy glycinin as food-grade Pickering stabilizers: Part. I. Structural characteristics, emulsifying properties and adsorption/arrangement at interface. *Food Hydrocoll.* **2016**, *60*, 606–619. [CrossRef]
15. Zhao, Y.; Wang, C.; Lu, W.; Sun, C.; Fang, Y. Evolution of physicochemical and antioxidant properties of whey protein isolate during fibrillization process. *Food Chem.* **2021**, *357*, 129751. [CrossRef] [PubMed]

16. Bolder, S.G.; Vasbinder, A.J.; Sagis, L.M.C.; Linden, E.V.D. Heat-induced whey protein isolate fibrils: Conversion, hydrolysis, and disulphide bond formation. *Int. Dairy J.* **2007**, *17*, 846–853. [CrossRef]

17. Loveday, S.M.; Su, J.; Rao, M.A.; Anema, S.G.; Singh, H.J.B. Effect of calcium on the morphology and functionality of whey protein nanofibrils. *Biomacromolecules* **2011**, *12*, 3780–3788. [CrossRef]

18. Oboroceanu, D.; Wang, L.; Magner, E.; Auty, M.A.E. Fibrillization of whey proteins improves foaming capacity and foam stability at low protein concentrations. *J. Food Eng.* **2014**, *121*, 102–111. [CrossRef]

19. Rodrigues, T.; Galindo-Rosales, F.J.; Campo-Deaño, L. Critical overlap concentration and intrinsic viscosity data of xanthan gum aqueous solutions in dimethyl sulfoxide. *Data Brief* **2020**, *33*, 106431. [CrossRef]

20. Serfert, Y.; Lamprecht, C.; Tan, C.P.; Keppler, J.K.; Appel, E.; Rossier-Miranda, F.J.; Schroen, K.; Boom, R.M.; Gorb, S.; Selhuber-Unkel, C.J.J. Characterisation and use of β-lactoglobulin fibrils for microencapsulation of lipophilic ingredients and oxidative stability thereof. *J. Food Eng.* **2014**, *143*, 53–61. [CrossRef]

21. Zhang, B.; Lei, M.; Huang, W.; Liu, G.; Jin, W. Improved Storage Properties and Cellular Uptake of Casticin-Loaded Nanoemulsions Stabilized by Whey Protein-Lactose Conjugate. *Foods* **2021**, *10*, 1640. [CrossRef] [PubMed]

22. Peng, D.; Jin, W.; Sagis, L.M.C.; Li, B. Adsorption of microgel aggregates formed by assembly of gliadin nanoparticles and a β-lactoglobulin fibril-peptide mixture at the air/water interface: Surface morphology and foaming behavior. *Food Hydrocoll.* **2022**, *122*, 1070399. [CrossRef]

23. Lai, C.S.; Li, S.; Chai, C.Y.; Wang, Y.J. Anti-inflammatory and antitumor promotional effects of a novel urinary metabolite, 3′,4′-didemethylnobiletin, derived from nobiletin. *Carcinogenesis* **2008**, *29*, 2415–2424. [CrossRef] [PubMed]

24. Yan, L.; Zheng, J.; Hang, X.; Mcclements, D. Nanoemulsion-based delivery systems for poorly water-soluble bioactive compounds: Influence of formulation parameters on polymethoxyflavone crystallization. *Food Hydrocoll.* **2012**, *27*, 517–528.

25. Li, S.; Wang, H.; Guo, L.; Zhao, H.; Ho, C.T. Chemistry and bioactivity of nobiletin and its metabolites. *J. Funct. Foods* **2014**, *6*, 2–10. [CrossRef]

26. Wagoner, T.B.; Cakir-Fuller, E.; Drake, M.A.; Foegeding, E.A. Sweetness perception in protein-polysaccharide beverages is not explained by viscosity or critical overlap concentration. *Food Hydrocoll.* **2019**, *94*, 229–237. [CrossRef]

27. Brodkorb, A.; Egger, L.; Alminger, M. INFOGEST static in vitro simulation of gastrointestinal food digestion. *Nat. Protoc.* **2019**, *14*, 991–1014. [CrossRef]

28. Ting, Y.; Jiang, Y.; Lan, Y.; Xia, C.; Lin, Z.; Rogers, M.A.; Huang, Q. Viscoelastic Emulsion Improved the Bioaccessibility and Oral Bioavailability of Crystalline Compound: A Mechanistic Study Using in Vitro and in Vivo Models. *Mol. Pharm.* **2015**, *12*, 2229. [CrossRef]

29. Heller, C. Capillary electrophoresis of proteins and nucleic acids in gels and entangled polymer solutions. *J. Chromatogr. A* **1995**, *698*, 19–31. [CrossRef]

30. Blanchet, C.E.; Svergun, D.I. Small-angle X-ray scattering on biological macromolecules and nanocomposites in solution. *Annu. Rev. Phys. Chem.* **2013**, *64*, 37–54. [CrossRef]

31. Shi, C.; Li, Y. Progress on the Application of Small-angle X-ray Scattering in the Study of Protein and Protein Complexes. *Acta Polym. Sin.* **2015**, *8*, 871–883.

32. Li, Y.; Xia, Q.; Shi, K.; Huang, Q. Scaling behaviors of α-zein in acetic acid solutions. *J. Phys. Chem. B* **2011**, *115*, 9695–9702. [CrossRef] [PubMed]

33. Zhou, B.; Tobin, J.T.; Drusch, S.; Hogan, S.A. Dynamic adsorption and interfacial rheology of whey protein isolate at oil-water interfaces: Effects of protein concentration, pH and heat treatment. *Food Hydrocoll.* **2021**, *116*, 106640. [CrossRef]

34. Cui, Z.; Chen, Y.; Kong, X.; Zhang, C.; Hua, Y. Emulsifying Properties and Oil/Water (O/W) Interface Adsorption Behavior of Heated Soy Proteins: Effects of Heating Concentration, Homogenizer Rotating Speed, and Salt Addition Level. *J. Agric. Food Chem.* **2014**, *62*, 1634–1642. [CrossRef]

35. Peng, D.; Yang, J.; Li, J.; Tang, C.; Li, B. Foams Stabilized by β-Lactoglobulin Amyloid Fibrils: Effect of pH. *J. Agric. Food Chem.* **2017**, *65*, 10658–10665. [CrossRef]

36. Silva, A.; Almeida, F.S.; Sato, A. Functional characterization of commercial plant proteins and their application on stabilization of emulsions. *J. Food Eng.* **2020**, *292*, 110277. [CrossRef]

37. Peng, J.; Simon, J.R.; Venema, P.; van der Linden, E. Protein Fibrils Induce Emulsion Stabilization. *Langmuir* **2016**, *32*, 2164–2174. [CrossRef]

38. Ebert, S.; Grossmann, L.; Hinrichs, J.; Weiss, J. Emulsifying properties of water-soluble proteins extracted from the microalgae Chlorella sorokiniana and Phaeodactylum tricornutum. *Food Funct.* **2019**, *10*, 754–764. [CrossRef]

39. Wu, J.; Shi, M.; Li, W.; Zhao, L.; Wang, Z.; Yan, X.; Norde, W.; Li, Y. Pickering emulsions stabilized by whey protein nanoparticles prepared by thermal cross-linking. *Colloids Surf. B Biointerfaces* **2015**, *127*, 96–104. [CrossRef]

40. Liu, G.; Li, W.; Qin, X.; Zhong, Q. Pickering emulsions stabilized by amphiphilic anisotropic nanofibrils of glycated whey proteins. *Food Hydrocoll.* **2019**, *101*, 105503. [CrossRef]

41. Lv, P.; Wang, D.; Chen, Y.; Yuan, F. Pickering emulsion gels stabilized by novel complex particles of high-pressure-induced WPI gel and chitosan: Fabrication, characterization and encapsulation. *Food Hydrocoll.* **2020**, *108*, 105992. [CrossRef]

42. Dong, Y.; Wei, Z.; Wang, Y.; Huang, Q. Oleogel-based Pickering emulsions stabilized by ovotransferrin–carboxymethyl chitosan nanoparticles for delivery of curcumin. *LWT-Food Sci. Technol.* **2022**, *157*, 113121. [CrossRef]

43. Lesmes, U.; Baudot, P.; Mcclements, D.J. Impact of Interfacial Composition on Physical Stability and In Vitro Lipase Digestibility of Triacylglycerol Oil Droplets Coated with Lactoferrin and/or Caseinate. *J. Agric. Food Chem.* **2010**, *58*, 7962–7969. [CrossRef] [PubMed]
44. Xiao, J.; Li, C.; Huang, Q. Kafirin Nanoparticle-Stabilized Pickering Emulsions as Oral Delivery Vehicles: Physicochemical Stability and in Vitro Digestion Profile. *J. Agric. Food Chem.* **2015**, *63*, 10263–10270. [CrossRef]
45. Infantes-Garcia, M.R.; Verkempinck, S.H.E.; Gonzalez-Fuentes, P.G.; Hendrickx, M.E.; Grauwet, T. Lipolysis products formation during in vitro gastric digestion is affected by the emulsion interfacial composition. *Food Hydrocoll.* **2021**, *110*, 106163. [CrossRef]
46. Li, Y.; Mcclements, D.J. New Mathematical Model for Interpreting pH-Stat Digestion Profiles: Impact of Lipid Droplet Characteristics on in Vitro Digestibility. *J. Agric. Food Chem.* **2010**, *58*, 8085–8092. [CrossRef]

Article

Whey Protein Isolate Nanofibers Prepared by Subcritical Water Stabilized High Internal Phase Pickering Emulsion to Deliver Curcumin

Xin Xu [1], Zhiyi Zhang [1], Junlong Zhu [1], Dan Wang [1], Guoyan Liu [1], Li Liang [1], Jixian Zhang [1], Xiaofang Liu [1], Youdong Li [1], Jiaoyan Ren [2], Qianchun Deng [3] and Chaoting Wen [1,*]

[1] College of Food Science and Engineering, Yangzhou University, Yangzhou 225127, China; xuxin@yzu.edu.cn (X.X.); zhzhyi0928@163.com (Z.Z.); z17502110240@163.com (J.Z.); wangdan683280@163.com (D.W.); yzufsff@163.com (G.L.); liangli0508@hotmail.com (L.L.); zjx@yzu.edu.cn (J.Z.); liuxf@yzu.edu.cn (X.L.); 008040@yzu.edu.cn (Y.L.)

[2] School of Food Science and Engineering, South China University of Technology, Guangzhou 510641, China; jyren@scut.edu.cn

[3] Oil Crops Research Institute (OCRI), Chinese Academy of Agricultural Sciences, Wuhan 430062, China; dengqianchun@caas.cn

* Correspondence: chaoting@yzu.edu.cn; Tel.: +86-514-9797-8002

Citation: Xu, X.; Zhang, Z.; Zhu, J.; Wang, D.; Liu, G.; Liang, L.; Zhang, J.; Liu, X.; Li, Y.; Ren, J.; et al. Whey Protein Isolate Nanofibers Prepared by Subcritical Water Stabilized High Internal Phase Pickering Emulsion to Deliver Curcumin. *Foods* **2022**, *11*, 1625. https://doi.org/10.3390/foods11111625

Academic Editor: Paul J.A. Sobral

Received: 26 April 2022
Accepted: 25 May 2022
Published: 31 May 2022

Publisher's Note: MDPI stays neutral with regard to jurisdictional claims in published maps and institutional affiliations.

Abstract: This study aimed to design a Pickering emulsion (PE) stabilized by whey protein isolate nanofibers (WPINs) prepared with subcritical water (SW) to encapsulate and prevent curcumin (Cur) degradation. Cur-loaded WPINs–SW stabilized PE (WPINs–SW–PE) and hydrothermally prepared WPINs stabilized PE (WPINs–H–PE) were characterized using the particle size, zeta potential, Congo Red, CD, and TEM. The results indicated that WPINs–SW–PE and WPINs–H–PE showed regular spherical shapes with average lengths of 26.88 ± 1.11 μm and 175.99 ± 2.31 μm, and zeta potential values were −38.00 ± 1.00 mV and −34.60 ± 2.03 mV, respectively. The encapsulation efficiencies of WPINs–SW–PE and WPINs–H–PE for Cur were 96.72 ± 1.05% and 94.07 ± 2.35%. The bioaccessibility of Cur of WPINs–SW–PE and WPINs–H–PE were 57.52 ± 1.24% and 21.94 ± 2.09%. In addition, WPINs–SW–PE had a better loading effect and antioxidant activities compared with WPINs–H–PE. SW could be a potential processing method to prepare a PE, laying the foundation for the subsequent production of functional foods.

Keywords: subcritical water; whey protein isolate nanofibers; high internal phase Pickering emulsion; curcumin deliver system; bioavailability

1. Introduction

Curcumin (Cur) is found in the rhizome of turmeric (*Curcuma longa*) and is a characteristic polyphenolic compound. Cur has antioxidant, anti-inflammatory, and anticancer properties, among others [1–3]. The U.S. Food and Drug Administration (FDA) generally considers Cur safe at low levels [4]. Cur is a hydrophobic molecule and practically insoluble in water, inhibiting its application in medicines and health products [5]. Therefore, a delivery system is needed to promote the water dispersibility, chemical stability, and bioavailability of Cur [6,7]. The nano-system enhances the biological effects of drug ingestion by protecting the drug from enzymatic degradation, providing a controlled release and altering residence time, among others [8]. Delivery systems for Cur nano-formulations include emulsion, liposomes, solid lipid nanoparticles, polymer nanoparticles, polymer micelles, etc. Emulsions are divided into two main categories, including traditional emulsion and Pickering emulsion (PE). Recently, PE has attracted much attention in delivery systems due to its use of solid particles to stabilize oil droplets, with higher thickness and surface loading than traditional emulsion [9]. PE can protect hydrophobic bioactive substances and deliver them to target sites due to their good chemical and physical stability [10]. In addition, PE includes traditional PE and high internal phase PE (HIPPE). The minimum internal ratio of HIPPE was 0.74. HIPPE can be prepared with only a few stabilizers, which have strong anti-coalescence ability and storage ability. HIPPE has significant advantages as a delivery system, with strong stability (chemical, environmental, gastrointestinal stability, etc.) and high bioavailability for delivering bioactive substances. Since the interface is occupied by most of the particles, the interfacial tension is reduced [11]. It is worth noting that food-grade macromolecules (polysaccharide, protein, polyphenol, etc.) were used to stabilize HIPPE with significant advantages, especially for animal-derived protein [12–15]. Therefore, the selection and development of suitable animal protein stabilizers to prepare HIPPE are of great significance for improving the utilization rate of biologically active substances.

Whey protein isolate (WPI) is a functional food with high nutritional value and contains more essential amino acids than plant protein [16]. Some studies have found that the heated protein is a good stabilizer for emulsion [17]. Thermal treatment techniques can alter the emulsification and functional properties of protein [18]. The tertiary structure of the protein is partially or fully unfolded upon heating, resulting in the exposure of hydrophobic groups, thereby increasing its flexibility [19]. Whey protein isolate nanofibers (WPINs) are nanofibers formed by heat-induced denaturation of protein [20]. Fibrosis usually begins with globular protein complete or partial unfolding and ends with an ordered fibrous structure [21]. Studies have shown that WPINs can be prepared via heating in a water bath at 80 °C for 10 h [22]. WPINs can also be prepared by using an oven reaction kettle at 110 °C for 4 h [23]. These two traditional methods (heat treatment method, oven reaction kettle treatment method) for preparing WPINs were time-consuming, energy-intensive, and difficult for industrial production. Therefore, researchers began to adopt new technologies, such as the subcritical water (SW) treatment method. In the SW treatment method, the liquid is heated to 100–374 °C under a certain pressure [24]. This method is very environmentally friendly and efficient. A previous study confirmed that the *Lycium barbarum* polysaccharide-protein conjugates obtained under SW conditions (120 °C) can be used for stabilized selenium nanoparticles and exhibit good activity [25]. Another study also explored the synthesis of nano-catalysts with SW at 100–500 °C and found that the optimal synthesis temperature of different chemical materials was different [26]. Compared with natural heat-treated soybean protein (90 °C), SW (>100 °C) extraction of protein from heat-denatured soybean exhibits excellent interfacial properties and higher surface activity [27]. Studies have shown that the high temperature and pressure in SW conditions are more conducive to the degradation of cellulose and the preparation of WPINs, resulting in a shorter preparation time [28]. As far as we know, the preparation of WPINs stabilized HIPPE from SW treatment has not been performed.

This study aimed to efficiently prepare WPINs in a short time using SW (pressure: 0.2 MPa). WPINs (c = 5 wt%) and corn oil (Φ = 0.74) were used to prepare the HIPPE. Cur was embedded in HIPPE to explore its bioavailability and antioxidant activity after simulated digestion in vitro.

2. Materials and Methods

2.1. Materials and Chemicals

Whey protein isolate (>80%) and curcumin (>95%) were provided by Yuanye Biotechnology Co., Ltd. (Shanghai, China). Corn oil was purchased from Suguo supermarket (Yangzhou, China) without further purification. Congo Red, Sodium dodecyl sulfate (SDS), 1,1-diphenyl-2-picryl-hydroxyl (DPPH), 2,2-Azino-bis (3-ethylbenzothiazoline-6-sulphonic acid) diammonium salt (ABTS), and 2,4,6-tripyridyl-s-triazine were purchased from Shanghai Aladdin Biochemical Technology Co., Ltd. (Shanghai, China). FITC, Nile red, and porcine pancreatic lipase (100–650 μ/mg) were obtained from Sigma-Aldrich (St. Louis, MO, USA). Bile salts and porcine pepsin (12 μ/mg) were purchased from Sunson (Beijing, China). All other reagents were analytical grade. All solutions were prepared using purified deionized (DI) water.

2.2. Preparation of Whey Protein Isolate Nanofibers by Subcritical Water

The pretreatment steps of the 5% WPI solution (w/w) refer to the experiment of Yang et al. [23]. In brief, the pH of the WPI solution (5%, w/w) was adjusted to 2.0 with 3 M hydrochloric acid, stirred at $25 \pm 1\,°C$ for 30 min, then centrifuged at $4\,°C$ and $10,000 \times g$ for 15 min (L550 Cence, Changsha, China). Undissolved protein was removed by vacuum filtration of the supernatant with a filter (0.22 μm pore size, Xinya, Shanghai, China). The filtrate was sent to an SW reactor, and the oven temperature was set to $110\,°C$. After heating for 5 min and setting 0.2 MPa, 0.5 MPa, 1.0 MPa, and 1.5 MPa pressure, the SW reaction kettle was cooled with an ice bat to obtain WPI nanofibers (WPINs–SW) and stored in a refrigerator at 4 °C. The WPI nanofibers (WPINs−H) prepared by the hydrothermal method were incubated in a water bath at 80 °C for 10 h as a control.

2.3. Structural Characterization of WPINs

2.3.1. Particle Size and Zeta Potential

The particle and zeta potential measurements of samples were based on the previous method with some modifications [7]. The particle size of WPINs was measured using a Malvern Nano sizer (Mastersizer 1000, Malvern Instruments Ltd., Malvern, UK). Samples were diluted 1000 times by the distilled water and 3200 rpm vortex vibrations for 1 min before measurement.

Zeta potential Analyzer (Mastersizer 1000, Malvern Instruments Ltd., Malvern, UK) was used to measure the zeta potential of the nanoparticles. The samples were diluted 100 times by the distilled water and were added to a cuvette equipped with an electrode.

2.3.2. Congo Red Binding Spectrum

The Congo Red binding capacity measurement of samples was in accordance with the method of Nilsson et al. [29]. After mixing 500 μL of the sample solution with 5 mL of the Congo Red solution (70 μg/mL, pH 7.0, 10 mM phosphate buffer), we let it stand for 15 min at room temperature. The spectra were then observed at 400–600 nm with a photometer (Cary 5000, Varian, Palo Alto, CA, USA).

2.3.3. Circular Dichroism (CD) Spectroscopy

The secondary structural changes of WPINs were measured according to the previously described methods using CD spectroscopy [30]. CD spectra were observed with a Jasco spectropolarimeter (Model J-810, Jasco, Tokyo, Japan) in the far ultraviolet (190–260 nm) region and at room temperature. The final protein concentration of the WPINs solution was diluted to 0.25 mg/mL. Each scan was repeated three times and averaged.

2.3.4. Transmission Electron Microscopy (TEM)

According to the method of Wang et al. [22], the concentration was diluted to 0.1%. Then, the WPINs solution was slowly dropped onto a specially made copper mesh (with a diameter of 3 mm and a thickness of 10–30 μm). After standing for 15 min, we gently blotted off excess liquid with filter paper. Then, 2% uranyl acetate was added dropwise to a dry mesh, and the copper was left to ventilate for 8 min. Before TEM measurement, the mesh was absorbed again with filter paper to remove the unwanted solution. Electron micrographs were taken by an HT7800 transmission electron microscope (Hitachi, Tokyo, Japan).

2.4. Preparation and Characterization of High Internal Phase Pickering Emulsion

HIPPE was prepared by homogenizing WPINs solution (5%, w/w) with a fixed oil–water ratio of 7.4:2.6 (mL/mL). The suspension (2.6 mL) with different WPINs concentrations was mixed with corn oil (7.4 mL) using a shear emulsifying machine (T18, IKA, Staufen im Breisgau, Germany) at 20,000 rpm for 5 min. WPINs–SW–PE was prepared with WPINs (SW treatment)-stabilized PE, while WPINs–H–PE was prepared with WPINs (hydrothermal treatment)-stabilized PE.

2.5. Structural Properties of Pickering Emulsion

2.5.1. Optical Microscopy

Photographs refer to the method of Ren et al. with modifications [31]. The morphology of the PE was observed by an optical microscope (Motic, Panthera LBA310-T, Hong Kong, China). We diluted HIPPE by a factor of 5 and obtained photographs using $100\times$, $400\times$, and $1000\times$ lenses.

2.5.2. Confocal Laser Scanning Microscopy (CLSM)

The adsorption of nanoparticles at the oil–water interface can be observed by confocal laser scanning microscopy (CLSM) (LSM 880 NLO, Carl Zeiss AG, Ober-kochen, Germany). PE was diluted by 100 times before shooting. The emulsion was further characterized with the method of Li et al. with minor modifications [32]. The oil phase was stained with Nile red, and the protein phase was stained with FITC during the emulsion preparation. Emulsions were observed under a $10\times$ objective. The laser excitation source for Nile red was 488 nm and the laser excitation source for FITC was 543 nm.

2.5.3. Particle Size and Zeta Potential

The PE size was measured by the Malvern Nano sizer (Mastersizer 3000, Malvern Instruments Ltd., Malvern, UK). The Zeta Potential Analyzer (Mastersizer 1000, Malvern Instruments Ltd., Malvern, UK) was used to measure the zeta potential of the nanoparticles. The samples were diluted 1000 times by the distilled water and were added to a cuvette equipped with an electrode.

2.5.4. Emulsifying Capacity (EAI) and Emulsifying Stability (ESI)

According to the previous method of de Souza et al., the evaluation of emulsifying performance mainly relied on emulsifying ability (EAI) and emulsifying stability (ESI) [33]. The PE preparation was performed using the same procedure as Section 2.4. The freshly prepared emulsion (50 µL) was diluted with 0.1% SDS, and the absorbance at 500 nm was measured (recorded as A_0). After standing at room temperature for 10 min, the absorbance of the diluted emulsion was measured again (denoted as A_{10}). EAI and ESI were calculated according to Equations (1) and (2).

$$\text{EAI} \left(\text{m}^2/\text{g} \right) = \frac{2 \times 2.303 \times A_0 \times \text{DF}}{C \times \varnothing \times \theta \times 10000} \tag{1}$$

$$\text{ESI (min)} = \frac{A_0 \times 10}{\Delta A} \tag{2}$$

DF, C, and $\varnothing$ represent the dilution factor, protein concentration, and oil volume fraction, respectively. θ represents the optical path (0.01 m). ΔA represents the difference between A_0 and A_{10}.

2.5.5. Centrifugation Stability

The measurement of centrifugal stability employed the method of Gond et al. [34]. After preparing the PE, an appropriate amount of the emulsion was immediately taken out and placed in a 10 mL centrifuge tube. After centrifugation at a centrifugal force of $10,000\times g$ for 10 min, the emulsion oil–water separation and particle precipitation were compared.

2.6. Preparation of Curcumin-Loaded Pickering Emulsion

Cur was added to corn oil at a concentration of 0.1 wt%. To ensure the maximum solubility of Cur in corn oil, the mixture was stirred overnight under magnetic stirring (800 rpm). The undissolved Cur in the mixture was then removed by centrifugation at $10,000\times g$ for 10 min. The supernatant was used as an oil phase in the SW prepared WPINs to stabilize PE (WPINs–SW–PE–Cur) and hydrothermal-method-prepared WPINs to stabilize PE (WPINs–H–PE–Cur), which was prepared using the same procedure as Section 2.4.

2.6.1. Embedding Rate of Curcumin

The embedding rate of Cur was measured according to the method of Han et al. [35] Freshly prepared Cur-loaded PE was centrifuged at $10,000\times g$ for 10 min to remove any large particles and nonencapsulated Cur crystals and dissolved in ethanol. The embedding rate of the PE was then calculated using Equation (3):

$$\text{Embedding rate} = \frac{\text{mass of Cur in PE}}{\text{total mass of Cur}} \times 100\%, \qquad (3)$$

The concentration of PE can be tested by measuring the absorbance at 425 nm with a UV spectrophotometer (TianMei UV100, Shanghai, China). We drew a standard curve by measuring the absorbance of a series of known concentrations of Cur in ethanol. We diluted the oil phase in ethanol and measured the absorbance at 425 nm recovery of Cur from PE by demulsification with ethanol. Briefly, 100 µL of the emulsion was added to 900 µL of ethanol, and the mixture was centrifuged at $10,000\times g$ for 1 min to pellet WPINs. After centrifugation, the supernatant was diluted 10 times with ethanol. We then converted the absorbance of the diluted ethanolic extract to the Cur concentration according to the standard curve.

2.7. Release of Curcumin from the Pickering Emulsion during Digestion In Vitro

Before performing in vitro digestion experiments, two mock digestion solutions were prepared. Artificially simulated gastric fluid (SGF; 34.22 mM NaCl, 226.11 mM hydrochloric acid) and simulated intestinal fluid (SIF; 3.75 M NaCl, 249.49 mM calcium chloride dihydrate) were prepared. Bile salts are difficult to dissolve completely, so they were dissolved in a buffer solution (pH 7.0, 53.57 g/L) 24 h before digestion in the small intestine. All mock digests were preheated to 37 °C before each digestion stage.

The lipid content of all samples was adjusted to 2%. Pepsin (0.064 g) was dissolved in 20 g of SGF to obtain the gastric phase electrolyte. Furthermore, 20 g of the emulsion was mixed with 20 g of SGF. The solution was preheated to 37 °C. We adjusted the pH of the mixture to 2.5 with HCl and spun it at 37 °C for 2 h at a stirring speed of 100 rpm.

Chyme samples (30 g) obtained from the final stage were placed in a constant temperature water bath at 37 °C and the pH was adjusted to 7.0. While stirring, we added 1.5 mL of SIF and 3.5 mL of the bile salt solution, prepared in advance, to the chyme sample while adjusting the pH of the mixture to 7.0 for a second time. We then added 2.5 mL of porcine pancreatic lipase under continuous stirring (24 mg/mL), the temperature was maintained at 37 °C, and it was rotated at a stirring speed of 100 rpm for 2 h [36].

2.7.1. Bio-Accessibility

The fraction dissolved in gastrointestinal fluids can generally be seen as the bioavailability of Cur in vitro [37]. Therefore, the amount of released Cur from various formulations was measured in the gastric and small intestine stages. Cur released from the gastric phase was measured after the samples were incubated in simulated gastric fluid for 2 h. In contrast, in the small intestine phase, it was measured after 2 h of incubation of the models in simulated small intestinal fluid. Finally, the samples were centrifuged at $5000\times g$ for 30 min, and the supernatant was collected and diluted in ethanol to determine the release of Cur (Section 2.6.1). The bio-accessibility of Cur was calculated in vitro using Equation (1):

$$\text{Bio} - \text{accessibility} = \frac{\text{Weight of solubilized Cur}}{\text{Weight of Cur before digestion}} \times 100\%, \qquad (4)$$

It should be noted that this expression for bio-accessibility depends on the fraction of Cur lost due to chemical degradation and the solubilized fraction within the micelle phase.

2.7.2. Curcumin Release

Samples were taken at 120 min in the simulated stomach and 240 min in the intestine, diluted with ethanol, and measurements were performed with a UV spectrophotometer with a wavelength of 425 nm.

2.8. Determination of the Antioxidant Capacity

The antioxidant capacity of different Cur formulations (encapsulated and free) was measured following simulated gastrointestinal digestion in vitro.

2.8.1. DPPH• Scavenging Capacity

This spectrophotometric assay uses the stable free radical 1,1-diphenyl-2-picryl-hydroxyl (DPPH) as a reagent. To evaluate the DPPH• scavenging ability of Cur, the method of Gulcin was used with slight modifications [38]. The DPPH• absorbs at 517 nm, but its absorption decreases after reduction by antioxidants or radical species. Based on this principle, we prepared a 0.1 mM DPPH solution in ethanol and added 0.5 mL to 1.5 mL of Cur ethanol solutions of various concentrations (15–45 ug/mL). The mixed solution was vortexed for 3 min and incubated in the absence of light for 30 min. The antioxidant capacity of the control group was determined by a blank sample containing no scavenger and then subtracted from the antioxidant capacity of the corresponding model containing Cur. Controls included free Cur (Free–Cur) simulated gastrointestinal fluid, Free–Cur biopolymer-coated WPINs (Cur for encapsulation) exposed to simulated gastrointestinal fluid, and a Free–Cur physical mixture exposed to simulated gastrointestinal fluid (for non-encapsulated Cur).

2.8.2. ABTS·+ Scavenging Capacity

The ABTS·+ scavenging assay is based on a previously described method [39]. ABTS was dissolved in deionized water to a concentration of 7 mM one day in advance, then 2.45 mM potassium persulfate was dissolved in the ABTS solution and stored at room temperature in the dark. The ABTS solution was then diluted with 10 mM phosphate buffer (pH 7.4) to an absorbance of 0.70 ± 0.01 cm^{-1} at 734 nm and equilibrated at 30 °C. The Cur preparation was centrifuged through simulated GIT at $5000 \times g$ for 30 min and the supernatant was collected. We adjusted the Cur concentration of the supernatant with phosphate buffer to achieve 20–80% inhibition of blank absorbance by pre-assay. We then dissolved 4.0 mL of sample in 40 mL of the ABTS solution and took absorbance readings at 30 °C after 7 min of initial mixing. Refer to 2.8.1 for the rest of the steps.

2.8.3. Determination of Reducing Power

The FRAP assay was performed with some modifications [40]. The FRAP reagent was prepared with $FeCl_3 \cdot 6H_2O$ (20 mmol/L), 2,4,6-tripyridyl-s-triazine (10 mmol/L), and acetate buffer (300 mmol/L, pH 3.6.) at a ratio of 1:1:10.

The stand curve was drawn using various concentrations of $FeSO_4 \cdot 7H_2O$ (100–800 μM). Cur solution (100 μL) was added to 3 mL of the FRAP reagent. The mixed reagents were placed in a constant temperature water bath at 37 °C for 30 min. The absorbance was measured at a wavelength of 539 nm using a UV spectrophotometer (TianMei UV100, China). FRAP values are expressed as the ratio of Fe^{2+} per μmol to sample per mg (μmol/mg).

2.9. Statistical Analysis

All experiments were performed in triplicate, and the results were reported as the mean ± standard. Statistical analysis was performed using the software IBM SPSS Statistics 26.0. Data were subjected to analysis of variance, and the Turkey HSD test was used to determine the significance of differences between data values ($p < 0.05$).

3. Results and Discussion

3.1. Effects of Subcritical Water on the Structural Properties of WPINs

3.1.1. Particle Size and Zeta Potential of WPINs

Numerous studies report that protein nanoparticles were widely used to stabilize PE due to their high stability. The particle size and zeta potential of the nanoparticles were the main indicators for assessing the structural properties of the nanoparticles. To study the effects of SW pressure on the fibrosis degree of WPI, the particle size and zeta potential of WPINs were characterized. As shown in Figure 1A, different SW pressures can affect the particle size of WPINs. The order of particle size from largest to smallest was WPINs–SW (1.5 MPa), WPINs–SW (0.5 MPa), WPINs–SW (1.0 MPa), and WPINs–SW (0.2 MPa), of which the values were 232.40 nm, 225.44 nm, 222.26 nm, and 190.90 nm, respectively. Compared with other SW pressures, the nanoparticles treated with 0.2 MPa had the smallest particle size, which showed that SW significantly promoted the degree of protein fibrosis, which was conducive to the formation of smaller particles, possibly related to the structural changes of the protein [41]. In addition, the particle size of nanofibers increased with the increase in pressure, which might be due to the excessive pressure promoting the aggregation of protein fibers, and it was not conducive to the preparation of emulsion. Similar results were also reported. For example, the increase in protein particle size was due to the aggregation of nanoparticles caused by excessive pressure [28]. Similarly, the zeta potential of the samples had similar trends, and there were no significant differences between the samples (Figure 1B). Therefore, WPINs–SW (0.2 MPa) was chosen as the optimum SW pressure for subsequent nanofiber preparation.

As shown in Figure 1C, the particle size of WPI, WPINs–SW, and WPINs–H was 243.50 nm, 190.90 nm, and 218.90 nm, respectively. Compared with WPI, the particle sizes of WPINs–SW and WPINs–H decreased by 52.60 nm and 24.60 nm, respectively. The zeta potential of WPI, WPINs–SW, and WPINs–H was 23.50 mV, 26.37 mV, and 27.20 mV, respectively. The zeta potential of WPINs–SW and WPINs–H was considerably higher than that of WPI. The PE prepared by WPINs–SW was more stable, and WPINs–H had more significant zeta potential.

3.1.2. Congo Red Binding Spectrum

Congo Red can specifically bind to nanofibers, and the absorption intensity and absorption peak of Congo Red was increased. The change in absorption intensity can characterize the degree of WPI fibrosis [42]. As shown in Figure 1D, the absorption curve of Congo Red is the black curve, and the absorption curves of Congo Red, which are bound to WPINs–H, WPINs–SW, and WPI, are the red, blue, and green curves. When Congo Red reacted with WPINs, the absorption peak shifted from 490 nm to 540 nm. The order of the absorbance magnitude was WPINs–SW > WPINs–H > WPI, and the absorbance increased with mounting pressure, likely because the pressure caused more nanofibers to be formed and aggregated. This indicated that pressure was a factor affecting the generation of WPINs. In addition, the absorption intensity of WPINs–SW for Congo Red was much higher than that of WPINs–H. It indicated that the appropriate pressure promoted the formation of WPINs. This result was similar to the previous report that the zein structure was altered by subcritical water [43]. The results show that the high temperature and acidic environment caused the protein secondary structure to open, the protein was denatured, and protein particles became nanofibers.

3.1.3. Circular Dichroism (CD) Spectroscopy

Far-ultraviolet CD spectra (190–240 nm) can reflect the secondary structure of WPINs, including α-helix, β-sheet, β-turn, and random coil conformation [44]. For WPINs and WPI, the CD spectra had negative peaks near 190–280 nm and partial α-helix structure peaks had negative peaks at 208 nm and 222 nm. The average residual ellipticity at 216 nm may represent the content of the β-sheet structure. In Figure 1E, the β-sheet structure of WPINs–SW and WPINs–H were significantly more than the β-sheet structure of WPI, indicating that the formation of the β-sheet secondary structure promoted the formation of WPINs. The β-sheet structure of WPINs–SW was more than the β-sheet structure of WPINs–H. Under acidic conditions, some proteins were hydrolyzed into polypeptide fragments due to high-temperature denaturation, and their secondary structures were expanded. Exposure of hydrophobic groups in proteins enhanced intermolecular hydrophobic interactions and assembled into fibers [45]. The formation of the β-sheet during SW was due to the reconstruction of a stable native secondary structure [46]. This result was consistent with the Congo Red analysis.

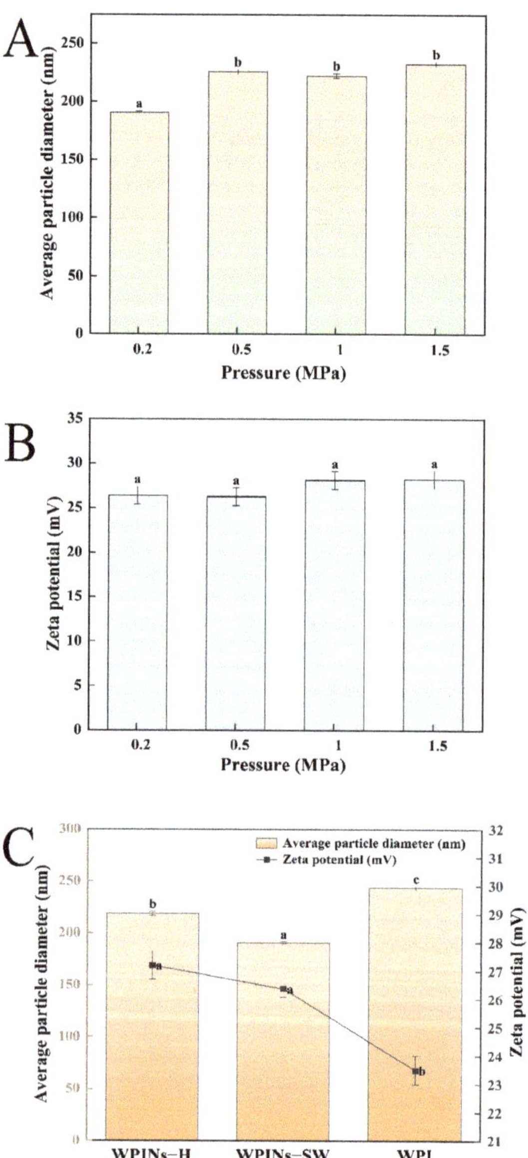

Figure 1. *Cont.*

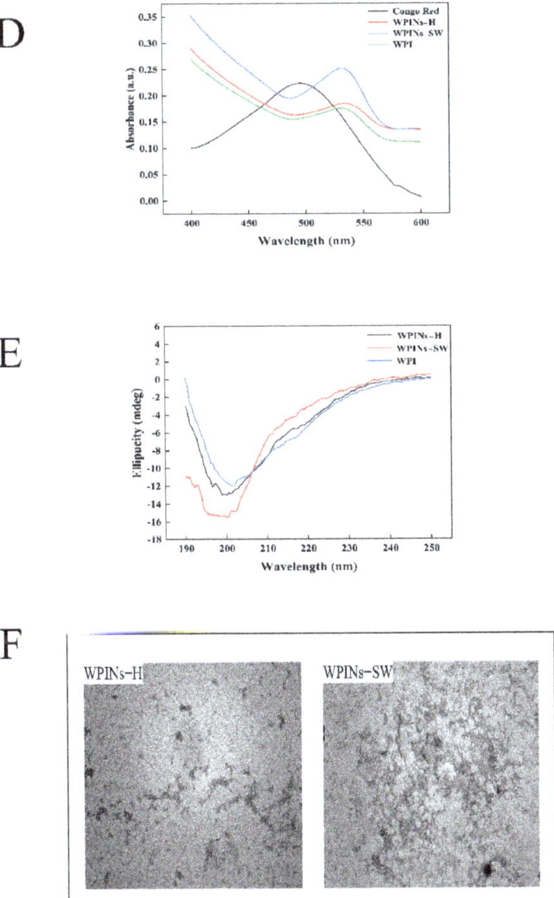

Figure 1. Effects of subcritical water on the structural properties of WPINs. (**A**) The impact of subcritical water pressure on the particle size of WPINs; (**B**) the effect of subcritical water pressure on the zeta potential of WPINs; (**C**) the effect of hydrothermal method and subcritical water method on the particle size and zeta potential of WPINs; (**D**) Congo Red analysis; (**E**) circular dichroism (CD) analysis; (**F**) transmission electron microscopy (TEM) analysis. The different lowercase letters mean that the variance of different samples is significant ($p < 0.05$). Note: WPINs–H represents whey protein isolate nanofibers prepared from hydrothermal method; WPINs–SW represents whey protein isolate nanofibers prepared from subcritical water; WPI represents whey protein isolate.

3.1.4. Transmission Electron Microscopy

WPI can form rough protein nanotube aggregates after a period of thermal induction treatment, and the physicochemical properties of the fibers changed significantly. The length and morphology of nanotube aggregates formed by different preparation methods were significantly different [47]. As shown in Figure 1F, WPINs–SW was a straight linear aggregate. It had many branches and individual protein particles. WPINs–H was sparse with short chapters. Fibrosis occurred in both WPINs–SW and WPINs–H. At pH 2.0, the surface of WPI had a higher charge, and there was still electrostatic repulsion between WPI molecules, resulting in the formation of WPINs [48]. SW denatured the protein, which unfolded the structure and exposed hydrophobic groups, and enhanced hydrophobic interactions between adjacent molecules [45]. It was shown that the average length of protein fiber aggregates increased with increasing pressure. This result proved that pressure is one of the reasons for the WPI fibrosis. Pressure might promote the fibrillation of WPI, causing PE prepared from WPINs to be more stable.

3.2. Effects of Subcritical Water on the Properties of High Internal Phase Pickering Emulsion Stabilized by WPINs

3.2.1. Optical Microscopy

In Figure 2A, the WPINs–SW–PE formed spherical aggregates with uniform size distribution. Since the WPINs–SW was closely arranged on the surface of the oil droplets, a large amount of electrostatic repulsion was provided between the PE particles, which made the WPINs–SW–PE very dense and oil droplets disperse independently [49]. However, the particle sizes of WPINs–H–PE were different, whereby one part was dispersed and the other part was aggregated, which meant that the oil droplets were loosely wrapped by WPINs–H, and the oil droplets were aggregated together. The particle size of WPINs–SW–PE was significantly smaller than that of WPINs–H–PE, and the result was consistent with the results of the particle size measurement. The WPINs–SW–PE under 0.2 MPa was more uniform. The emulsion had good stability and a good effect on the subsequent embedding of Cur.

A

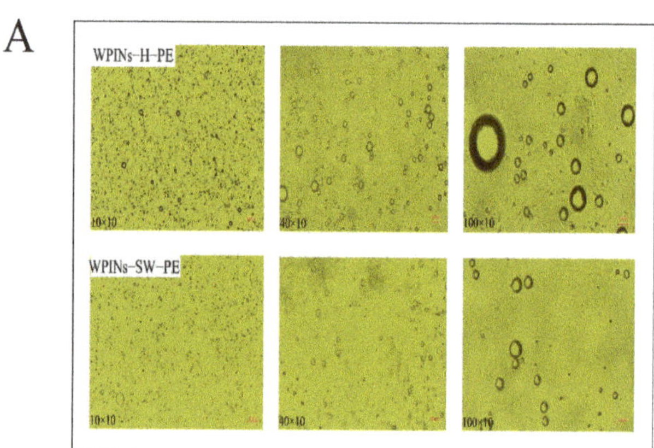

B

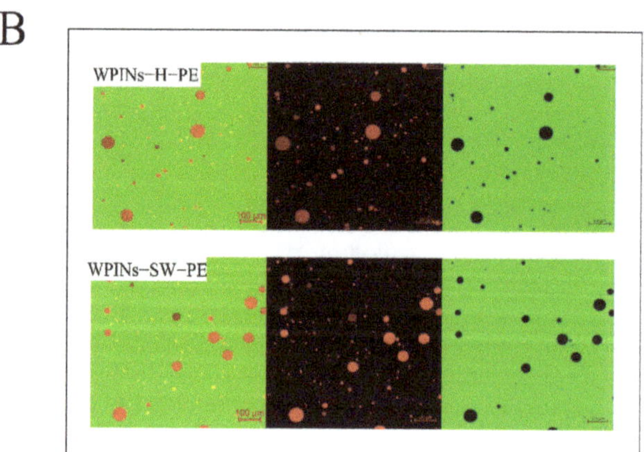

Figure 2. *Cont.*

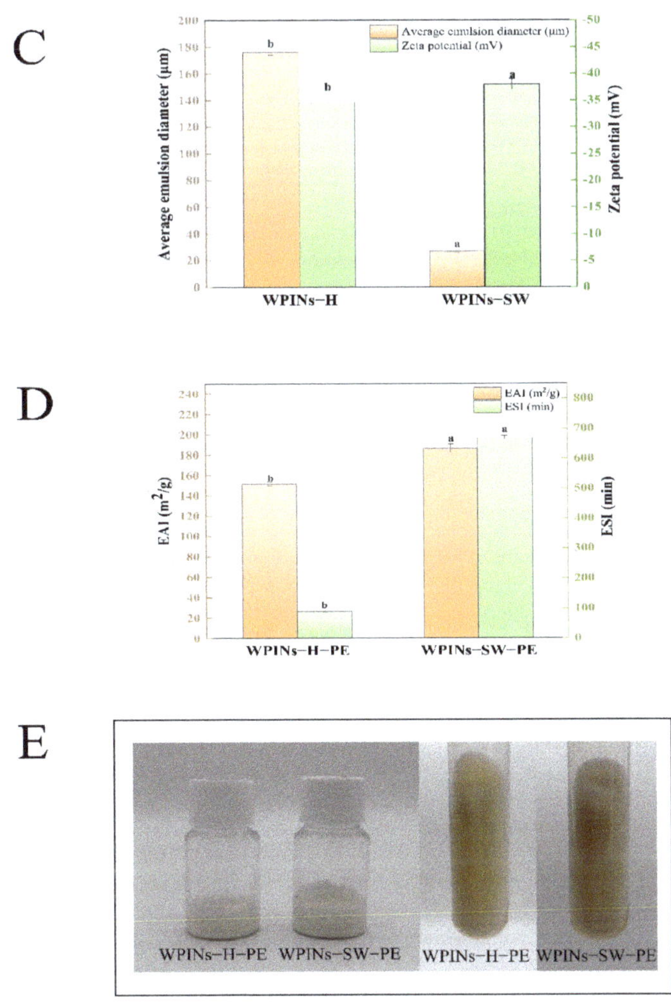

Figure 2. Effects of subcritical water on the properties of high internal phase Pickering emulsion stabilized by WPINs. (**A**) The optical microscope characterization of Pickering emulsion; (**B**) the CLSM imaging of Pickering emulsion; (**C**) the particle size and zeta potential of Pickering emulsion; (**D**) the EAI and ESI of Pickering emulsion; (**E**) the centrifugal stability of Pickering emulsion. The different lowercase letters mean that the variance of different samples is significant ($p < 0.05$). Note. WPINs–H–PE represents whey protein isolate nanofibers prepared from hydrothermal method stabilized Pickering emulsion; WPINs–SW–PE represents whey protein isolate nanofibers prepared from subcritical water stabilized Pickering emulsion.

3.2.2. CLSM

The mechanism of PE stability can be observed by CLSM photography. Under the fluorescent field, the protein bound specifically to FITC and emitted green light. The oil phase was combined explicitly with Nile red and glowed red. As shown in Figure 2B, the red oil phase particles were covered by green nanoparticles, indicating that the emulsion type was an oil-in-water (O/W) emulsion. A red ring layer protected the surface of oil particles, and FITC-stained protein was adsorbed at the oil–water interface and provided a barrier for the coalescence of oil particles [50]. This finding provided the most intuitive evidence for the mechanism as WPINs stabilized for PE. The emulsion did not show significant agglomeration. The oil–water interface presented a yellow ring layer caused by the superposition of the green color of FITC and the red color of Nile red during CLSM imaging, indicating the presence of green-stained protein at the oil–water interface. The CLSM images of WPINs–SW–PE showed that WPINs–SW was completely adsorbed around the oil droplets and dispersed in the spaces between the oil droplets. The photos of WPINs–SW–PE showed no aggregation and flocculation of oil droplets, and the number of particles at the interface was sufficient for protein adsorption. There was a slight aggregation of oil droplets in WPINs–H–PE, indicating that WPINs–H did not wrap oil droplets well and WPINs–H did not emulsify as well as WPINs–SW.

3.2.3. Particle Size and Zeta Potential

Particle size and zeta potential were important parameters for judging the stability of particles to the PE. The average particle size and zeta potential of the WPINs stabilized HIPPE are shown in Figure 2C. The particle size of WPINs–H–PE was 175.99 μm. The particle size of WPINs–SW–PE was 26.88 μm. SW led to a reduction in the average particle size. In addition, the zeta potential was used to represent the electrostatic repulsion between charged particles. The absolute value of the high zeta potential indicated the high stability of PE [51]. The absolute value of WPINs–SW–PE on the zeta potential was greater than that of WPINs–H–PE on the zeta potential. Interestingly, both of them were greater than 30 mV, indicating that WPINs treated with both SW and hydrothermal methods could enhance the stability of PE [52]. Based on these results, SW treatment may reduce the average particle size and increase the zeta potential of PE, which was attributed to the increase in the surface charge and distribution of WPINs during SW. WPINs–SW–PE will act as a higher and more stable encapsulation rate during the subsequent encapsulation of Cur.

3.2.4. Emulsification Activity Index (EAI) and Emulsification Stability Index (ESI)

The emulsification activity index (EAI) and emulsification stability index (ESI) of PE were determined. The EAI value was the maximum interfacial area per unit weight of protein in a stable solution. The ESI indicated the stability of a diluted emulsion over a specified period [53]. As shown in Figure 2D, the EAI of WPINs–SW–PE was 186.63 m^2/g, and the EAI of the WPINs–H–PE was 151.79 m^2/g. The hydrophobic and electrostatic interactions between droplets were increased because the protein mass ratio was 5%, so the WPINs exhibited the best emulsification activity [49]. The fibrosis of WPI became more complete under the pressure of SW, and the emulsifying activity of PE was increased. Meanwhile, the ESI of the WPINs–H–PE was 89.05 min, and the ESI of WPINs–SW–PE was 670.30 min. PE can maintain a relatively stable state because of the strong electrostatic repulsion between molecules, which was conducive to the adsorption of WPINs on the oil–water interface. Besides, the emulsion showed no demulsification phenomenon, indicating that the particles adsorbed on the interface had higher emulsification stability [54]. Significant emulsification activity and emulsification stability of the PE were a function of the excellence of the water phase particle, which provided a more stable emulsion option for the embedded Cur below.

3.2.5. Centrifugation Stability

As shown in Figure 2E, the water–oil ratio of the HIPPE was 2.6:7.4 after centrifugation at 10,000× *g* for 10 min. The oil phase was not precipitated, and no sediment particles were observed in the lower water phase. It may be that HIPPE was highly cohesive and able to form a dense network structure, while PE was hard and did not flow easily. The particles can be firmly adsorbed by the interface to prevent the emulsion from merging, resulting in ultra-high stability of the emulsion. Most PE prepared with protein-based particles was difficult to maintain under centrifugal forces of more than 5000× *g* [34,55,56].

3.3. Effects of High Internal Phase Pickering Emulsion on Curcumin Bioavailability

3.3.1. Embedding Rate of Curcumin

As shown in Figure 3A, the embedding rate of WPINs–SW–PE–Cur was 96.72%, and the embedding rate of WPINs–H–PE–Cur was 94.07%. The results showed that the PE was stabilized by WPINs, and the stabilized particles were adsorbed and arranged at the oil–water interface, which helped improve the embedding rate of Cur. The embedding rate of Cur in WPINs–SW–PE was higher than that of WPINs–H–PE because the embedding of oil droplets by WPINs–SW was stronger than that of WPINs–H, and the structure of WPINs prepared by SW was more stable.

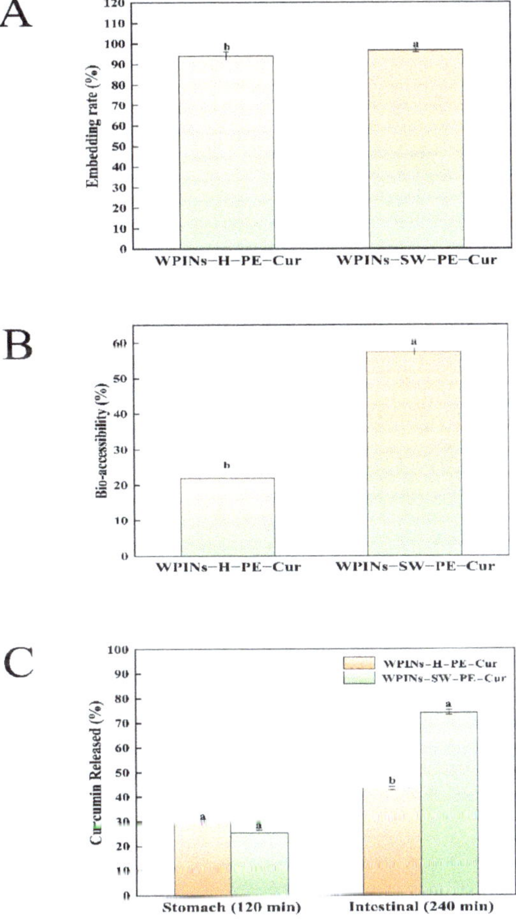

Figure 3. Effects of high internal phase Pickering emulsion on curcumin bioavailability. (**A**) The effect of Pickering emulsion on the embedding rate of curcumin; (**B**) the bio-accessibility of curcumin in Pickering emulsion during gastrointestinal digestion; (**C**) the release rate of curcumin during gastrointestinal digestion. The different lowercase letters mean that the variance of different samples is significant ($p < 0.05$). Note: WPINs–H–PE–Cur represents curcumin-loaded Pickering emulsion prepared from hydrothermal method, WPINs–SW–PE–Cur represents curcumin-loaded Pickering emulsion prepared from subcritical water.

3.3.2. Bio-Accessibility

The release of Cur occurred in the small intestinal stage, which was a prerequisite for exploring bio-accessibility. The bio-accessibility of Cur-loaded HIPPE after in vitro digestion was explored, and the results are presented in Figure 3B. The bio-accessibility of WPINs–SW–PE–Cur was 57.52%, and the bio-accessibility of WPINs–H–PE–Cur was 21.94%. The structure of WPINs–SW was more compact to protect Cur from releasing during simulated gastric digestion in vitro. In addition, the particle size of WPINs–SW–PE was smaller, and the specific surface area in contact with lipase during the simulated small intestinal digestion process in vitro was larger, which caused more Cur to be released and then used by cells, thus causing the higher bio-accessibility of WPINs–SW–PE–Cur [57,58]. However, the particle size of WPINs–H–PE was large, so a large amount of oil cannot be digested by the gastrointestinal tract. Cur was a lipophilic substance, which was dissolved in corn oil, so Cur cannot be released at the target site. The bio-accessibility of WPINs–H–PE–Cur was low.

3.3.3. Curcumin Release

Cur was released in simulated gastric fluid (SGF) and simulated intestinal fluid (SIF). As shown in Figure 3C, after incubation in SGF for 2 h, the Cur release of WPINs–SW–PE–Cur and WPINs–H–PE–Cur reached 25.25% and 29.37%, respectively. After 2 h in SIF, the Cur release of WPINs–SW–PE–Cur and WPINs–H–PE–Cur reached 74.25% and 43.45%, respectively. In the SGF environment, the release of WPINs–H–PE–Cur was higher than that of WPINs–SW–PE–Cur, while the results were reversed in SIF. Perhaps the pH 2.0 gastric acid environment and the presence of pepsin cause the WPINs to be encapsulated in PE hydrolyzed by pepsin, and the electrostatic repulsion between adjacent particles was weakened and the included oil phase was exposed. Since Cur was heavily exposed in the stomach, the amount of Cur reaching the small intestine was reduced [57,59,60]. However, WPINs–SW–PE had good stability, which enabled Cur to be protected in the stomach and achieved a large amount of targeted release in the small intestine. In conclusion, WPINs–SW–PE–Cur significantly improved the bioavailability of Cur. A previous study investigated the release of Cur from stabilized PE encapsulated in a chitosan/gum Arabic antiparticle and noted that approximately 36% of Cur was released after 3 h incubation in SIF [35]. As a comparison, a much higher rate of Cur release can be observed in the current study, which demonstrated that the WPINs–SW stabilized PE had a considerable effect on the targeted release in SIF.

3.4. Effects of Simulated Gastrointestinal Digestion on the Antioxidant Capacity of Pickering Emulsion

3.4.1. DPPH• Scavenging Capacity

The DPPH• scavenging capacity of Cur encapsulated in PE was compared to that of Free–Cur after in vitro digestion to understand the effect of protective nano and PE on the antioxidant activity of Cur [61]. As shown in Figure 4A, the DPPH• scavenging capacity of WPINs–SW–PE–Cur, WPINs–H–PE–Cur, and Free–Cur was 48.67%, 38.45%, and 18.43%, respectively. DPPH• scavenging capacity after digestion in vitro was 22.56%, 18.74%, and 7.34%, respectively. It can be seen that the antioxidant activity of encapsulated Cur was significantly higher than that of Free–Cur. The antioxidant activity of WPINs–SW–PE–Cur was significantly better than that of WPINs–H–PE–Cur, which may be related to the solubility of Cur. The solubility of Cur in WPINs–SW–PE was higher, which was more conducive to the full contact between Cur and oxidized substances. The protective effect of PE on the antioxidant activity of Cur was confirmed, thus helping to increase its antioxidant capacity [62].

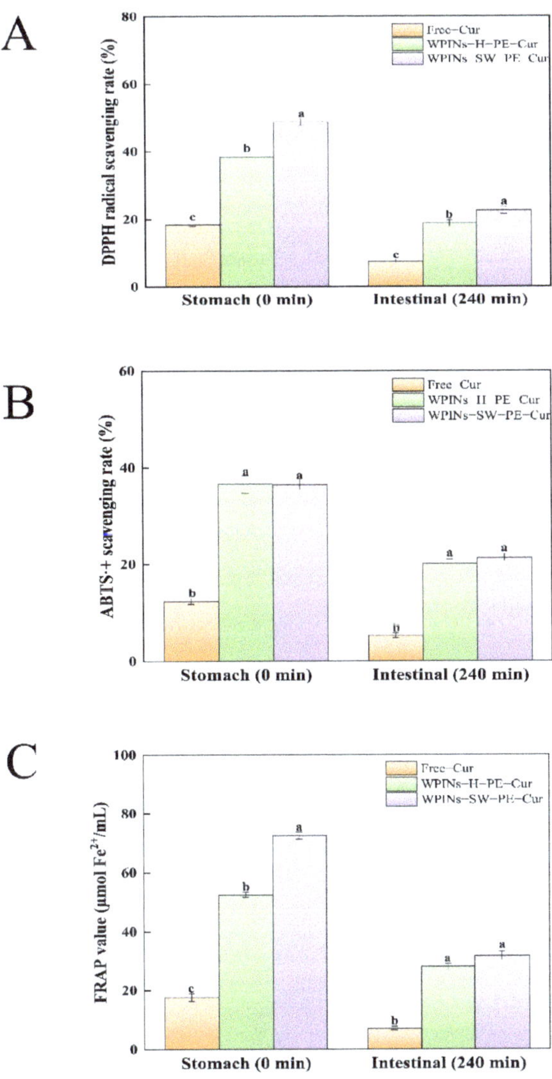

Figure 4. Effects of simulated gastrointestinal digestion on the antioxidant capacity of Pickering emulsion. (**A**) DPPH• scavenging capacity on the Pickering emulsion; (**B**) ABTS·+ scavenging capacity on the Pickering emulsion; (**C**) reducing power on the Pickering emulsion. The different lowercase letters mean that the variance of different samples is significant ($p < 0.05$). Note: WPINs–H–PE–Cur represents curcumin-loaded Pickering emulsion prepared by hydrothermal method; WPINs–SW–PE–Cur represents curcumin-loaded Pickering emulsion prepared by subcritical water.

3.4.2. ABTS·+ Scavenging Capacity

As shown in Figure 4B, the antioxidant activities of WPINs–SW–PE–Cur, WPINs–H–PE–Cur, and Free–Cur were determined. The ABTS·+ scavenging abilities of the WPINs–SW–PE–Cur, WPINs–H–PE–Cur, and Free–Cur were 36.50%, 36.53%, and 12.38%, respectively. After digestion in vitro, the ABTS·+ scavenging abilities were 21.27%, 20.02%, and 5.30%, respectively. The antioxidant activity of Free–Cur was always significantly lower than that of encapsulated Cur, and the ABTS·+ scavenging ability of WPINs–SW–PE–Cur was stronger than that of WPINs–H–PE–Cur. The antioxidant groups of Cur were exposed in the system and combined with ABTS·+ in large quantities because PE improved the solubility of Cur and expanded the dispersibility of Cur in the solution. This result was consistent with the DPPH• scavenging ability.

3.4.3. Reducing Power

Figure 4C displays the FRAP iron-reducing ability of WPINs–SW–PE–Cur, WPINs–H–PE–Cur, and Free–Cur before and after simulated digestion in vitro. In the reaction system of the FRAP method, the FRAP iron-reducing ability of the sample was positively correlated with the color and absorbance changes of the final reaction product, which can be used as an essential indicator to measure the antioxidant activity of the sample [63]. Before simulated digestion in vitro, the FRAP iron-reducing ability of Free–Cur was 17.60 μmol/mL, and the FRAP iron-reducing ability of the WPINs–SW–PE–Cur and WPINs–H–PE–Cur was 72.49 μmol/mL and 52.53 μmol/mL. After digestion in vitro, the FRAP iron-reducing ability of the WPINs–SW–PE–Cur and WPINs–H–PE–Cur was 31.52 μmol/mL and 27.91 μmol/mL. The FRAP iron-reducing ability of the Free–Cur was 7.04 μmol/mL, indicating that the PE prepared by WPINs had a good encapsulation effect on Cur and promoted the water solubility of Cur, and the WPINs–SW–PE was more stable. The previous result reported that gelatin-encapsulated Cur showed significantly higher antioxidant activity [64]. In fact, the encapsulation of Cur increased the water solubility of Cur [65].

4. Conclusions

This study successfully prepared WPINs–SW at 110 °C and 0.2 MPa in 5 min. TEM images showed that WPINs–SW is composed of fibrous nanoparticles. Short-term high temperature and pressure can allow complete protein modification and acts as a stabilizer for HIPPE. CLSM, optical microscopy, particle size, zeta potential, and EAI and ESI results indicated that the WPINs–SW–PE had a smaller particle size and better stability. The results of simulated intestinal digestion clearly showed that the WPINs–SW–PE had higher Cur release than that of WPINs–H–PE. These findings suggested that WPINs–SW–PE–Cur might be a promising encapsulating agent to protect the loaded Cur and achieve sustainable release under intestinal conditions. However, the absorption and transport functions of active substances delivered by HIPPE remain to be further explored.

Author Contributions: Conceptualization and methodology, X.X.; validation, Z.Z., investigation, J.Z. (Junlong Zhu) and D.W.; resources, G.L. and L.L.; writing—original draft preparation, J.Z. (Jixian Zhang), Z.Z. and C.W.; writing—review and editing, X.L., J.R., Y.L. and Q.D.; visualization, Z.Z.; supervision, C.W.; project administration, X.X.; funding acquisition, L.L. All authors have read and agreed to the published version of the manuscript.

Funding: This research was funded by the National Natural Science Foundation of China (31901651).

Institutional Review Board Statement: Not applicable.

Informed Consent Statement: Not applicable.

Data Availability Statement: Data are contained within the article.

Conflicts of Interest: The authors declare no conflict of interest.

References

1. Kunnumakkara, A.B.; Anand, P.; Aggarwal, B.B. Curcumin inhibits proliferation, invasion, angiogenesis and metastasis of different cancers through interaction with multiple cell signaling proteins. *Cancer Lett.* **2008**, *269*, 199–225. [CrossRef] [PubMed]
2. Abadi, A.J.; Mirzaei, S.; Mahabady, M.K.; Hashemi, F.; Zabolian, A.; Hashemi, F.; Raee, P.; Aghamiri, S.; Ashrafizadeh, M.; Aref, A.R.; et al. Curcumin and its derivatives in cancer therapy: Potentiating antitumor activity of cisplatin and reducing side effects. *Phytother. Res.* **2022**, *36*, 189–213. [CrossRef] [PubMed]
3. Soleimani, V.; Sahebkar, A.; Hosseinzadeh, H. Turmeric (*Curcuma longa*) and its major constituent (curcumin) as nontoxic and safe substances: Review. *Phyther. Res.* **2018**, *32*, 985–995. [CrossRef] [PubMed]

4. Kocaadam, B.; Şanlier, N. Curcumin, an active component of turmeric (*Curcuma longa*), and its effects on health. *Crit. Rev. Food Sci. Nutr.* **2017**, *57*, 2889–2895. [CrossRef] [PubMed]

5. Chang, C.; Wang, T.; Hu, Q.; Zhou, M.; Xue, J.; Luo, Y. Pectin coating improves physicochemical properties of caseinate/zein nanoparticles as oral delivery vehicles for curcumin. *Food Hydrocoll.* **2017**, *70*, 143–151. [CrossRef]

6. Patel, A.R.; Velikov, K.P. Colloidal delivery systems in foods: A general comparison with oral drug delivery. *LWT-Food Sci. Technol.* **2011**, *44*, 1958–1964. [CrossRef]

7. Zou, L.; Zheng, B.; Liu, W.; Liu, C.; Xiao, H.; McClements, D.J. Enhancing nutraceutical bioavailability using excipient emul-sions: Influence of lipid droplet size on solubility and bioaccessibility of powdered curcumin. *J. Funct. Foods* **2015**, *15*, 72–83. [CrossRef]

8. Jeetah, R.; Bhaw-Luximon, A.; Jhurry, D. Nanopharmaceutics: Phytochemical-Based Controlled or Sustained Drug-Delivery Systems for Cancer Treatment. *J. Biomed. Nanotechnol.* **2014**, *10*, 1810–1840. [CrossRef]

9. Jafari, S.M.; Sedaghat Doost, A.; Nikbakht Nasrabadi, M.; Boostani, S.; Van der Meeren, P. Phytoparticles for the stabilization of Pickering emulsions in the formulation of novel food colloidal dispersions. *Trends Food Sci. Technol.* **2020**, *98*, 117–128. [CrossRef]

10. Tavasoli, S.; Liu, Q.; Jafari, S.M. Development of Pickering emulsions stabilized by hybrid biopolymeric particles/nanoparticles for nutraceutical delivery. *Food Hydrocoll.* **2022**, *124*, 107280. [CrossRef]

11. Shi, A.; Feng, X.; Wang, Q.; Adhikari, B. Pickering and high internal phase Pickering emulsions stabilized by protein-based particles: A review of synthesis, application and prospective. *Food Hydrocoll.* **2020**, *109*, 106117. [CrossRef]

12. Zeng, T.; Wu, Z.L.; Zhu, J.Y.; Yin, S.W.; Tang, C.H.; Wu, L.Y.; Yang, X.Q. Development of antioxidant Pickering high inter-nal phase emulsions (HIPEs) stabilized by protein/polysaccharide hybrid particles as potential alternative for PHOs. *Food Chem.* **2017**, *231*, 122–130. [CrossRef] [PubMed]

13. Zembyla, M.; Murray, B.S.; Sarkar, A. Water-In-Oil Pickering Emulsions Stabilized by Water-Insoluble Polyphenol Crystals. *Langmuir* **2018**, *34*, 10001–10011. [CrossRef] [PubMed]

14. Xu, Y.-T.; Liu, T.-X.; Tang, C.-H. Novel pickering high internal phase emulsion gels stabilized solely by soy β-conglycinin. *Food Hydrocoll.* **2019**, *88*, 21–30. [CrossRef]

15. Hao, Z.-Z.; Peng, X.-Q.; Tang, C.-H. Edible pickering high internal phase emulsions stabilized by soy glycinin: Improvement of emulsification performance and pickering stabilization by glycation with soy polysaccharide. *Food Hydrocoll.* **2020**, *103*, 105672. [CrossRef]

16. Hulmi, J.J.; Lockwood, C.M.; Stout, J.R. Effect of protein/essential amino acids and resistance training on skeletal muscle hyper-trophy: A case for whey protein. *Nutr. Metab.* **2010**, *7*, 1–11. [CrossRef]

17. Thaiphanit, S.; Anprung, P. Physicochemical and emulsion properties of edible protein concentrate from coconut (*Cocos nucifera* L.) processing by-products and the influence of heat treatment. *Food Hydrocoll.* **2016**, *52*, 756–765. [CrossRef]

18. Li, X.; Mao, L.; He, X.; Ma, P.; Gao, Y.; Yuan, F. Characterization of β-lactoglobulin gels induced by high pressure processing. *Innov. Food Sci. Emerg. Technol.* **2018**, *47*, 335–345. [CrossRef]

19. Raikos, V. Effect of heat treatment on milk protein functionality at emulsion interfaces. A review. *Food Hydrocoll.* **2010**, *24*, 259–265. [CrossRef]

20. Mantovani, R.A.; de Figueiredo Furtado, G.; Netto, F.M.; Cunha, R.L. Assessing the potential of whey protein fibril as emulsifier. *J. Food Eng.* **2018**, *223*, 99–108. [CrossRef]

21. Adamcik, J.; Mezzenga, R. Proteins Fibrils from a Polymer Physics Perspective. *Macromolecules* **2012**, *45*, 1137–1150. [CrossRef]

22. Wang, Q.; Yu, H.; Tian, B.; Jiang, B.; Xu, J.; Li, D.; Feng, Z.; Liu, C. Novel Edible Coating with Antioxidant and Antimicrobial Activities Based on Whey Protein Isolate Nanofibrils and Carvacrol and Its Application on Fresh-Cut Cheese. *Coatings* **2019**, *9*, 583. [CrossRef]

23. Yang, Y.; Jiao, Q.; Wang, L.; Zhang, Y.; Jiang, B.; Li, D.; Feng, Z.; Liu, C. Preparation and evaluation of a novel high internal phase Pickering emulsion based on whey protein isolate nanofibrils derived by hydrothermal method. *Food Hydrocoll.* **2022**, *123*, 107180. [CrossRef]

24. Zhang, J.; Wen, C.; Zhang, H.; Duan, Y.; Ma, H. Recent advances in the extraction of bioactive compounds with subcritical water: A review. *Trends Food Sci. Technol.* **2020**, *95*, 183–195. [CrossRef]

25. Zhang, J.; Ji, T.; Yang, X.; Liu, G.; Liang, L.; Liu, X.; Wen, C.; Ye, Z.; Wu, M.; Xu, X. Properties of selenium nanoparticles stabilized by *Lycium burburum* polysaccharide-protein conjugates obtained with subcritical water. *Int. J. Biol. Macromol.* **2022**, *205*, 672–681. [CrossRef]

26. Yoko, A.; Seong, G.; Tomai, T.; Adschiri, T. Utilization of Sub-and Supercritical Water for Nano-Catalyst Synthesis and Waste and Bio-mass Processing. *Waste Biomass Valorizat.* **2022**, *13*, 51–66. [CrossRef]

27. Wang, J.-M.; Xia, N.; Yang, X.-Q.; Yin, S.-W.; Qi, J.-R.; He, X.-T.; Yuan, D.-B.; Wang, L.-J. Adsorption and Dilatational Rheology of Heat-Treated Soy Protein at the Oil–Water Interface: Relationship to Structural Properties. *J. Agric. Food Chem.* **2012**, *60*, 3302–3310. [CrossRef]

28. Novo, L.P.; Bras, J.; García, A.; Belgacem, N.; Curvelo, A.A.D.S. A study of the production of cellulose nanocrystals through subcritical water hydrolysis. *Ind. Crops Prod.* **2016**, *93*, 88–95. [CrossRef]

29. Nilsson, M.R. Techniques to study amyloid fibril formation in vitro. *Methods* **2004**, *34*, 151–160. [CrossRef]

30. Feng, Z.; Li, L.; Zhang, Y.; Li, X.; Liu, C.; Jiang, B.; Xu, J.; Sun, Z. Formation of whey protein isolate nanofibrils by endoproteinase GluC and their emulsifying properties. *Food Hydrocoll.* **2019**, *94*, 71–79. [CrossRef]

31. Ren, Z.; Chen, Z.; Zhang, Y.; Lin, X.; Li, Z.; Weng, W.; Yang, H.; Li, B. Effect of heat-treated tea water-insoluble protein nanoparticles on the characteristics of *Pickering emulsions*. *LWT* **2021**, *149*, 111999. [CrossRef]
32. Li, J.; Xu, X.; Chen, Z.; Wang, T.; Lu, Z.; Hu, W.; Wang, L. Zein/gum Arabic nanoparticle-stabilized Pickering emulsion with thymol as an antibacterial delivery system. *Carbohydr. Polym.* **2018**, *200*, 416–426. [CrossRef] [PubMed]
33. de Souza, T.S.P.; Dias, F.F.G.; Koblitz, M.G.B.; de Moura Bell, J.M. Effects of enzymatic extraction of oil and protein from almond cake on the physicochemical and functional properties of protein extracts. *Food Bioprod. Processing* **2020**, *122*, 280–290. [CrossRef]
34. Gong, X.; Wang, Y.; Chen, L. Enhanced emulsifying properties of wood-based cellulose nanocrystals as Pickering emulsion stabilizer. *Carbohydr. Polym.* **2017**, *169*, 295–303. [CrossRef]
35. Han, J.; Chen, F.; Gao, C.; Zhang, Y.; Tang, X. Environmental stability and curcumin release properties of Pickering emulsion stabilized by chitosan/gum arabic nanoparticles. *Int. J. Biol. Macromol.* **2020**, *157*, 202–211. [CrossRef]
36. Liang, L.; Zhang, X.; Wang, X.; Jin, Q.; McClements, D.J. Influence of Dairy Emulsifier Type and Lipid Droplet Size on Gastrointestinal Fate of Model emulsions: In Vitro Digestion Study. *J. Agric. Food Chem.* **2018**, *66*, 9761–9769. [CrossRef]
37. Zhou, H.; Tan, Y.; Lv, S.; Liu, J.; Mundo, J.L.M.; Bai, L.; Rojas, O.J.; McClements, D.J. Nanochitin-stabilized pickering emulsions: Influence of nanochitin on lipid digestibility and vitamin bioaccessibility. *Food Hydrocoll.* **2020**, *106*, 105878. [CrossRef]
38. Ak, T.; Gülçin, I. Antioxidant and radical scavenging properties of curcumin. *Chem. Interact.* **2008**, *174*, 27–37. [CrossRef]
39. Wen, C.; Zhang, J.; Feng, Y.; Duan, Y.; Ma, H.; Zhang, H. Purification and identification of novel antioxidant peptides from watermelon seed protein hydrolysates and their cytoprotective effects on H_2O_2-induced oxidative stress. *Food Chem.* **2020**, *327*, 127059. [CrossRef]
40. Chen, Y.; Huang, F.; Xie, B.; Sun, Z.; McClements, D.J.; Deng, Q. Fabrication and characterization of whey protein isolates-lotus seedpod proanthocyanin conjugate: Its potential application in oxidizable emulsions. *Food Chem.* **2021**, *346*, 128680. [CrossRef] [PubMed]
41. Mo, H.; Li, Q.; Liang, J.; Ou, J.; Jin, B. Investigation of physical stability of Pickering emulsion based on soy protein/β-glucan/coumarin ternary complexes under subcritical water condition. *Int. J. Food Sci. Technol.* **2021**, *56*, 5017–5024. [CrossRef]
42. Mohammadian, M.; Salami, M.; Emam-Djomeh, Z. Characterization of hydrogels formed by non-toxic chemical cross-linking of mixed nanofibrillated/heat-denatured whey proteins. *J. Iran. Chem. Soc.* **2019**, *16*, 2731–2741. [CrossRef]
43. Zhang, J.; Wen, C.; Zhang, H.; Zandile, M.; Luo, X.; Duan, Y.; Ma, H. Structure of the zein protein as treated with subcritical water. *Int. J. Food Prop.* **2018**, *21*, 128–138. [CrossRef]
44. Wei, G.; Su, Z.; Reynolds, N.P.; Arosio, P.; Hamley, I.W.; Gazit, E.; Mezzenga, R. Self-assembling peptide and protein amyloids: From structure to tailored function in nanotechnology. *Chem. Soc. Rev.* **2017**, *46*, 4661–4708. [CrossRef]
45. Chanamai, R.; McClements, D.J. Comparison of gum arabic, modified starch, and whey protein isolate as emulsifiers: Influ-ence of pH, CaCl2 and temperature. *J. Food Sci.* **2002**, *67*, 120–125. [CrossRef]
46. Haque, M.A.; Kaur, P.; Islam, A.; Hassan, M.I. Application of circular dichroism spectroscopy in studying protein folding, stability, and interaction. In *Advances in Protein Molecular and Structural Biology Methods*; Academic Press: Cambridge, MA, USA, 2022; pp. 213–224.
47. Mahardika, M.; Abral, H.; Kasim, A.; Arief, S.; Asrofi, M. Production of Nanocellulose from Pineapple Leaf Fibers via High-Shear Homogenization and Ultrasonication. *Fibers* **2018**, *6*, 28. [CrossRef]
48. Chung, C.; Rojanasasithara, T.; Mutilangi, W.; McClements, D.J. Stability improvement of natural food colors: Impact of amino acid and peptide addition on anthocyanin stability in model beverages. *Food Chem.* **2017**, *218*, 277–284. [CrossRef]
49. Wei, Y.; Liu, Z.; Guo, A.; Mackie, A.; Zhang, L.; Liao, W.; Mao, L.; Yuan, F.; Gao, Y. Zein Colloidal Particles and Cellulose Nanocrystals Synergistic Stabilization of Pickering Emulsions for Delivery of β-Carotene. *J. Agric. Food Chem.* **2021**, *69*, 12278–12294. [CrossRef]
50. Xiao, J.; Wang, X.; Gonzalez, A.J.P.; Huang, Q. Kafirin nanoparticles-stabilized Pickering emulsions: Microstructure and rheological behavior. *Food Hydrocoll.* **2016**, *54*, 30–39. [CrossRef]
51. Shah, B.R.; Xu, W.; Mráz, J. Formulation and characterization of zein/chitosan complex particles stabilized Pickering emulsion with the encapsulation and delivery of vitamin D3. *J. Sci. Food Agric.* **2021**, *101*, 5419–5428. [CrossRef]
52. Yu, G.; Zhao, J.; Wei, Y.; Huang, L.; Li, F.; Zhang, Y.; Li, Q. Physicochemical Properties and Antioxidant Activity of Pumpkin Polysaccharide (*Cucurbita moschata* Duchesne ex Poiret) Modified by Subcritical Water. *Foods* **2021**, *10*, 197. [CrossRef] [PubMed]
53. De Almeida, N.M.; de Moura Bell, J.M.L.N.; Johnson, L.A. Properties of Soy Protein Produced by Countercurrent, Two-Stage, Enzyme-Assisted Aqueous Extraction. *J. Am. Oil Chem. Soc.* **2014**, *91*, 1077–1085. [CrossRef]
54. Feng, X.; Ma, L.; Dai, H.; Fu, Y.; Fu, Y.; Zhu, K.; Wang, H.; Zhang, Y. The study on stability of food-grade Pickering emulsion and the loading of β-carotene. *Food Ferment. Ind.* **2021**, *47*, 18–25.
55. Marefati, A.; Matos, M.; Wiege, B.; Haase, N.U.; Rayner, M. Pickering emulsifiers based on hydrophobically modified small granular starches Part II—Effects of modification on emulsifying capacity. *Carbohydr. Polym.* **2018**, *201*, 416–424. [CrossRef] [PubMed]
56. Silva, C.E.P.; Tam, K.C.; Bernardes, J.S.; Loh, W. Double stabilization mechanism of O/W Pickering emulsions using cationic nanofibrillated cellulose. *J. Colloid Interface Sci.* **2020**, *574*, 207–216. [CrossRef] [PubMed]
57. Salvia-Trujillo, L.; Qian, C.; Martin-Belloso, O.; McClements, D. Influence of particle size on lipid digestion and β-carotene bioaccessibility in emulsions and nanoemulsions. *Food Chem.* **2013**, *141*, 1472–1480. [CrossRef]

58. Khan, M.A.; Chen, L.; Liang, L. Improvement in storage stability and resveratrol retention by fabrication of hollow ze-in-chitosan composite particles. *Food Hydrocoll.* **2021**, *113*, 106477. [CrossRef]
59. Gao, H.; Ma, L.; Cheng, C.; Liu, J.; Liang, R.; Zou, L.; Liu, W.; McClements, D.J. Review of recent advances in the preparation, properties, and applications of high internal phase emulsions. *Trends Food Sci. Technol.* **2021**, *112*, 36–49. [CrossRef]
60. Han, L.; Lu, K.; Zhou, S.; Zhang, S.; Xie, F.; Qi, B.; Li, Y. Development of an oil-in-water emulsion stabilized by a black bean protein-based nanocomplex for co-delivery of quercetin and perilla oil. *LWT* **2020**, *138*, 110644. [CrossRef]
61. Shah, B.R.; Zhang, C.; Li, Y.; Li, B. Bioaccessibility and antioxidant activity of curcumin after encapsulated by nano and Pickering emulsion based on chitosan-tripolyphosphate nanoparticles. *Food Res. Int.* **2016**, *89*, 399–407. [CrossRef]
62. Malik, P.; Ameta, R.; Singh, M. Preparation and characterization of bionanoemulsions for improving and modulating the antioxidant efficacy of natural phenolic antioxidant curcumin. *Chem. Interact.* **2014**, *222*, 77–86. [CrossRef] [PubMed]
63. Ou, B.; Huang, D.; Hampsch-Woodill, M.; Flanagan, J.A.; Deemer, E.K. Analysis of antioxidant activities of common vegetables employing oxygen radical absorbance capacity (ORAC) and ferric reducing antioxidant power (FRAP) assays: A comparative study. *J. Agric. Food Chem.* **2002**, *50*, 3122–3128. [CrossRef]
64. Gómez-Estaca, J.; Balaguer, M.P.; López-Carballo, G.; Gavara, R.; Hernández-Muñoz, P. Improving antioxidant and antimicrobial properties of curcumin by means of encapsulation in gelatin through electrohydrodynamic atomization. *Food Hydrocoll.* **2017**, *70*, 313–320. [CrossRef]
65. Pan, K.; Zhong, Q.; Baek, S.J. Enhanced Dispersibility and Bioactivity of Curcumin by Encapsulation in Casein Nanocapsules. *J. Agric. Food Chem.* **2013**, *61*, 6036–6043. [CrossRef] [PubMed]

Article

Effects of Soybean Oil Body as a Milk Fat Substitute on Ice Cream: Physicochemical, Sensory and Digestive Properties

Wan Wang, Min Wang, Cong Xu, Zhijing Liu, Liya Gu, Jiage Ma, Lianzhou Jiang, Zhanmei Jiang and Juncai Hou *

College of Food Science, Northeast Agricultural University, Harbin 150030, China; 13159806631@163.com (W.W.); 18246037810@163.com (M.W.); 15636116265@163.com (C.X.); liuzhijing22@126.com (Z.L.); andrea_hh@163.com (L.G.); jiage_ma@neau.edu.cn (J.M.); jlzname@neau.edu.cn (L.J.); zhanmeijiang@neau.edu.cn (Z.J.)
* Correspondence: jchou@neau.edu.cn; Tel.: +86-451-55190710

Abstract: Soybean oil body (SOB) has potential as a milk fat substitute due to its ideal emulsification, stability and potential biological activity. In this study, SOB was used as a milk fat substitute to prepare ice cream, expecting to reduce the content of saturated fatty acid and improve the quality defects of ice cream products caused by the poor stability of milk fat at low temperatures. This study investigated the effect of SOB as a milk fat substitute (the substitution amount was 10–50%) on ice cream through apparent viscosity, particle size, overrun, melting, texture, sensory and digestive properties. The results show SOB substitution for milk fat significantly increased the apparent viscosity and droplet uniformity and decreased the particle size of the ice cream mixes, indicating that there were lots of intermolecular interactions to improve ice cream stability. In addition, ice cream with 30% to 50% SOB substitution had better melting properties and texture characteristics. The ice cream with 40% SOB substitution had the highest overall acceptability. Furthermore, SOB substitution for milk fat increased unsaturated fatty acid content in ice cream and fatty acid release during digestion, which had potential health benefits for consumers. Therefore, SOB as a milk fat substitute may be an effective way to improve the nutritional value and quality characteristics of dairy products.

Keywords: soybean oil body; ice cream; melting properties; texture; sensory evaluation; digestive properties

Citation: Wang, W.; Wang, M.; Xu, C.; Liu, Z.; Gu, L.; Ma, J.; Jiang, L.; Jiang, Z.; Hou, J. Effects of Soybean Oil Body as a Milk Fat Substitute on Ice Cream: Physicochemical, Sensory and Digestive Properties. *Foods* **2022**, *11*, 1504. https://doi.org/10.3390/foods11101504

Academic Editors: Qianchun Deng, Ruijie Liu and Xin Xu

Received: 28 April 2022
Accepted: 20 May 2022
Published: 22 May 2022

Publisher's Note: MDPI stays neutral with regard to jurisdictional claims in published maps and institutional affiliations.

1. Introduction

Ice cream has good flavor and texture as a consumer-friendly frozen dairy product. Ice cream, as an oil-in-water frozen aerated emulsion, contains partially agglomerated fat globules, unfrozen viscous whey, ice crystals and air bubbles [1]. Generally, ice cream contains 10–16% fat, which is an important ingredient in ice cream and affects the melting resistance, shape retention and smoothness of ice cream after the freezing process [2]. Milk is usually the main raw material for preparing ice cream, containing about 3.5–5.0% fat, which mainly contains 98% triacylglycerol and a small amount of phospholipids, free fatty acids and cholesterol [3,4]. Interestingly, milk fat is triacylglycerol present in milk as an emulsion (oil-in-water), commonly known as fat globules. The particle sizes of fat globules range from 0.1 to 15 μm. Due to the wider particle size distribution, milk fat stored at low temperatures could cause aggregation of fat globules as fat crystals protrude from the globule surface and damage the fat globule membrane, resulting in poor performance defects in whole milk ice cream stability [5]. In addition, milk fat is associated with many negative health effects, mainly due to the relatively high content of saturated fatty acids, which could lead to the increase of cholesterol and low-density lipoprotein, thereby increasing the risk of cardiovascular disease [3]. Furthermore, the increasing consumer preference for natural and healthy functional foods has prompted ice cream makers to look for innovations in health-beneficial ingredients to meet consumer demand [6–8].

Soybean oil body (SOB) is lipid-storing organelles in soybean with a particle size in the approximate range of 0.4–2.0 µm [9]. The purified SOB consists mainly of neutral lipid droplets (87–91.89%) surrounded by natural emulsifiers, mainly phospholipids and basic proteins. SOB is endowed with unique stability and emulsifying properties owing to the charge of the surface proteins of SOB and the polarity of the phospholipids that increase the mutual repulsion between the SOBs to prevent aggregation [10]. In addition, SOB is rich in nutrients such as vitamin E, unsaturated fatty acids and phytosterols [10,11]. Among them, palmitic acid, oleic acid, α-linolenic acid and linoleic acid are the main unsaturated fatty acids in SOB, amounting to around 96% of the total fatty acids [12]. Typically, SOB is extracted and concentrated into a white cream that can be diluted to form a natural oil-in-water emulsion, and SOB retains its natural function and physical stability under processing conditions. SOB is rich in a variety of bioactive components, and ideal emulsification and stability, therefore it has the potential to be used as a milk fat substitute in various foods (e.g., ice cream, mayonnaise) [13].

In this study, SOB was served as a milk fat substitute to prepare ice cream, expecting to reduce the content of saturated fatty acid and improve the quality defects of ice cream products caused by the poor stability of milk fat at low temperatures. The apparent viscosity, particle size, melting properties, chemical composition, microstructure and digestive properties of ice cream were characterized to explore the consequence of SOB as a milk fat substitute on ice cream quality. While improving the beneficial ingredients and quality of ice cream to further determine the application value of SOB in ice cream, this study provides a theoretical underpinning for expanding the application of SOB in the food industry and also supplies innovative ideas for the development of healthy frozen foods.

2. Materials and Methods

2.1. Materials

Soybeans were obtained from the Soybean Institute of Northeast Agricultural University (Harbin, China). Skimmed milk powder was purchased from Inner Mongolia Yili Industrial Group Co., Ltd. (Hohhot, China). Vast cream was purchased from Qingdao Nestle Co., Ltd. (Qingdao, China).

2.2. Preparation of SOB

The preparation method of SOB referred to the description of Zhou et al., [9]. Briefly, after soaking clean soybeans in deionized water for 12 h, a 20% sucrose solution was added. The mixture was ground (18,000 r/min 120 s) and filtered to obtain soybean milk. Then, the soybean milk was centrifuged ($8000\times g$, 4 °C, 20 min) to collect the precipitation. The precipitation was resuspended in 20% sucrose solution and centrifuged ($8000\times g$, 4 °C, 20 min). This step was repeated three times to obtain SOB and sterilized at 100 °C for 20 min.

2.3. Preparation of Ice Cream

Skimmed milk powder (14%, w/w), granulated sugar (10%, w/w), egg yolk (5%, w/w), cream (12%, w/w) and water were mixed and filtered. Among them, SOB was used to replace cream according to 0% (control group), 10%, 20%, 30%, 40% and 50% of the amount of cream. Then, the above mixture was homogenized (15 MPa, 65 °C), pasteurized (75 °C, 20 min) and aged (4 °C, 12 h) to obtain the ice cream mixes. Finally, the ice cream mixes were whipped and frozen for 30 min and hardened (−18 °C, 24 h) to obtain ice cream samples.

2.4. Steady Shear Rheological Properties of Ice Cream Mixes

The method for the determination of the steady shear rheological properties of ice cream mixes referred to the description of Jiang et al., and was modified [14]. The steady shear rheological properties of the ice cream mixes were determined at 4 °C by the rotational rheometer (MARS40, Thermo, Waltham, MA, USA). The ice cream mixes were placed on the plate system (60 mm in diameter) of the rheometer. The strain was set to 0.6%. The range of the shear rate was 1–100 s^{-1} in 120 s to measure the apparent viscosity, consistency index (K) and flow behavior index (n).

2.5. Particle Size of Ice Cream Mixes

The measuring method of average particle diameter referred to the report of Zhao et al., [15]. The ice cream mixes were diluted 100-fold with SDS solution (1%, w/v) and measured by laser particle size distribution analyzer (HYL-1076, Haoyu Technology, Dandong, China). Droplet size measurements were reported as surface average diameter ($D_{[3,2]}$), volume average diameter ($D_{[4,3]}$) and median diameter (D_{50}).

2.6. Overrun of Ice Cream

The overrun of ice cream was measured according to the method of Pon et al., [16]. Equal volumes of ice cream mixes and ice cream samples were weighed to calculate ice cream overrun. The formula for calculating the overrun was as follows, where the mass of the ice cream mixes was recorded as W_0 and the mass of the ice cream sample was recorded as W_1:

$$\text{Overrun (\%)} = 100 \times (W_0 - W_1)/W_1 \qquad (1)$$

2.7. Melting Properties of Ice Cream

The determination method of the melting properties referred to the description of Kurt et al., [1]. The ice cream samples were placed on a metal mesh screen at 37 °C for 60 min to determine the first dripping time (min) and the ice cream melting rate (%).

2.8. Texture of Ice Cream

The hardness, adhesiveness, springiness and chewiness of ice cream samples were determined by texture analyzer (TA-XT Plus, SMATA, Godalming, England). The specific parameters are as follows: the texture analyzer probe was P/5, the probe diameter was 5 mm, the pre-measurement rate was 8 mm/s, the mid- and post-measurement rate was 2 mm/s, the penetration depth was 15 mm and the trigger force was 20 g.

2.9. Physicochemical Properties of Ice Cream

The chemical composition of the ice cream was determined according to AOAC [17]. The content of fatty acids was determined by gas chromatography–mass spectrometry (Nexis GC-2030, Shimadzu, Kyoto, Japan) [18]. The chromaticity of ice cream samples was measured by color difference meter (ZE6000, Nippon Denshoku, Tokyo, Japan), in which L^* value represented the light dark value, a^* represented the red green value and b^* represented the yellow blue value. The flavor of the ice cream samples was determined by electronic nose.

2.10. Microstructure of Ice Cream

The observation method of ice cream microstructure referred to the report of Zhou et al., and was modified [19]. The microstructures of ice cream were observed using ultra-high distraction microscopy (Deltavision OMX SR, GE, Boston, MA, USA). An amount of 2 mL of diluted 5 times ice cream samples were added with 40 μL of Nile red (0.1%) and 50 μL Nile blue (0.1%) (Amresco, Washington, DC, USA) and then avoided light reaction for 30 min. An amount of 15 μL of the reaction liquid was placed in a slide and then the microstructure and the distribution of fat and protein were observed at the excitation wavelength of 488 nm.

2.11. Sensory Evaluation of Ice Cream

The sensory evaluation system is established by using the fuzzy mathematical model. The sensory scoring standard was shown in Table S1 in the Supplementary Materials.

Then, according to the sensory evaluation index, the index set was established. $U = \{u_1, u_2, u_3, u_4\}$, where u_1, u_2, u_3 and u_4 represented the color, taste, texture and flavor of ice cream, respectively. The sensory evaluation comment set was established $V = \{v_1, v_2, v_3, v_4\}$, where v_1, v_2, v_3 and v_4 represented excellent, good, medium and poor, respectively. The middle score of each grade was selected as the final sensory score, $V = \{9, 7, 5, 3\}$.

In addition, 10 professionally trained sensory panelists (half male and half female) performed a weighted analysis on the color, taste, texture and flavor of ice cream samples. The frequency statistics method was adopted to determine the weight of each evaluation index. The weight set was $X = \{x_1, x_2, x_3, x_4\}$.

Furthermore, the fuzzy matrix was established. The 10 sensory panelists scored ice cream samples based on color, taste, texture and flavor. The final votes for each grade (excellent, good, medium, poor) were counted. Fuzzy matrix R was the number of votes at each grade divided by the total number of sensory panelists. Fuzzy relation evaluation set was $Y = X \times R$. Sensory score was $T = Y \times V$.

2.12. Simulated Digestion of Ice Cream In Vitro

Preparation of artificial saliva: 15.1 mmol KCl, 3.7 mmol KH_2PO_4, 13.6 mmol $NaHCO_3$, 0.15 mmol $MgCl_2(H_2O)_6$, 0.06 mmol $(NH_4)_2CO_3$, 0.075 mmol $CaCl_2(H_2O)_2$ and 50 mg ptyalin (Shanghai Yuanye Biotechnology Co., Ltd., Shanghai, China) were dissolved in deionized water with constant volume to 500 mL and adjusted pH to 6.8.

Preparation of artificial gastric juice: 6.9 mmol KCl, 0.9 mmol KH_2PO_4, 25 mmol $NaHCO_3$, 47.2 mmol NaCl, 0.1 mmol $MgCl_2(H_2O)_6$, 0.5 mmol $(NH_4)_2CO_3$, 0.075 mmol $CaCl_2(H_2O)_2$ and 10^6 U pepsin were dissolved in deionized water with constant volume to 500 mL and adjusted pH to 2.0.

Preparation of artificial intestinal juice: 6.8 mmol KCl, 0.8 mmol KH_2PO_4, 85 mmol $NaHCO_3$, 38.4 mmol NaCl, 0.33 mmol $MgCl_2(H_2O)_6$, 0.3 mmol $CaCl_2(H_2O)_2$ 160 mmol bile salt and 5×10^4 U trypsin were dissolved in deionized water with constant volume to 500 mL and adjusted pH to 7.0.

The digestion model in vitro was constructed according to the method of Ma et al., [20]. Briefly, ice cream samples were mixed with artificial saliva (1:1, v/v) and digested for 5 min at 100 rpm/min on a 37 °C incubator shaker. Then, the artificial saliva digesta was mixed with artificial gastric juice (1:1, v/v) and digested for 60 min at 100 rpm/min on a 37 °C incubator shaker. Finally, the artificial gastric digesta was mixed with artificial intestinal juice (1:1, v/v) and digested for 120 min at 100 rpm/min on a 37 °C incubator shaker. Finally, the digesta were placed at -80 °C to terminate the reaction and stored. In addition, the artificial saliva, artificial gastric juice and artificial intestinal juice were replaced with equal volumes of deionized water in the control sample (undigested ice cream dilution sample).

2.13. Determination of Free Fatty Acid

The pH-stat method was used to measure the content of the free fatty acids released in simulated intestinal juice digestion (20, 40, 60, 80, 120 min) according to the method of Hageman et al., [21]. Changes in the pH of the digesta were due to the release of free fatty acids during digestion, so the content of free fatty acids can be determined by recording the consumption of NaOH to neutralize the pH of the digesta. The formula for calculating the concentration of free fatty acid was as follows, where the volume of NaOH consumed when the pH of the digesta reached 7.0 was recorded as V_1 (μL), the concentration of NaOH was recorded as C_1 (mol/L) and the digest volume was recorded as V_2 (mL):

$$\text{Free fatty acid (μmol/mL)} = (V_1 \times C_1)/V_2 \tag{2}$$

2.14. Determination of Protein Digestibility

The determination method of protein digestibility followed the previous report of Li et al., [22]. An amount of 1 mL of trichloroacetic acid (10%) was added to 1 mL of gastric digesta, intestinal digesta and the control sample (undigested ice cream dilution sample), respectively. After mixing, the mixture was centrifuged at 20 °C for 15 min with a centrifugal force of $8000 \times g$ and the supernatant was taken. The protein content of each supernatant was then determined using the BCA protein assay kit. The protein content remaining in the digesta was recorded as C_1 and the protein content in the control sample (undigested ice cream dilution sample) was recorded as C_2. The formula for calculating the protein digestibility was as follows:

$$\text{Protein digestibility (\%)} = 100 \times (1 - C_1/C_2) \tag{3}$$

2.15. Sodium Dodecyl Sulphate–Polyacrylamide Gel Electrophoresis

The experiment method referred to the description of Nikiforidis et al., and was improved [23]. Briefly, ice cream samples were treated with Tris-HCL buffer solution (6.25 mmol/L) and heated by water bath (100 °C, 10 min). After the electrophoretic, the gel was stained by Komas Blue R-250 and was discolored by glacial acetic acid

2.16. Statistical Analysis

All results were the mean of three independent replicates and expressed as mean $\pm$ standard deviation. One-way analysis of variance (ANOVA) was used with SPSS statistics software (SPSS 26.0, Chicago, IL, USA). $p < 0.05$ represented significant difference with Duncan's test.

3. Results

3.1. Apparent Viscosity of Ice Cream Mixes

The apparent viscosity is a crucial indicator reflecting the flow behavior and the taste of ice cream. The sample exhibited shear thinning behavior shown in Figure 1A, as the apparent viscosity of the ice cream mixes decreased with the increase of shear rate [1]. This was because the shearing of the ice cream mixes would produce a destructive effect, which increased the arrangement of constituent molecules in its structure, resulting in the non-Newtonian pseudoplastic flow of the ice cream mixes [24]. In addition, the consistency index (K) and flow behavior index (n) of the ice cream mixes were shown in Table 1. The n values of ice cream mixes were less than 1 and ranged from 0.54 to 0.65, indicating that the ice cream mixes were pseudoplastic fluids. Meanwhile, the K value of ice cream mixes increased significantly with the increase of SOB substitution amount, while the n value decreased significantly. When the SOB substitution amount was 40%, the K value reached the maximum and showed the lowest n value. The smaller value of n reflected a departure from the Newtonian flow, indicating higher pseudoplasticity of the ice cream mixes with 40% SOB substitution [25]. Moreover, the apparent viscosity of ice cream mixes increased significantly with the increase of SOB substitution amount. Possibly due to the excellent emulsion stability of SOB, high concentration of SOB could act as a stabilizer to maintain the emulsion system of ice cream, thereby increasing the apparent viscosity of SOB ice cream [26]. Furthermore, the increase of apparent viscosity and consistency index of the ice cream mixes with SOB as a milk fat substitute might be due to the differences in intermolecular interactions. Variations in interactions were caused by differences in biopolymer types and differences in the chemical composition of the surfaces of SOB and milk fat globules. As the substitution amount of SOB increased, the changes in aggregation degree of droplets and the chemical composition of fat globules in ice cream mixes led to altered intermolecular interactions. Biopolymers generate new ordered and improved structures through hydrogen bonding or hydrophobic interactions between non-polar side segments of the carbon backbone to cause differences in rheological behavior [27,28].

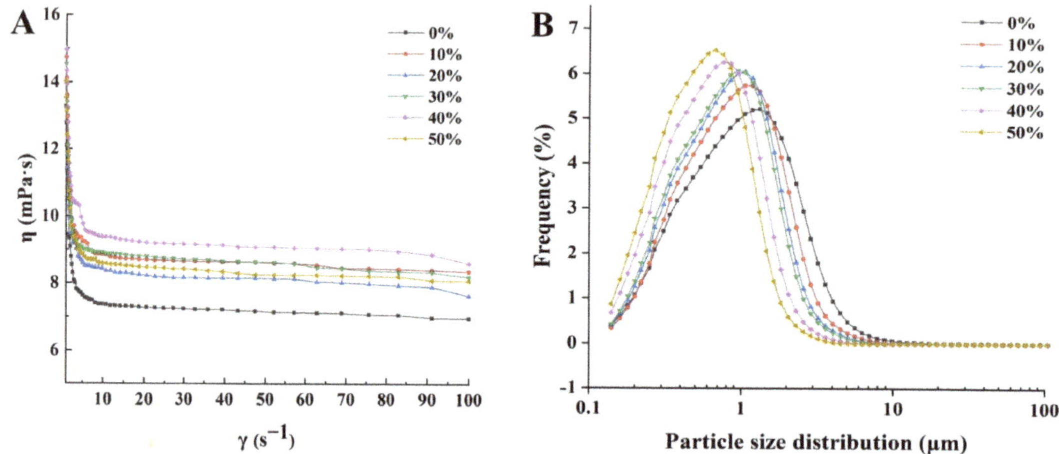

Figure 1. Effects of different SOB substitution amounts on (**A**) apparent viscosity and (**B**) particle size distribution of ice cream mixes.

Table 1. Effects of different SOB substitution amounts on consistency index, flow behavior index, particle size and uniformity of ice cream mixes.

		SOB Substitution Amounts (%)					
		0	10	20	30	40	50
Rheological	K (Pa·s^n)	0.22 ± 0.01 [e]	0.32 ± 0.02 [b]	0.26 ± 0.01 [d]	0.31 ± 0.01 [b]	0.36 ± 0.01 [a]	0.29 ± 0.01 [c]
	n	0.65 ± 0.02 [a]	0.56 ± 0.01 [d]	0.61 ± 0.02 [b]	0.58 ± 0.01 [c]	0.54 ± 0.01 [e]	0.59 ± 0.02 [bc]
	R^2	0.9372	0.9396	0.9452	0.9403	0.941	0.9308
Particle size	D$_{50}$ (µm)	1.04 ± 0.02 [a]	0.92 ± 0.02 [b]	0.84 ± 0.01 [c]	0.80 ± 0.01 [d]	0.63 ± 0.01 [e]	0.58 ± 0.01 [f]
	D$_{[4,3]}$ (µm)	1.36 ± 0.10 [a]	1.15 ± 0.03 [b]	1.02 ± 0.01 [c]	0.97 ± 0.01 [d]	0.77 ± 0.02 [e]	0.69 ± 0.03 [f]
	D$_{[3,2]}$ (µm)	0.72 ± 0.01 [a]	0.67 ± 0.03 [b]	0.63 ± 0.01 [c]	0.60 ± 0.01 [d]	0.50 ± 0.01 [e]	0.46 ± 0.01 [f]
	PDI	0.84 ± 0.03 [a]	0.68 ± 0.01 [b]	0.64 ± 0.00 [c]	0.61 ± 0.01 [d]	0.58 ± 0.01 [e]	0.53 ± 0.02 [f]

K—consistency index; n—flow behavior index; R^2—correlation coefficient; PDI—particle dispersion index. Results are mean ± SD of three determinations. Different alphabet represents significant difference in the row ($p < 0.05$).

3.2. Particle Size of Ice Cream Mixes

The volume-weighted average particle size (D$_{[4,3]}$) and particle size distribution of droplets are effective indicators for evaluating the stability of ice cream mixes and the degree of droplet coalescence and are important for improving the sensory properties of food products [29]. The particle size of the ice cream mixes showed a unimodal distribution and the particle size peak distribution of the ice cream mixes without SOB substitution was at 1.32 µm (Figure 1B). With the increase of SOB substitution amount, the particle size of ice cream slurry gradually decreased, especially when SOB replacement amount was 50%, the particle size peak distribution of ice cream slurry was at 0.67 µm. The SOB ice cream mixes had a smaller particle size compared with full milk fat ice cream mixes, probably because the particle size of SOB was less than the particle size of the milk fat globules. Therefore, as the SOB substitution increased, the particle size of the ice cream mixes gradually decreased [9]. On the other hand, as a natural emulsifier, SOB had an excellent stabilizing effect on the dispersion of droplets in ice cream mixes [30]. In addition, due to the surface protein and polarity phospholipids, the SOB surface could form a tight and stable charged layer, thereby increasing the spatial resistance between SOBs to avoid aggregation [10]. Therefore, the ice cream mixes with SOB substitution had a smaller particle size and particle dispersion index than the ice cream mixes without SOB substitution, indicating that the ice cream mixes with SOB substitution had a more stable emulsion system and showed a trend of increasing dependence on the SOB substitution amount [31]. In addition, the smaller particle size could enhance the creamy flavor characteristics of ice cream.

3.3. Overrun of Ice Cream

The overrun represents the air content in ice cream and is a crucial indicator affecting the melting, texture, and sensory characteristics of ice cream [1]. According to Figure 2A, the overrun of ice cream without SOB substitution was (18.37 ± 0.99)%, while when the SOB substitution amount was 40%, the overrun of ice cream reached the maximum, which was (29.74 ± 0.47)%. The overrun of ice cream with SOB substitution increased significantly ($p < 0.05$), which may be correlated with significantly higher foamability of SOB than milk fat globules. In addition, it may also be attributed to the increase in the apparent viscosity of the ice cream mix with increasing SOB substitution. According to reports, the apparent viscosity of the emulsion system had a crucial impact on the overrun of ice cream, because a certain apparent viscosity was required to produce moderate overrun [32]. The spherical bubbles in ice cream were usually surrounded by a network of partially coalesced fat droplets, which were also surrounded by whey protein, casein and emulsifiers; therefore, as a milk fat substitute, the addition of SOB increased the apparent viscosity of the ice cream mixes, which stabilized the formation of air bubbles. Meanwhile, the higher apparent viscosity allowed the air to form many smaller air chambers in the ice cream, giving the ice cream a higher overrun [33]. Furthermore, the increase in the apparent viscosity of the ice cream mixture caused by the increase in SOB substitution was conducive to preventing the collapse and coagulation of air bubbles in the ice cream [34].

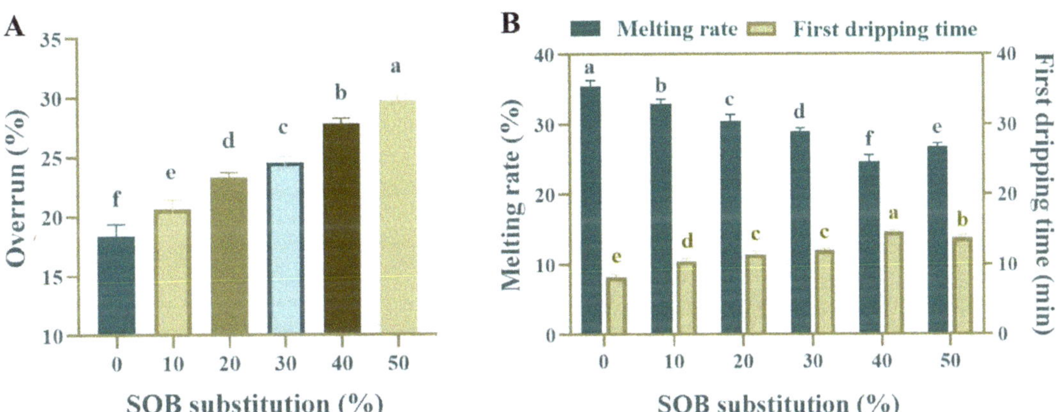

Figure 2. Effects of different SOB substitution amounts on (**A**) overrun and (**B**) melting properties of ice cream. Different alphabet represents significant difference in the same indicator ($p < 0.05$).

3.4. Melting Properties of Ice Cream

The effects of different SOB substitution amounts on the melting rate and the first dripping time of ice cream were shown in Figure 2B. The melting rate of ice cream without SOB substitution was (35.50 ± 2.12)%, while when the SOB substitution amount was 40%, the melting rate of ice cream decreased to (24.57 ± 1.24)% ($p < 0.05$). Meanwhile, compared with the ice cream without SOB substitution for milk fat, the first dripping time of the ice cream with SOB substitution was significantly extended ($p < 0.05$). This may be because the overrun of the ice cream increased with the amount of SOB substitution, meaning there were more air chambers in the ice cream, which reduced the melt rate. Air is a bad conductor of heat, which will reduce the heat diffusion rate of ice cream, so the melting rate of ice cream will be decreased [35]. In addition, the increase in SOB substitution increased the apparent viscosity of the ice cream mixes, which could prevent the migration of water molecules in the ice cream at ambient temperature to decrease the fluidity of ice cream, thereby increasing the anti-meltdown of ice cream [27]. Furthermore, it has been reported that the degree of aggregation of the droplets in the ice cream was also significantly correlated with the melting rate of ice cream [36]. As the amount of SOB substituted for milk fat increased, the particle size and the particle distribution index of the ice cream mixes decreased, which could improve the stability of the ice cream emulsion system, thereby preventing the migration of water molecules and decreasing the melting rate of SOB ice cream [36]. In contrast, for the full milk fat ice cream with lower overrun and stability there was not enough network space structure to prevent melting and complete collapse. Therefore, the substitution of milk fat with SOB could improve the melting properties of full milk fat ice cream and the decrease of the melting rate was dependent on the increase of the substitution amount of SOB.

3.5. Texture Properties of Ice Cream

The effects of different SOB substitution amounts on the texture properties of ice cream were shown in Table 2. Compared with ice cream without SOB substitution, the hardness of SOB ice cream decreased significantly with the increase of SOB substitution amount, but the adhesiveness, springiness and chewiness increased significantly with the increase of SOB substitution amount ($p < 0.05$).

Table 2. Effects of different SOB substitution amounts on the texture properties of ice cream.

	SOB Substitution Amounts (%)					
	0	10	20	30	40	50
Hardness (g)	5899.93 ± 56.69 [a]	5506.79 ± 92.58 [b]	5042.31 ± 130.08 [c]	4740.50 ± 96.28 [d]	4326.72 ± 79.66 [e]	4223.43 ± 47.14 [f]
Adhesiveness (g·s)	138.97 ± 2.80 [f]	165.94 ± 4.59 [e]	221.68 ± 1.78 [d]	256.21 ± 6.72 [c]	270.37 ± 9.10 [b]	315.96 ± 4.35 [a]
Springiness	0.73 ± 0.02 [d]	0.78 ± 0.01 [c]	0.80 ± 0.01 [bc]	0.81 ± 0.01 [b]	0.82 ± 0.01 [ab]	0.84 ± 0.04 [a]
Chewiness	243.18 ± 5.15 [e]	263.56 ± 10.98 [d]	314.82 ± 4.80 [c]	346.46 ± 6.96 [b]	384.99 ± 4.10 [a]	392.91 ± 2.14 [a]

Results are mean ± SD of three determinations. Different alphabet represents significant difference in the row ($p < 0.05$).

The hardness distinction between ice cream with SOB substituted and full milk fat ice cream may be that SOB substitution increased the apparent viscosity of the ice cream mixes. The increased viscosity of the ice cream network structure could decrease the formation of bulky and large-sized ice crystals during freezing, which would decrease the hardness of the ice cream [37]. This result was similar to the report of Kurt et al., that the addition of quince seed increased the viscosity of ice cream mixes, which led to a decrease in hardness of ice cream [1]. In addition, the ice cream with 50% SOB substitution had high adhesiveness, springiness and chewiness, which indicated that the addition of SOB improved the recovery rate of ice cream after deformation and signified that the yield stress of ice cream increased [1]. In general, there is a significant correlation between oral processing parameters and textural properties (i.e., adhesiveness, springiness and chewiness) [38]. The ice cream with SOB substitution will have a smoother mouthfeel during tasting due to the increase in ice cream adhesiveness and the decrease in ice crystal size [39]. In addition, the food with high springiness has reversible deformation during chewing, and chewiness is generally positively correlated with the number of chews before swallowing [38]. Therefore, the chewiness and springiness of ice cream increased with SOB substitution, which could prolong the residence time of ice cream in the mouth, which was important for increasing the oral perception of milk and soy flavors.

3.6. Physicochemical Analysis of Ice Cream

The effects of different SOB substitution amounts on the physicochemical analysis of ice cream were shown in Table 3. The contents of total soluble solids (TSS), protein and carbohydrate in ice cream with SOB substitution increased significantly and the content of fat decreased significantly ($p < 0.05$) compared with ice cream without SOB substitution, indicating that there were differences in the chemical composition of SOB and milk fat. This is because milk fat globules contain 98% triacylglycerols and less than 2% protein, while SOB contains 87–91.89% neutral lipid droplets and 5.42–13% basic protein [3,10]. In addition, the content of saturated fatty acids in ice cream with SOB substitution for milk fat decreased significantly and the content of unsaturated fatty acids increased significantly compared with full milk fat ice cream ($p < 0.05$), therefore the contents of saturated fatty acids and unsaturated fatty acids in SOB and milk fat globule was different. Milk fat mainly contained saturated fatty acids, while the SOB mainly contained unsaturated fatty acids [40]. More evidence shows that a large amount of myristic acid intake would increase the risk of plasma cholesterol and cardiovascular disease [41]. In contrast, a diet rich in unsaturated fatty acids has proven to improve high density lipoprotein function that protects patients with cardiovascular disease [42]. Meanwhile, unsaturated fatty acids play a crucial part in improving immune function and reducing systemic inflammation by regulating patterns on immune cells [43]. Therefore, the ice cream prepared with SOB substitution for milk fat had potential and healthy functions.

Table 3. Effects of different SOB substitution amounts on the physicochemical parameters of ice cream.

		SOB Substitution Amounts (%)					
		0	10	20	30	40	50
Physicochemical properties	TSS °Bx	30.65 ± 0.03 [f]	30.81 ± 0.04 [e]	30.87 ± 0.02 [d]	30.97 ± 0.03 [c]	31.04 ± 0.02 [b]	31.08 ± 0.01 [a]
	Fat %	5.97 ± 0.04 [a]	5.88 ± 0.03 [b]	5.78 ± 0.04 [c]	5.71 ± 0.05 [d]	5.63 ± 0.02 [e]	5.53 ± 0.03 [f]
	Protein %	5.57 ± 0.03 [f]	5.68 ± 0.04 [e]	5.76 ± 0.03 [d]	5.84 ± 0.05 [c]	5.92 ± 0.04 [b]	6.01 ± 0.02 [a]
	Carbohydrate %	18.14 ± 0.06 [f]	18.25 ± 0.05 [e]	18.34 ± 0.06 [d]	18.42 ± 0.03 [c]	18.48 ± 0.02 [b]	18.53 ± 0.04 [a]
Fatty acid content	Myristic acid %	5.92 ± 0.01 [a]	5.34 ± 0.03 [b]	4.96 ± 0.08 [c]	4.51 ± 0.10 [d]	4.31 ± 0.04 [e]	3.79 ± 0.03 [f]
	Hyperic acid %	27.25 ± 0.11 [a]	26.40 ± 0.03 [b]	25.61 ± 0.02 [c]	24.25 ± 0.60 [d]	23.93 ± 0.22 [e]	22.54 ± 0.17 [f]
	Glycolic acid %	9.08 ± 0.08 [a]	8.64 ± 0.00 [b]	8.36 ± 0.10 [b]	8.06 ± 0.19 [c]	7.95 ± 0.18 [c]	7.40 ± 0.04 [d]
	Oleic acid %	25.65 ± 0.06 [a]	25.90 ± 0.01 [a]	25.93 ± 0.32 [a]	25.98 ± 0.51 [a]	26.10 ± 0.33 [a]	26.20 ± 0.08 [a]
	Linoleic acid %	4.78 ± 0.03 [f]	7.90 ± 0.04 [e]	11.13 ± 0.12 [d]	14.48 ± 0.20 [c]	15.97 ± 0.01 [b]	17.95 ± 0.13 [a]
	Linolenic acid %	0.34 ± 0.01 [f]	0.96 ± 0.03 [e]	1.40 ± 0.04 [d]	1.85 ± 0.01 [c]	2.05 ± 0.02 [b]	2.31 ± 0.01 [a]
Color properties	L^*	95.95 ± 2.10 [a]	97.40 ± 0.59 [a]	96.60 ± 0.56 [a]	96.86 ± 2.02 [a]	97.90 ± 1.31 [a]	98.83 ± 3.87 [a]
	a^*	0.64 ± 0.02 [a]	0.56 ± 0.02 [b]	0.41 ± 0.05 [c]	0.45 ± 0.03 [c]	0.35 ± 0.03 [d]	0.30 ± 0.02 [e]
	b^*	21.31 ± 0.27 [ab]	20.65 ± 0.05 [b]	21.42 ± 0.08 [a]	20.97 ± 0.17 [ab]	21.32 ± 0.33 [ab]	21.52 ± 0.79 [a]
	ΔE	-	1.60	0.70	0.99	1.97	2.91

TSS—total soluble solids; L^*—lightness; a^*—redness; b^*—yellowness; ΔE—chromatic aberration. Results are mean ± SD of three determinations. Different alphabet represents significant difference in the row ($p < 0.05$).

3.7. Color Properties of Ice Cream

The effects of different SOB substitution amounts on the color properties of ice cream were shown in Table 3. The a^* value in ice cream with SOB substitution significantly decreased with the increase of SOB substitution amount compared with ice cream without SOB substitution ($p < 0.05$), which was due to the difference in the color properties of SOB and milk fat itself, indicating that SOB has less red than the milk fat. Additionally, it could be due to the decreased particle size of the droplets, which scattered the light, making the ice cream whiter [44]. However, the difference between the L^* and b^* values of ice cream with different SOB substitution amounts was not significant ($p > 0.05$). In addition, the ΔE value of the ice cream with different SOB substitution amounts was 1.60, 0.70, 0.99, 1.97 and 2.91. Bayram et al., [45] reported that when the ΔE value reached 3.7, the naked eye could perceive the difference in color. Therefore, the ice cream with SOB substitution for milk fat cannot bring unpleasant color properties to the naked eye.

3.8. Microstructure of Ice Cream

Ultra-high-resolution optical microscope was used to observe the microstructure of ice cream as shown in Figure 3 (green represented fat and red represented protein). The droplets of ice cream without SOB substitution for milk fat aggregated and formed larger particles. However, the droplets of ice cream with SOB substitution were dispersed and uniform, and the particle size of the droplets in the ice cream gradually decreased with the increase of the SOB substitution amounts. This may be due to the small particle size and the high stability of SOB, so there are few droplet aggregates formed [36]. In addition, due to the surface protein and polarity phospholipids, the SOB surface could form a tight and stable charged layer, thereby increasing the spatial resistance between SOBs to avoid aggregation [10]. The above results indicated that SOB has an excellent stabilizing effect on the dispersion of droplets in ice cream. This result was consistent with the decrease in particle size and particle distribution index of the ice cream mixes with the increase in SOB substitution.

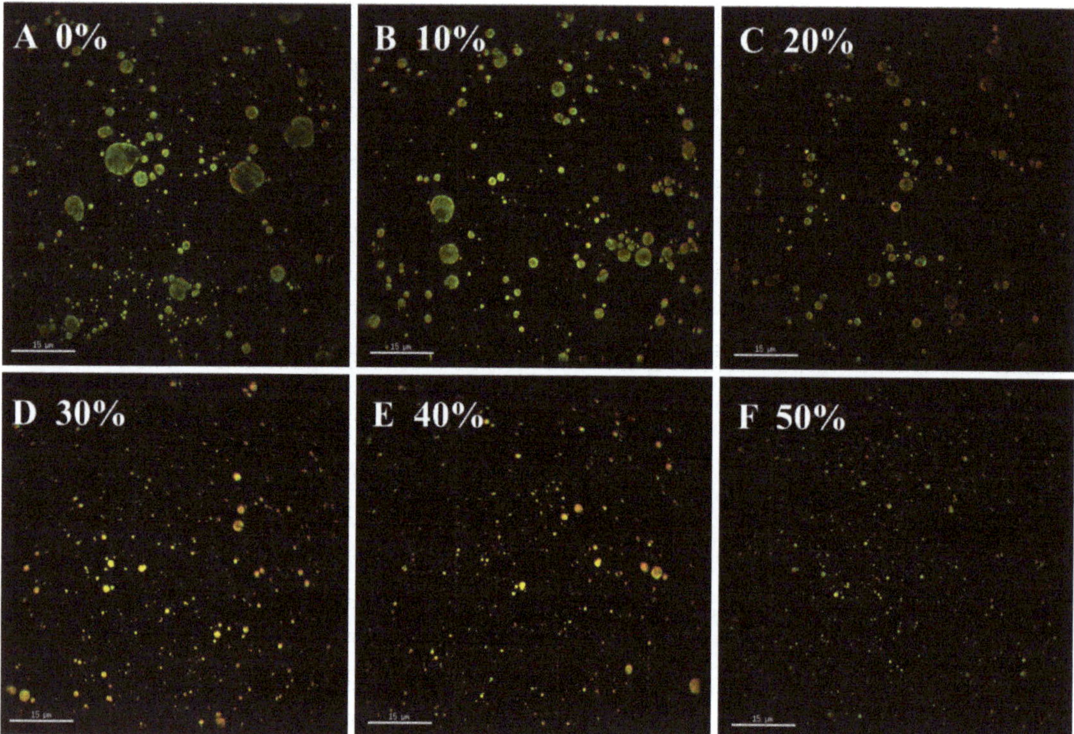

Figure 3. Effects of different SOB substitution amounts on microstructure of ice cream. (**A–F**) represent the ice cream with 0%, 10%, 20%, 30%,40% and 50% of SOB substitution for milk fat, respectively.

3.9. Flavor Properties of Ice Cream

The electronic nose signals of ice cream with different SOB substitution amounts were analyzed to distinguish the flavor differences. The principal component analysis (PCA) chart (Figure 4A) showed that the contribution rate of the first principal component of the ice cream sample was 75.94%, the contribution rate of the second principal component was 13.46% and the cumulative contribution rate was 89.40%, indicating that the sample information was sufficient [46]. In addition, the radar chart (Figure 4B) showed W1S (methyls), W2S (alcohols, aldehydes and ketones), W3S (long-chain alkanes) and W6S (hydrides) in the volatile flavor compounds of ice cream majored contribution. Among them, the response intensity of W2S in ice cream with SOB substitution for milk fat was higher than that in ice cream without SOB substitution and the response intensity of W2W (sulfides) was lower than that in ice cream without SOB substitution. This may be due to the difference in flavor compounds in SOB and milk fat, illustrating that ice cream with SOB substitution had higher abundances of alcohols and aromatics compounds and lower abundances of sulfides compounds than full milk fat ice cream. However, except for the ice cream with 50% SOB substitution, the PCA chart of the ice cream samples with a different SOB substitution amount were overlapped, which was due to the smaller difference between the response values of the electronic nose sensor for different ice cream samples. The reason for the difference in flavor properties between the ice cream with 50% SOB substitution and the other ice cream samples may be that the soy flavor of the 50% SOB significantly masked the creamy flavor. Furthermore, except for W2W and W2S, the response values of other sensors were similar between the full milk fat ice creams and the ice cream with SOB substitution. The results indicated that the SOB substitution of less than 50% could maintain the original flavor of full milk fat ice cream while imparting proper bean flavor.

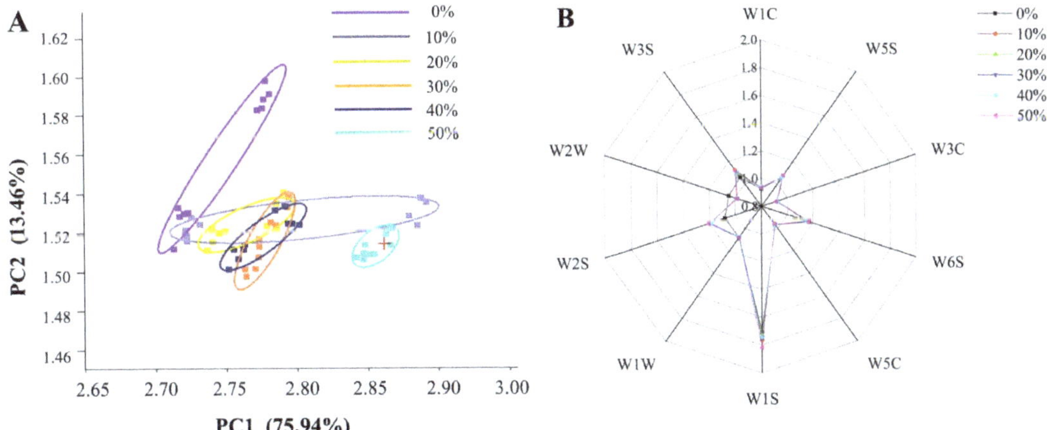

Figure 4. Effects of different SOB substitution amounts on flavor properties of ice cream, they should be listed as: (**A**) principal component analysis plot; (**B**) radar chart.

3.10. Sensory Score of Ice Cream

Table 4 showed the scoring of ice cream with different SOB substitution amounts for 10 sensory panelists. Then, based on the fuzzy mathematical sensory evaluation model, the comprehensive score matrix was established to calculate the sensory score. According to the sensory panelists, the ice cream with SOB substituted for milk fat had a distinct soybean flavor and this flavor increased with SOB substituted amounts without negatively affecting sensory scores. The results showed that 40% SOB substitution improved the sensory score (8.46) of ice cream, indicating improved acceptability. In addition, although the sensory scores of ice cream with 50% SOB substitution and with 40% SOB substitution were not significantly different, but, according to the sensory evaluation panelists, the bean flavor of ice cream with 50% SOB substitution obviously masked the creamy flavor, so the creamy flavor was not strong enough to make ice cream with 50% SOB substitution less acceptable than ice cream with 40% SOB substitution. Moreover, according to the sensory evaluation panelists, ice cream substituted with 30–50% SOB had a smooth texture and uniform melting compared with full milk fat ice cream, which was consistent with the texture property results in this study. This may be due to the particle size and particle distribution index of ice cream gradually decreasing with the increase of SOB substitution, indicating that the droplets in the ice cream were uniformly dispersed and easily melted uniformly in the mouth. Furthermore, the smooth mouthfeel indicated that SOB substitution for milk fat eliminated the roughness of the full milk fat ice cream, suggesting that the increase in SOB substitution decreased the ice crystals detectable in the ice cream during melting [37]. This result was consistent with the change trends of ice cream hardness. The sensory evaluation results showed that the ice cream with 40% SOB substitution for milk fat had the highest sensory score, even more generally acceptable than the full milk fat ice cream.

Table 4. Effects of different SOB substitution amounts on the sensory score of ice cream.

SOB Substitution Amounts (%)	Color				Taste				Texture				Flavor				Sensory Score
	E	G	M	P	E	G	M	P	E	G	M	P	E	G	M	P	
0	7	3	0	0	5	3	2	0	6	3	1	0	7	3	0	0	8.06 ± 0.19 [bc]
10	7	3	0	0	2	5	2	1	4	4	1	1	6	3	1	0	7.44 ± 0.19 [d]
20	7	3	0	0	3	5	2	0	5	4	1	0	8	1	1	0	7.89 ± 0.34 [c]
30	8	2	0	0	3	6	1	0	8	2	0	0	8	2	0	0	8.26 ± 0.24 [ab]
40	9	1	0	0	5	4	1	0	8	2	0	0	9	1	0	0	8.46 ± 0.20 [a]
50	8	2	0	0	5	5	0	0	8	2	0	0	7	3	0	0	8.38 ± 0.13 [a]

E—excellent; G—good; M—moderate; P—poor. Different alphabet represents significant difference in the column ($p < 0.05$).

3.11. Particle Size of Ice Cream In Vitro Digestion

The particle size of ice cream in vitro digestion was shown in Figure 5A. With the extension of digestion time, the particle size of ice cream digesta showed a trend of first increasing and then decreasing. During simulated saliva digestion, the $D_{[4,3]}$ of ice cream increased significantly, which may be due to the binding of anionic mucins in saliva to the positive charges of droplet surface proteins, depleting the interaction between droplets and causing aggregation, resulting in the increase of $D_{[4,3]}$ [47]. In addition, the $D_{[4,3]}$ of ice cream digesta increased significantly when digested in simulated gastric juice for 20 min. This result was because the gastric juice was acidic and contained various protein hydrolase, thus the casein micelles and fat globule surface proteins in ice cream samples were denaturation, hydrolyzed and aggregated [47]. Meanwhile, the interface protein of the droplet was partially hydrated, resulting in a decrease in electrostatic exclusion, i.e., flocculation, and the $D_{[4,3]}$ of the digesta increased [48]. In particular, in simulated gastric juice for 20 min, the $D_{[4,3]}$ of the full milk fat ice cream digesta was significantly higher than that of the ice cream with SOB substitution, which may be because the surface protein of milk fat globules was more easily hydrolyzed by pepsin than the surface protein of SOB, thereby destabilizing the ice cream emulsion system, resulting in digesta particle size increase. Furthermore, with the increase of digestion time in gastric and intestinal juice, the flocculated protein was decomposed by protease, resulting in the decrease of $D_{[4,3]}$ of ice cream digesta [47]. Among them, the $D_{[4,3]}$ of ice cream digesta with SOB substitution was always smaller than that of full milk fat ice cream digesta and the $D_{[4,3]}$ of digesta decreased with the increase of SOB substitution amount. This meant that the ice cream with SOB substitution was easier to digest and absorb than full milk fat ice cream.

3.12. Free Fatty Acids of Ice Cream In Vitro Digestion

The intestine is the main place for fat digestion and release, which transforms fat into absorption forms through interaction with pancreatic and bile secretions [3]. The free fatty acid release of ice cream in vitro digestion was shown in Figure 5B. With the movement of digestion time, the free fatty acid releases showed the trend of increased first and then gentle. This is because the triacylglycerols were converted into free fatty acids under the action of trypsin and bile salts when the ice cream gastric digesta was digested in simulated intestinal juice for 20 min, thus significantly increasing the release of free fatty acids. Furthermore, at the same digestion time, the release of free fatty acids from the ice cream digesta increased significantly with increasing SOB substitution. This may be due to the smaller particle size of droplets of ice cream with SOB substitution, so the same volume of fat had a larger surface, which could promote lipase to enter the oil–water interface to convert triglycerides into free fatty acids and glycerol, resulting in free fatty acids release in digestive juices increased [49,50]. The results indicated that the ice cream prepared by substitution SOB for milk fat was easier to digest and release fatty acids than the whole milk fat ice cream, which was more conducive to the intake of nutrients by the human body.

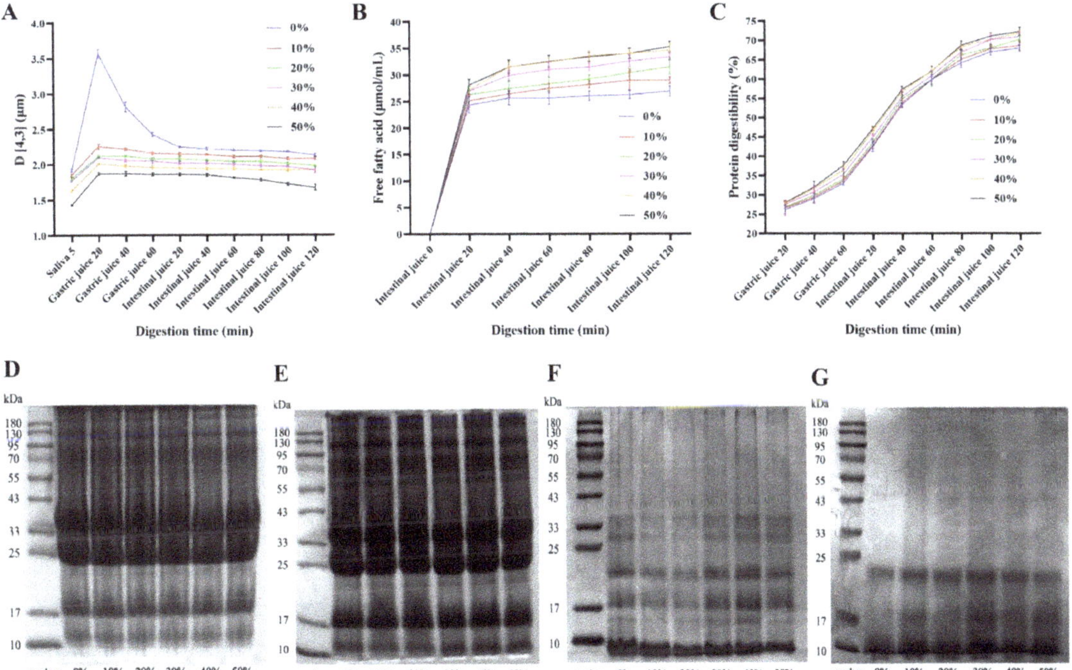

Figure 5. Effects of different SOB substitution amounts on digestive properties of ice cream, they should be listed as: (**A**) average particle size; (**B**) free fatty acid; (**C**) protein digestibility; (**D**) not digested; (**E**) digestive in artificial saliva for 5 min; (**F**) digestive in artificial gastric juice for 60 min; (**G**) digestive in artificial intestinal juice for 120 min.

3.13. Protein Digestibility of Ice Cream In Vitro Digestion

The protein digestibility of ice cream in vitro digestion was shown in Figure 5C. With the increase of digestion time, the digestibility of protein increased significantly and the protein digestibility of samples in intestinal juice was significantly higher than that in gastric juice. This may be due to the fact that trypsin in intestinal juice is more likely to break peptide bonds than pepsin in gastric juice [51]. This result was consistent with the decreased protein molecular weight after simulated gastric and intestinal digestion shown in Figure 5F,G. Furthermore, in addition, the protein digestibility of ice cream samples increased significantly with the SOB substitution and they were higher than that of full milk fat ice cream, which may be due to the difference in the composition of SOB and milk fat globule surface proteins, that is, SOB surface proteins are more easily hydrolyzed by protease than milk fat globule surface proteins. Moreover, it may also be because the ice cream with SOB substitution had a smaller particle size, so the droplet surface protein had a larger contact surface area with the protease, thus improving the protein digestibility [49]. Therefore, the results showed that the ice cream prepared with SOB as a milk fat substitute was easier to digest than the protein in full milk fat ice cream, which had a positive contribution to the digestion and absorption of ice cream by the human body.

4. Conclusions

This study provided new insights into ice cream prepared with SOB as a milk fat substitute. SOB substitution for milk fat not only decreased the content of saturated fatty acids in ice cream, but also improved the quality defects of ice cream products caused by the poor stability of milk fat at low temperatures. SOB substitution for 40–50% milk fat could improve the stability of ice cream by increasing the apparent viscosity and decreasing the particle size of the ice cream mixes, which made the ice cream form the desired quality. Therefore, ice cream prepared substitution 40–50% SOB for milk fat had ideal physicochemical properties and potential biological activity. Among them, the ice cream prepared substitution 40% SOB for milk fat had the best sensory acceptability and had the potential for practical application. This study provided a significant theoretical underpinning for the application of SOB in dairy products.

Supplementary Materials: The following supporting information can be downloaded at: https://www.mdpi.com/article/10.3390/foods11101504/s1, Table S1: Sensory scoring standard.

Author Contributions: Conceptualization, W.W. and M.W.; methodology, M.W.; software, C.X.; validation, Z.L. and L.G.; formal analysis, Z.L.; investigation, L.G.; resources, J.M.; data curation, J.M.; writing—original draft preparation, W.W.; writing—review and editing, W.W.; visualization, M.W.; supervision, L.J.; project administration, J.H. and Z.J.; funding acquisition, J.H. All authors have read and agreed to the published version of the manuscript.

Funding: This research was funded by National Natural Science Foundation of China, grant number 31871727.

Institutional Review Board Statement: Not applicable.

Informed Consent Statement: Not applicable.

Data Availability Statement: Data is contained within the article and Supplementary Material.

Conflicts of Interest: The authors declare no conflict of interest.

References

1. Kurt, A.; Atalar, I. Effects of quince seed on the rheological, structural and sensory characteristics of ice cream. *Food Hydrocoll.* **2018**, *82*, 186–195. [CrossRef]
2. Akbari, M.; Eskandari, M.H.; Davoudi, Z. Application and functions of fat replacers in low-fat ice cream: A review. *Trends Food Sci. Technol.* **2019**, *86*, 34–40. [CrossRef]
3. Singh, H. Symposium review: Fat globules in milk and their structural modifications during gastrointestinal digestion. *J. Dairy Sci.* **2019**, *102*, 2749–2759. [CrossRef] [PubMed]
4. Zhang, W.; Zhao, P.; Li, J.; Wang, X.; Hou, J.; Jiang, Z. Effects of ultrasound synergized with microwave on structure and functional properties of transglutaminase-crosslinked whey protein isolate. *Ultrason. Sonochem.* **2022**, *83*, 105935. [CrossRef] [PubMed]
5. Huppertz, T.; Kelly, A. Physical chemistry of milk fat globules. In *Advanced Dairy Chemistry*, 3rd ed.; Fox, P., McSweeney, P., Eds.; Springer: New York, NY, USA, 2006; Volume 2, pp. 173–212.
6. Wang, W.; Liu, F.; Xu, C.; Liu, Z.; Ma, J.; Gu, L.; Jiang, Z.; Hou, J. *Lactobacillus plantarum* 69-2 combined with galacto-oligosaccharides alleviates d-galactose-induced aging by regulating the AMPK/SIRT1 signaling pathway and gut microbiota in mice. *J. Agric. Food Chem.* **2021**, *69*, 2745–2757. [CrossRef] [PubMed]
7. Liu, Z.; Zhao, J.; Sun, R.; Wang, M.; Wang, K.; Li, Y.; Shang, H.; Hou, J.; Jiang, Z. Lactobacillus plantarum 23-1 improves intestinal inflammation and barrier function through the TLR4/NF-κB signaling pathway in obese mice. *Food Funct.* **2022**. [CrossRef]
8. Xu, C.; Fu, Y.; Liu, F.; Liu, Z.; Ma, J.; Jiang, R.; Song, C.; Jiang, Z.; Hou, J. Purification and antimicrobial mechanism of a novel bacteriocin produced by Lactobacillus rhamnosus 1.0320. *LWT* **2021**, *137*, 110338. [CrossRef]
9. Zhou, X.; Zhao, J.; Zhao, X.; Sun, R.; Sun, C.; Hou, D.; Zhang, X.; Jiang, L.; Hou, J.; Jiang, Z. Oil bodies extracted from high-oil soybeans (Glycine max) exhibited higher oxidative and physical stability than oil bodies from high-protein soybeans. *Food Funct.* **2022**, *13*, 3271–3282. [CrossRef]
10. Zaaboul, F.; Zhao, Q.; Xu, Y.; Liu, Y. Soybean oil bodies: A review on composition, properties, food applications, and future research aspects. *Food Hydrocoll.* **2022**, *124*, 107296. [CrossRef]
11. Wang, W.; Cui, C.; Wang, Q.; Sun, C.; Jiang, L.; Hou, J. Effect of pH on physicochemical properties of oil bodies from different oil crops. *J. Food Sci. Technol.* **2019**, *56*, 49–58. [CrossRef]
12. Chen, Y.; Cao, Y.; Zhao, L.; Kong, X.; Hua, Y. Macronutrients and micronutrients of soybean oil bodies extracted at different pH. *J. Food Sci.* **2014**, *79*, C1285–C1291. [CrossRef]

13. Nikiforidis, C.V.; Matsakidou, A.; Kiosseoglou, V. Composition, properties and potential food applications of natural emulsions and cream materials based on oil bodies. *RSC Adv.* **2014**, *4*, 25067–25078. [CrossRef]

14. Jiang, Z.; Mu, S.; Ma, C.; Liu, Y.; Ma, Y.; Zhang, M.; Li, H.; Liu, X.; Hou, J.; Tian, B. Consequences of ball milling combined with high-pressure homogenization on structure, physicochemical and rheological properties of citrus fiber. *Food Hydrocoll.* **2022**, *127*, 107515. [CrossRef]

15. Zhao, X.; Wang, K.; Zhao, J.; Sun, R.; Shang, H.; Sun, C.; Liu, L.; Hou, J.; Jiang, Z. Physical and oxidative stability of astaxanthin microcapsules prepared with liposomes. *J. Sci. Food Agric.* **2022**. [CrossRef] [PubMed]

16. Pon, S.Y.; Lee, W.J.; Chong, G.h. Textural and rheological properties of stevia ice cream. *Int. Food Res. J.* **2015**, *22*, 1544–1549.

17. AOAC International. Official method 2000.18. In *Official Methods of Analysis of AOAC International*, 21st ed.; Horwitz, W., Latimer, Jr., Eds.; AOAC International: Gaithersburg, MD, USA, 2019.

18. Silva, J.M.; Klososki, S.J.; Silva, R.; Raices, R.S.L.; Silva, M.C.; Freitas, M.Q.; Barão, C.E.; Pimentel, T.C. Passion fruit-flavored ice cream processed with water-soluble extract of rice by-product: What is the impact of the addition of different prebiotic components? *LWT* **2020**, *128*, 109472. [CrossRef]

19. Zhou, X.; Liu, Z.; Wang, W.; Miao, Y.; Gu, L.; Li, Y.; Liu, X.; Jiang, L.; Hou, J.; Jiang, Z. NaCl induces flocculation and lipid oxidation of soybean oil body emulsions recovered by neutral aqueous extraction. *J. Sci. Food Agric.* **2021**. [CrossRef]

20. Ma, Y.; Liu, Y.; Yu, H.; Mu, S.; Li, H.; Liu, X.; Zhang, M.; Jiang, Z.; Hou, J. Biological activities and in vitro digestion characteristics of glycosylated α-lactalbumin prepared by microwave heating: Impacts of ultrasonication. *LWT* **2022**, *158*, 113141. [CrossRef]

21. Hageman, J.H.J.; Keijer, J.; Dalsgaard, T.K.; Zeper, L.W.; Carrière, F.; Feitsma, A.L.; Nieuwenhuizen, A.G. Free fatty acid release from vegetable and bovine milk fat-based infant formulas and human milk during two-phase in vitro digestion. *Food Funct.* **2019**, *10*, 2102–2113. [CrossRef]

22. Li, J.; Liu, Y.; Li, T.; Gantumur, M.-A.; Qayum, A.; Bilawal, A.; Jiang, Z.; Wang, L. Non-covalent interaction and digestive characteristics between α-lactalbumin and safflower yellow. Impacts of microwave heating temperature. *LWT* **2022**, *159*, 113206. [CrossRef]

23. Nikiforidis, C.V.; Kiosseoglou, V. Aqueous extraction of oil bodies from maize germ (Zea mays) and characterization of the resulting natural oil-in-water emulsion. *J. Agric. Food Chem.* **2009**, *57*, 5591–5596. [CrossRef] [PubMed]

24. Dogan, M.; Kayacier, A.; Toker, Ö.S.; Yilmaz, M.T.; Karaman, S. Steady, dynamic, creep, and recovery analysis of ice cream mixes added with different concentrations of xanthan gum. *Food Bioproc. Tech.* **2013**, *6*, 1420–1433. [CrossRef]

25. Mostafavi, F.S.; Tehrani, M.M.; Mohebbi, M. Rheological and sensory properties of fat reduced vanilla ice creams containing milk protein concentrate (MPC). *J. Food Meas. Charact.* **2017**, *11*, 567–575. [CrossRef]

26. Jirapeangtong, K.; Siriwatanayothin, S.; Chiewchan, N. Effects of coconut sugar and stabilizing agents on stability and apparent viscosity of high-fat coconut milk. *J. Food Eng.* **2008**, *87*, 422–427. [CrossRef]

27. Bahramparvar, M.; Tehrani, M.M. Application and functions of stabilizers in ice cream. *Food Rev. Int.* **2011**, *27*, 389–407. [CrossRef]

28. Soukoulis, C.; Tzia, C. Grape, raisin and sugarcane molasses as potential partial sucrose substitutes in chocolate ice cream: A feasibility study. *Int. Dairy J.* **2018**, *76*, 18–29. [CrossRef]

29. Sun, C.; Wu, T.; Liu, R.; Liang, B.; Tian, Z.; Zhang, E.; Zhang, M. Effects of superfine grinding and microparticulation on the surface hydrophobicity of whey protein concentrate and its relation to emulsions stability. *Food Hydrocoll.* **2015**, *51*, 512–518. [CrossRef]

30. Qian, C.; McClements, D.J. Formation of nanoemulsions stabilized by model food-grade emulsifiers using high-pressure homogenization: Factors affecting particle size. *Food Hydrocoll.* **2011**, *25*, 1000–1008. [CrossRef]

31. Aboulfazli, F.; Baba, A.S.; Misran, M. Effects of fermentation by *Bifidobacterium* bifidum on the rheology and physical properties of ice cream mixes made with cow and vegetable milks. *Int. J. Food Sci. Technol.* **2015**, *50*, 942–949. [CrossRef]

32. Liu, R.; Wang, L.; Liu, Y.; Wu, T.; Zhang, M. Fabricating soy protein hydrolysate/xanthan gum as fat replacer in ice cream by combined enzymatic and heat-shearing treatment. *Food Hydrocoll.* **2018**, *81*, 39–47. [CrossRef]

33. Muzammil, H.S.; Rasco, B.; Sablani, S. Effect of inulin and glycerol supplementation on physicochemical properties of probiotic frozen yogurt. *Food Nutr. Res.* **2017**, *61*, 1290314. [CrossRef] [PubMed]

34. Seo, C.W.; Oh, N.S. Functional application of Maillard conjugate derived from a κ-carrageenan/milk protein isolate mixture as a stabilizer in ice cream. *LWT* **2022**, *161*, 113406. [CrossRef]

35. Lomolino, G.; Zannoni, S.; Zabara, A.; Da Lio, M.; De Iseppi, A. Ice recrystallisation and melting in ice cream with different proteins levels and subjected to thermal fluctuation. *Int. Dairy J.* **2020**, *100*, 104557. [CrossRef]

36. Chen, W.; Liang, G.; Li, X.; He, Z.; Zeng, M.; Gao, D.; Qin, F.; Goff, H.D.; Chen, J. Effects of soy proteins and hydrolysates on fat globule coalescence and meltdown properties of ice cream. *Food Hydrocoll.* **2019**, *94*, 279–286. [CrossRef]

37. Javidi, F.; Razavi, S.M.A.; Behrouzian, F.; Alghooneh, A. The influence of basil seed gum, guar gum and their blend on the rheological, physical and sensory properties of low fat ice cream. *Food Hydrocoll.* **2016**, *52*, 625–633. [CrossRef]

38. Wee, M.S.M.; Goh, A.T.; Stieger, M.; Forde, C.G. Correlation of instrumental texture properties from textural profile analysis (TPA) with eating behaviours and macronutrient composition for a wide range of solid foods. *Food Funct.* **2018**, *9*, 5301–5312. [CrossRef] [PubMed]

39. Akbari, M.; Eskandari, M.H.; Niakosari, M.; Bedeltavana, A. The effect of inulin on the physicochemical properties and sensory attributes of low-fat ice cream. *Int. Dairy J.* **2016**, *57*, 52–55. [CrossRef]

40. Kapchie, V.N.; Yao, L.; Hauck, C.C.; Wang, T.; Murphy, P.A. Oxidative stability of soybean oil in oleosomes as affected by pH and iron. *Food Chem.* **2013**, *141*, 2286–2293. [CrossRef]
41. Speziali, G.; Liesinger, L.; Gindlhuber, J.; Leopold, C.; Pucher, B.; Brandi, J.; Castagna, A.; Tomin, T.; Krenn, P.; Thallinger, G.G.; et al. Myristic acid induces proteomic and secretomic changes associated with steatosis, cytoskeleton remodeling, endoplasmic reticulum stress, protein turnover and exosome release in HepG2 cells. *J. Proteom.* **2018**, *181*, 118–130. [CrossRef]
42. Hernáez, Á.; Castañer, O.; Elosua, R.; Pintó, X.; Estruch, R.; Salas-Salvadó, J.; Corella, D.; Arós, F.; Serra-Majem, L.; Fiol, M. Mediterranean diet improves high-density lipoprotein function in high-cardiovascular-risk individuals: A randomized controlled trial. *Circulation* **2017**, *135*, 633–643. [CrossRef]
43. Kumar, N.G.; Contaifer, D.; Madurantakam, P.; Carbone, S.; Price, E.T.; Van Tassell, B.; Brophy, D.F.; Wijesinghe, D.S. Dietary bioactive fatty acids as modulators of immune function: Implications on human health. *Nutrients* **2019**, *11*, 2974. [CrossRef] [PubMed]
44. Kenari, R.E.; Razavi, R. Effect of sonication conditions: Time, temperature and amplitude on physicochemical, textural and sensory properties of yoghurt. *Int. J. Dairy Technol.* **2021**, *74*, 332–343. [CrossRef]
45. Çörekçi, B.; Toy, E.; ÖZTÜRK, F.; Malkoc, S.; Öztürk, B. Effects of contemporary orthodontic composites on tooth color following short-term fixed orthodontic treatment: A controlled clinical study. *Turk. J. Med. Sci.* **2015**, *45*, 1421–1428. [CrossRef] [PubMed]
46. Yang, C.J.; Ding, W.; Ma, L.J.; Jia, R. Discrimination and characterization of different intensities of goaty flavor in goat milk by means of an electronic nose. *J. Dairy Sci.* **2015**, *98*, 55–67. [CrossRef]
47. Silletti, E.; Vingerhoeds, M.H.; Norde, W.; van Aken, G.A. The role of electrostatics in saliva-induced emulsion flocculation. *Food Hydrocoll.* **2007**, *21*, 596–606. [CrossRef]
48. Sarkar, A.; Goh, K.K.T.; Singh, R.P.; Singh, H. Behaviour of an oil-in-water emulsion stabilized by β-lactoglobulin in an in vitro gastric model. *Food Hydrocoll.* **2009**, *23*, 1563–1569. [CrossRef]
49. Clulow, A.J.; Salim, M.; Hawley, A.; Boyd, B.J. A closer look at the behaviour of milk lipids during digestion. *Chem. Phys. Lipids* **2018**, *211*, 107–116. [CrossRef]
50. Li, M.; Liu, Y.; Zhao, J.; Yu, R.; Hussain, M.A.; Qayum, A.; Jiang, Z.; Qu, B. Glycosylated whey protein isolate enhances digestion behaviors and stabilities of conjugated linoleic acid oil in water emulsions. *Food Chem.* **2022**, *383*, 132402. [CrossRef]
51. Dai, C.; Zhang, W.; He, R.; Xiong, F.; Ma, H. Protein breakdown and release of antioxidant peptides during simulated gastrointestinal digestion and the absorption by everted intestinal sac of rapeseed proteins. *LWT* **2017**, *86*, 424–429. [CrossRef]

Article

Diverse Krill Lipid Fractions Differentially Reduce LPS-Induced Inflammatory Markers in RAW264.7 Macrophages In Vitro

Dan Xie [1,2], Fangyuan He [1], Xiaosan Wang [2], Xingguo Wang [2], Qingzhe Jin [2] and Jun Jin [2,*]

[1] College of Biology and Food Engineering, Anhui Polytechnic University, Beijing Zhong Road, Wuhu 241000, China; xdwawj@163.com (D.X.); Hfy1124900348@163.com (F.H.)
[2] Collaborative Innovation Center of Food Safety and Quality Control in Jiangsu Province, International Joint Research Laboratory for Lipid Nutrition and Safety, School of Food Science and Technology, Jiangnan University, 1800 Lihu Road, Wuxi 214122, China; wxstongxue@163.com (X.W.); wangxg1002@gmail.com (X.W.); jqzwuxi@163.com (Q.J.)
* Correspondence: junjin@jiangnan.edu.cn; Tel.: +86-0510-85876799

Abstract: Antarctic krill oil is an emerging marine lipid and expected to be a potential functional food due to its diverse nutrients, such as eicosapentaenoic acid (EPA), docosahexaenoic acid (DHA), phospholipids, astaxanthin and tocopherols. Although krill oil has been previously proved to have anti-inflammatory activity, there is little information about the relationship between its chemical compositions and anti-inflammatory activity. In this study, the RAW264.7 macrophages model was used to elucidate and compare the anti-inflammatory potential of different krill lipid fractions: KLF-A, KLF-H and KLF-E, which have increasing phospholipids, EPA and DHA contents but decreasing astaxanthin and tocopherols levels. Results showed that all the krill lipid fractions alleviated the inflammatory reaction by inhibition of production of nitric oxide (NO), release of tumor necrosis factor-α (TNF-α), interleukin (IL)-1β and IL-6 and gene expression of proinflammatory mediators including TNF-α, IL-1β, IL-6, cyclooxygenase-2 (COX-2) and inducible nitric oxide synthase (iNOS). In addition, KLF-E with the highest phospholipids, EPA and DHA contents showed the strongest inhibition effect on the LPS-induced proinflammatory mediator release and their gene expressions. The results would be helpful to provide powerful insights into the underlying anti-inflammatory mechanism of krill lipid and guiding the production of krill oil products with tailor-made anti-inflammatory activity.

Keywords: krill oil; anti-inflammatory effect; RAW 264.7 cell; chemical composition

Citation: Xie, D.; He, F.; Wang, X.; Wang, X.; Jin, Q.; Jin, J. Diverse Krill Lipid Fractions Differentially Reduce LPS-Induced Inflammatory Markers in RAW264.7 Macrophages In Vitro. *Foods* **2021**, *10*, 2887. https://doi.org/10.3390/foods10112887

Academic Editor: Shinichi Kitamura

Received: 4 October 2021
Accepted: 17 November 2021
Published: 22 November 2021

Publisher's Note: MDPI stays neutral with regard to jurisdictional claims in published maps and institutional affiliations.

1. Introduction

Chronic inflammation is a common pathological process that is closely associated with the pathogenesis of various diseases, such as obesity, atherosclerosis, cachexia, asthma, metabolic disorders and diabetes [1–3]. Regulation of inflammation is particularly important for maintaining human health. Recently, there is an increasing concerning about the side effects and high cost of anti-inflammatory drugs that are clinically used [4]. Thus, the exploration of a natural food resource with potential anti-inflammatory activities has been gaining great interests in treatment of inflammation [5].

Antarctic krill oil is an emerging marine lipid and expected to be a potential functional food due to its diverse nutrients, such as eicosapentaenoic acid (EPA), docosahexaenoic acid (DHA), phospholipids, astaxanthin and tocopherols [6]. Krill oil manufacturers claimed that arthritic patients could benefit from supplementation of the lipid products [7]. Many studies have also investigated the anti-inflammatory effect of krill oil from the perspective of cell experiments [8–10], animal models [11,12] and human trials [7,13]. Results of these studies have preliminarily confirmed that krill oil showed anti-inflammatory activity, but there is no sufficient information on the contribution of different components to the anti-inflammatory activity. Some studies indicated the anti-inflammatory activity of krill oil was attributed to the EPA and DHA [11,14]. Indeed, dietary EPA and DHA have been extensively confirmed to effectively reduce proinflammatory responses by modulating nuclear factor-κB and regulating achidonic acid cyclooxygenase-derived eicosanoids, primarily prostaglandin E2-dependent signaling [15,16]. As is known, most EPA and DHA in krill lipids are bound to phospholipids [6], and marine phospholipids rich in n-3 PUFA were shown to have better bioavailability and anti-inflammatory properties than those of neutral n-3 PUFA [17]. However, in addition to EPA and DHA, astaxanthin and tocopherols are also known as physiologically active substances that have shown anti-inflammatory properties in some studies [18–23]. For example, Kimble et al. [20] found that pre-treatment with astaxanthin reduced transcriptional activation of nuclear factor-κB and activator protein-1, resulting in the downregulation of the production of inflammatory cytokines and mediators in SW-1353 human chondrosarcoma cells. An inhibition effect of phorbol myristate acetate-induced proinflammatory IL-1β expression was also observed when treated with tocopherols in human monocyte leukemic cell line THP-1 [21] as well as LPS-induced activation of rat Kupffer cells [22]. Unfortunately, most of current studies on the anti-inflammation of krill oil have ignored the contribution of these minor components to the overall anti-inflammatory effects of krill lipid products.

In addition, some studies have indicated that the extraction method could significantly affect the composition of krill oil, including the contents of phospholipids, EPA, DHA, astaxanthin and tocopherols [24–26]. This raised the question of whether krill oils with different lipid fractions would cause differences in their anti-inflammatory effects. In our previous study, a three-step extraction method was adopted to selectively extract three lipid fractions with different compositions, and they were confirmed to show significantly different antioxidant capacities [27]. Li et al. (2013) also pointed out that krill oil containing more phospholipids was more effective in decreasing plasma total cholesterol and low-density lipoprotein cholesterol levels in rats fed a high cholesterol diet [28]. However, there is a lack of literature focused on the anti-inflammatory activity discrepancy between krill oils with different lipid compositions.

Inflammatory responses induced by lipopolysaccharide (LPS) in macrophages have been commonly used as a classical model to evaluate the anti-inflammatory activity of active substances [29,30]. In this study, three krill lipid fractions with different compositions were prepared according to our previous study [27], and their anti-inflammatory properties were compared in LPS-activated RAW264.7 macrophages by analyzing the secretion of cytokines and gene expression associated with inflammatory responses. The results would be helpful to provide powerful insights into the underlying anti-inflammatory mechanism of krill lipid and guide the production of krill oil products with tailor-made anti-inflammatory activity.

2. Materials and Methods

2.1. Materials and Reagents

Krill meal was bought from Antarctic Farm Biotechnology Co., Ltd. (Jinan, China) and stored at −40 °C until used. Dulbecco's modified Eagle's medium (DMEM), fetal bovine serum (FBS), trypsin-EDTA solution and penicillin-streptomycin (P/S) were obtained from Gibco (Gaithersburg, MD, USA). Phosphate-buffered saline (PBS) was obtained from Hyclone (Logan, UT, USA). Lipopolysaccharide (LPS) (*Escherichia coli*, serotype 0111:04) was purchased from Solarbio (Beijing, China) and 3-(4, 5-dimethylthiazol-2-yl)-2,5-diphenyltetrazolium bromide (MTT) was bought from Sigma-Aldrich (St. Louis, MO, USA). The nitric oxide (NO) detection kit and R0026 RNAeasy kit were purchased from Beyotime (Nanjing, China). The enzyme immunoassay (EIA) kits for tumor necrosis factor (TNF)-α, interleukin (IL)-1β and IL-6 were obtained from MultiSciences (Hangzhou, China). TNF-α, IL-1β and IL-6, as well as inducible nitric oxide synthase (iNOS), cyclooxygenase-2 (COX-2) and GAPDH oligonucleotide primers were purchased from Sangon (Shanghai, China). The FastQuant cDNA kit was purchased from Qiagen (Gaithersburg, MD, USA). TaKaRa

Premix Taq was purchased from TaKaRa (Otsu, Shiga, Japan), and iTaqTM SYBR® Green SuperMix was obtained from Bio-Rad (Hercules, CA, USA). RAW 264.7 macrophages were purchased from American Type Culture Collection (Manassas, VA, USA). A Milli-Q apparatus (Billerica, MA, USA) was used to produce ultrapure water. Chromatographic grade solvents were bought from Sigma-Aldrich (St. Louis, MO, USA). All other analytical reagents were purchased from Sinopharm Medicine (Shanghai, China).

2.2. Preparation and Analysis of Three Krill Lipid Fractions with Different Compositions

Three fractions of krill lipid, namely KLF-A, KLF-H and KLF-E, were extracted using a three-step extraction method (sequentially using acetone, hexane and ethanol as the extraction solvent in each step) according to our previous study [27]. Briefly, in step 1, 100 g of krill meal was mixed with 200 mL of precooled acetone, and lipids were extracted at 4 °C for 15 min. The mixture was then filtered using a Buchner funnel. The KLF-A was recovered from the filtrate by removing acetone at 30 °C using a vacuum rotary evaporator. The residual krill meal (KM) was then dried and weighed as the extraction material for the next step (KM-H). In step 2, the lipid was extracted from the KM-H using hexane at 30 °C for 15 min. The ratio of KM-H to hexane was 1:2. KLF-H was obtained in similar manner to step 1, and the residual krill meal obtained in step 2 was also dried and weighed as the extraction material for the next step (KM-E). In step 3, we extracted KLF-E from KM-E using ethanol at 30 °C for 20 min at a 1:3 ratio (KM-E/ethanol, w:v). The amount of the lipid extracted in each step was recorded. The total lipid amount was also extracted from the same amount of krill meal by the Folch method [31] as a reference to calculate the lipid extraction efficiency in each step. The compositions of krill lipid fractions including the contents of phospholipids, astaxanthin and tocopherols, as well as the fatty acids profile, were analyzed referring to our previous studies [25,27]. Specifically, the determination of phospholipids' content and composition was performed on a high-performance liquid chromatographic system (HPLC) (1260 Infinity, Agilent, Santa Clara, CA, USA) equipped with an evaporative light-scattering detector (ELSD). The amount of astaxanthin was determined by an HPLC (LC-20AT, Shimadzu, Kyoto, Japan) equipped with an ultraviolet detector (SPD-20A, Shimadzu, Kyoto, Japan) and a C18 column (5 μm, 4.6 × 250 mm; Hanbon, Huaian, Jiangsu, China) by comparing the peak area of the standard astaxanthin. The fatty acid composition was determined as fatty acid methyl esters (FAME) prepared according to our previous study [25] with a gas chromatographic system (GC) (7820A, Agilent, Santa Clara, CA, USA) equipped with a hydrogen flame ionization detector (FID) and a Trace TR-FAME capillary column (0.25 μm, 60 m × 0.25 mm; Thermo Fisher, Waltham, MA, USA), and the EPA and DHA contents were calculated as mg/g fraction by an external standard method.

The obtained krill lipid fraction was dissolved in DMSO with 2% tween 80 as reported by Kim et al. [32], and then the mixture was homogenized using an Ultra Turrax T25 blender (IKA, Staufen, Germany) at 12,000 rpm for 5 min. The mixture was used for cell experiments.

2.3. Cell Culture

RAW 264.7 was purchased from American Type Culture Collection (ATCC, Rockville, MD, USA). These cells were cultured in DMEM medium added with 10% heat-inactivated fetal bovine serum (FBS) and antibiotics composed of 100 U/mL of penicillin and 100 μg/mL of streptomycin and incubated in a 5% CO_2 humidified environment at 37 °C (CO_2 incubator, Heal Force).

The cells at 10–20 passages were normally cultured for 24 h, and then separately pretreated with the diverse krill lipid fraction at different concentrations. After 2 h, LPS (1 μg/mL) was directly added to stimulate inflammatory response for 24 h. The krill lipid fraction dissolved in DMSO with 2% tween 80 was diluted with DMEM to ensure the final concentration of DMSO with 2% tween 80 in culture medium less than 0.1%. In this study, the blank group refers to the cells cultured in the medium with no treatment, and the control group is those in the only LPS-stimulated medium. Both of the groups were treated with DMSO with 2% tween 80 of the same volume.

2.4. Cell Viability Assay

Cell viability treated with krill lipid fraction at different concentrations was measured by the methylthiazole tetrazolium (MTT) assay according to Mosmann et al. [33]. Briefly, after incubation with different concentrations (25–500 μg/mL) of krill lipid fraction and LPS stimulation (1 μg/mL) according to Section 2.3, 5 mg/mL of MTT working solution was added to the medium to dye the alive cells. After 4 h incubation, the supernatant was removed carefully, and DMSO was added into the well. Then, a Multiskan Go microplate reader (Thermo Scientific, Waltham, MA, USA) was used to determine the absorbance of each well at 490 nm. Cells treated with LPS alone were considered as the control group. The blank group was also set as the cells untreated, and its viability was taken as a reference. Cell viability was calculated as the following formula:

$$\text{cell viability} = (A_2 - A_0)/(A_1 - A_0) \times 100\% \tag{1}$$

where A_0 represents the absorbance of the group that did not contain cells, the sample and LPS; A_1 represents the absorbance of the group that only contained untreated cells, without the sample or LPS; and A_2 represents the absorbance of the sample group and control group.

2.5. Determination of NO Production

The NO content in the medium was measured using the Griess reaction by NO detection kit. After treatment with various concentrations of krill lipid fractions, RAW 264.7 cells were treated with or without LPS (1 μg/mL) for 24 h. The amounts of nitrite released in the cell cultures were determined as an indicator of NO release. Briefly, 100 μL of supernatant was reacted with 100 μL of Griess reagent, and the absorbance was determined at 540 nm using a Multiskan Go microplate reader (Thermo Scientific, Waltham, MA, USA). The NO production was calculated based on a standard sodium nitrite curve.

2.6. Cytokine Determinations

RAW 264.7 cells were cultured with different dosages of krill lipid fraction and then stimulated with LPS. Cytokine releases including IL-6, IL-1β and TNF-α in the culture medium were determined with EIA kits based on the manufacturer's instructions.

2.7. Quantitative Real-Time Fluorescent PCR (RT-PCR) Analysis
2.7.1. General

The levels of iNOS, COX-2, TNF-α, IL-6 and IL-1β mRNA expression were determined by RT-PCR.

2.7.2. RNA Extraction and cDNA Synthesis

Total cellular RNA was extracted using R0026 RNAeasy kit (Beyotime, Nanjing, China) based on the manufacturer's instructions. The RNA quality was evaluated by measuring the RNA quality index (RQI) on an Experion system (BioRad). All the samples reached an RQI higher than 7.0. The purity of extracted RNA was measured with a micro spectrophotometer (NanoDrop 2000, Thermo Scientific) and 260/280 indexes of all the samples were between 1.8 and 2.1. The quantity of total RNA was also determined with the micro spectrophotometer. Aliquots of isolated RNA (1 μg) from the samples were reverse transcribed into cDNA using the FastQuant cDNA kit (Gaithersburg, MD, USA) in a 25 μL reaction volume. One microliter of the obtained cDNA served as template for quantitative RT-PCR to quantify the relative mRNA content, and GAPDH was used as an internal control.

2.7.3. Quantitative RT-PCR

Table 1 exhibited the oligonucleotide primers used in this study. Quantitative RT-PCR transcript levels were analyzed on a Bio-Rad CFX-Connect PCR instrument (Hercules, CA, USA) using iTaqTM Universal SYBR® Green Supermix (Hercules, CA, USA). The relative mRNA expression levels of iNOS, COX-2, TNF-α, IL-6 and IL-1β transcripts were measured and calculated by the $2^{-\Delta\Delta CT}$ method with GAPDH mRNA as the invariant control.

Table 1. Primer sequences of the tested genes.

Gene	Forward (5$'$-3$'$)	Reverse (5$'$-3$'$)
GAPDH	ATG TAC GTA GCC ATC CAG GC	AGG AAG GAA GGC TGG AAG AG
IL-1β	CTG TCC TGC GTG TTG AAA	TTC TGC TTG AGA GGT GCT GA
IL-6	AGG AGA CTT GCC TGG TGA AA	CAG GGG TGG TTA TTG CAT CT
TNF-α	AGG CCT TGT GTT GTG TTT CCA	TGG GGG ACA GCT TCC TTC TT
iNOS	CAC CTT GGA GTT CAC CCA GT	ACC ACT CGT ACT TGG GAT GC
COX-2	TGA AAC CCA CTC CAA ACA CA	GAG AAG GCT TCC CAG CTT TT

2.8. Statistics

All the experiments were conducted in triplicate in the same way. The data were displayed as mean $\pm$ standard deviation of replicated measurements. The significant differences ($p < 0.05$) were analyzed with SPSS software (version 19.0, SPSS, Inc., Chicago, IL, USA) by one-way analysis of variance (ANOVA) combined with Duncan's multiple-range test (equal variances were assumed for the data) or the Games-Howell test (the heterogeneity for error variance was assumed for the data). All graphs were plotted with Origin 8.0.

3. Results and Discussion

3.1. Compositions of Three Krill Lipid Fractions

The three-step method adopted in this study resulted in different distributions of phospholipids and minor components in the three krill lipid fractions. As shown in Table 2, KLF-A extracted in the first step contained the lowest level of phospholipids but the highest concentration of astaxanthin and tocopherols. KLF-E was characterized by the highest phospholipids content and the least minor components. The levels of phospholipids and minor components of KLF-H fell between KLF-A and KLF-E. Additionally, the fractionation with high phospholipids content contained high EPA and DHA contents, as most of EPA and DHA are bonded to phospholipids [25]. The selective fractionation of krill lipids made it possible to illustrate the contributions of the diverse lipid components in krill to its anti-inflammatory activity.

Table 2. Compositions of the three krill lipid fractions used in this study [1].

Analytical Determination	KLF-A	KLF-H	KLF-E
Lipid yield (% d.b.)	5.23 ± 0.25 [b]	5.01 ± 0.36 [b]	8.97 ± 0.40 [a]
Lipid extraction efficiency (% total lipid)	25.37 ± 1.21 [b]	24.21 ± 1.72 [b]	45.94 ± 2.03 [a]
Phospholipids (g/100 g)	2.39 ± 0.11 [c]	35.02 ± 2.06 [b]	62.79 ± 2.45 [a]
Astaxanthin (mg/kg)	519.80 ± 23.56 [a]	30.03 ± 0.68 [b]	9.50 ± 0.06 [b]
Tocopherols (mg/100 g)	29.65 ± 0.52 [a]	$11.57 + 0.45$ [b]	3.73 ± 0.35 [c]
EPA (mg/g)	74.24 ± 4.31 [c]	132.57 ± 8.97 [b]	224.01 ± 9.97 [d]
DHA(mg/g)	25.51 ± 3.17 [c]	94.79 ± 7.24 [b]	134.04 ± 7.34 [a]

[1] Abbreviations are: KLF-A, the krill lipid fraction extracted with acetone in the first step; KLF-H, the krill lipid fraction extracted with hexane in the second step; KLF-E, the krill lipid fraction extracted with ethanol in the third step; EPA, eicosapentaenoic acid; DHA, docosahexaenoic acid. Values are means $\pm$ standard deviation. d.b., dry basis of initial krill meal. The significant differences ($p < 0.05$) were analyzed with SPSS software (version 19.0, SPSS) by one-way analysis of variance (ANOVA). Different superscript letters in a row indicate significant differences ($p < 0.05$).

3.2. Effect of Three Krill Lipid Fractions on the Viability of RAW264.7 Cells with LPS Stimulation

To evaluate and compare the potential anti-inflammatory effects of the three krill lipid fractions, RAW 264.7 macrophages stimulated with LPS were selected as the cell model. Firstly, appropriate dosage levels at which all the krill lipid fractions showed no cytotoxic need were determined to ensure enough cell viability. The effects of KLF-A, KLF-H and KLF-E at different concentrations in the culture (10, 25, 50, 100, 200 and 500 µg/mL) on the viability of RAW264.7 cells stimulated with LPS were investigated quantitatively, and the results are shown in Figure 1.

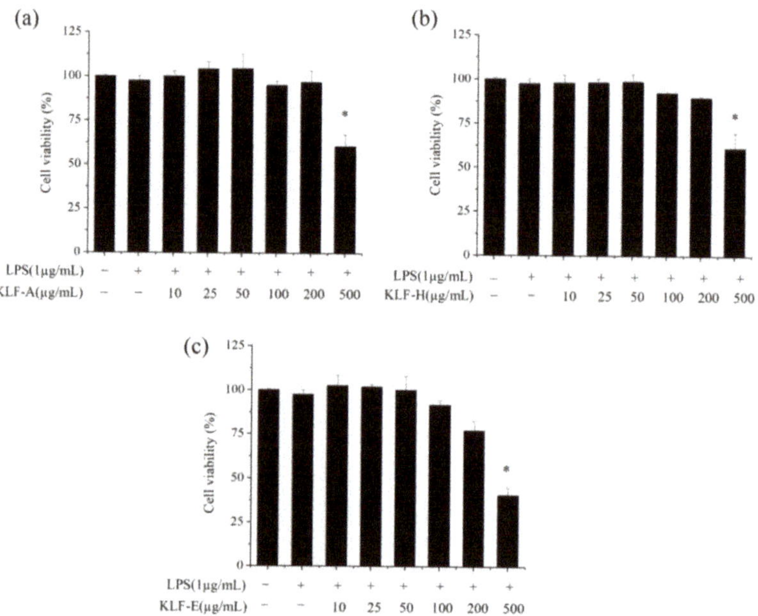

Figure 1. Effect of three krill lipid fractions with different concentrations on the cell viability of RAW264.7: (**a**) KLF-A, (**b**) KLF-H and (**c**) KLF-E. Blank group refers to the untreated cells and its viability was taken as reference (100%); cells treated with LPS alone were considered as the control group. * $p < 0.05$ versus the control group; one-way analysis of variance (ANOVA) combined with Games–Howell test was used to analyze the significance. Abbreviations: −, no addition; +, addition; LPS, lipopolysaccharide; KLF-A, the krill lipid fraction extracted with acetone in the first step; KLF-H, the krill lipid fraction extracted with hexane in the second step; KLF-E, the krill lipid fraction extracted with ethanol in the third step.

The results of MTT assay showed that RAW264.7 cell viabilities were reduced after LPS stimulation. Pre-incubation of all the three krill lipid fractions at 10~100 µg/mL before LPS stimulation could maintain better cell viability, but RAW264.7 cells showed the lowest tolerance to KLF-E. This may be attributed to the fact that the higher content of phospholipids in KLF-E increased the uptake of various components in krill lipid fractions by cells [34,35].

Based on the results of cell viability, the final concentrations administered to RAW 264.7 cells of three types of krill oils were set at 25, 50 and 100 µg/mL in the medium for subsequent experiments.

3.3. Effect of Three Krill Lipid Fractions on NO Production in RAW264.7 Cells with LPS Stimulation

NO is one ubiquitous cellular molecule that is associated with many physiological and pathological processes [36]. Overproduction of NO may cause DNA damage, mitochondrial respiratory depression or react with superoxide anions to produce highly oxidized peroxynitrite anions, thereby affecting cell survival, forming inflammatory cascade waterfalls, promoting inflammation and leading to various diseases [37,38].

The effects of KLF-A, KLF-H and KLF-E on NO production in LPS-stimulated RAW 264.7 cells are illustrated in Figure 2. As is shown, NO release in the control group significantly ($p < 0.05$) increased by up to six times in comparison to the blank group. Reductions of NO release induced by LPS were observed in different degrees when the cells were administrated to diverse krill lipid fractions at different dosages. For the specific treated group, the inhibition degree of NO production was positively correlated with the concentration of krill lipid fraction. However, different krill lipid fraction showed significant different in the inhibition of NO induced by LPS stimulation. It can be seen from Figure 2 that the NO production of LPS-induced inflammatory cells pretreated with 100 µg/mL of KLF-A, KLF-H and KLF-E decreased from 38.64 µM in the control group to 19.49, 14.83 and 8.56 µM, respectively. The NO production in the KLF-E group (8.56 µM) was almost the same as that in the blank group (6.78 µM) without LPS stimulation. Based on the results, KLF-E with the highest phospholipids content showed the strongest inhibitory effect on the NO production.

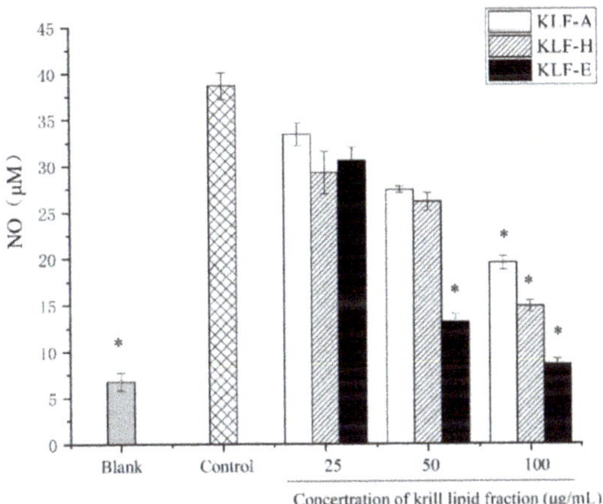

Figure 2. Effect of different concertrations of three krill lipid fractions on nitric oxide (NO) production in RAW264.7 cells with LPS stimulation. Blank group referred to the untreated cells; cells treated with LPS alone were considered as the control group. * $p < 0.05$ versus the control group; one-way analysis of variance (ANOVA) combined with Duncan's multiple-range test was used to analyze the significance. Abbreviations: LPS, lipopolysaccharide; KLF-A, the krill lipid fraction extracted with acetone in the first step; KLF-H, the krill lipid fraction extracted with hexane in the second step; KLF-E, the krill lipid fraction extracted with ethanol in the third step.

3.4. Effect of Three Krill Lipid Fractions on Cytokine Release and Related Gene Expression in RAW264.7 Cells with LPS Stimulation

When inflammation occurs, a series of cell pathways are activated and a variety of proinflammatory cytokines are releases. TNF-α, IL-1β and IL-6 are the most important proinflammatory cytokines produced by monocytes and macrophages, and their secretion levels can reflect the degree of inflammatory response [39–44]. In order to further explore the differences in the anti-inflammatory activities of the three krill lipid fractions, cytokine levels and gene expression including TNF-α, IL-6 and IL-1β were determined. The results are shown in Figure 3.

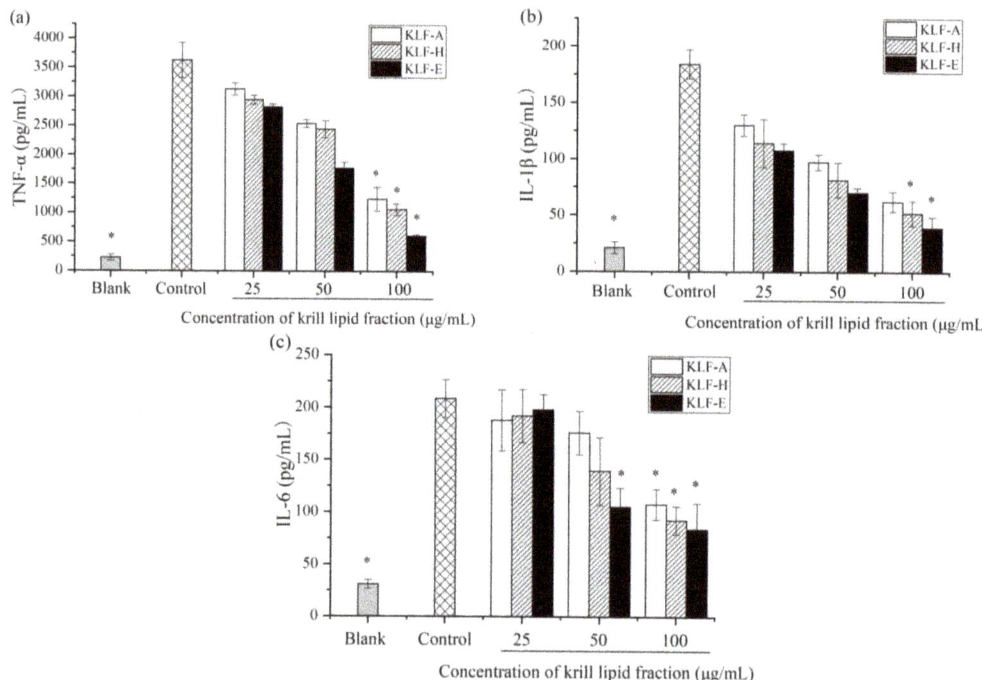

Figure 3. Effect of three krill lipid fractions on cytokine release in RAW264.7 cells with LPS stimulation: (**a**) TNF-α, (**b**) IL-1β and (**c**) IL-6. Blank group referred to the untreated cells; cells treated with LPS alone were considered as the control group. * $p < 0.05$ versus the control group; the one-way analysis of variance (ANOVA) combined with Games–Howell test was used to analyze the significance. Abbreviations: LPS, lipopolysaccharide; KLF-A, the krill lipid fraction extracted with acetone in the first step; KLF-H, the krill lipid fraction extracted with hexane in the second step; KLF-E, the krill lipid fraction extracted with ethanol in the third step.

Figure 3a demonstrated the effect of three krill lipid fractions on TNF-α secretion in LPS-induced inflammatory cells. Stimulation with LPS in RAW 264.7 macrophages resulted in an order of magnitude higher TNF-α production (3623.89 pg/mL) compared with the blank group (226.11 pg/mL). It can be also seen that all of the groups treated with krill lipid fraction exhibited lower TNF-α production induced by LPS. For one specific treated group, a higher concentration of krill lipid fraction was used, and a higher reduction of TNF-α level was observed. At the highest dosage of 100 μg/mL, TNF-α secretion levels of the cells treated with KLF-A, KLF-H and KLF-E before LPS stimulation decreased to 1237.78, 1062.78 and 607.22 pg/mL, respectively. Bonaterra et al. has confirmed that krill oil attenuated the release of TNF-α by inhibiting the binding of LPS to toll-like receptor 4 using differentiated THP-1 macrophages activated by LPS [9]. A downregulation effect of TNF-α mRNA expression was also observed in inflammatory human colorectal adenocarcinoma cell lines (Caco2, HT29) [10]. Our results were consistent with these studies. Notably, the KLF-E group showed the most significant inhibitory effect on the TNF-α production, followed by the KLF-H and KLF-A group at the same concentration. Figure 3b,c shows the cytokine production of IL-1β and IL-6 in RAW 264.7 macrophages, respectively. Similar to TNF-α, levels of IL-1β and IL-6 in the control group were remarkably increased by 8.7- and 6.7-fold more than those in blank group. However, these levels were significantly decreased by krill lipid fraction treatment. Additionally, higher dosages of krill lipid fractions administrated to RAW 264.7 exhibited stronger inhibition effects on IL-1β and IL-6 production. There was a slight difference with the results of TNF-α that, at the lowest level of 25 µg/mL, the three krill lipid fractions provided insignificantly different IL-1β and IL-6 production levels, while KLF-E showed the strongest inhibition effect on IL-1β and IL-6 release among the three fractions at both 50 and 100 µg/mL.

Quantitative RT-PCR results of TNF-α, IL-1β and IL-6 mRNA expression are exhibited in Figure 4. Compared with blank group, the mRNA expression of TNF-α, IL-1β and IL-6 was significantly increased in LPS-stimulated cells, similar with their protein production in the cell culture shown in Figure 3. In contrast, cells preincubated with KLF-A, KLF-H and KLF-E had significantly decreased mRNA expression of the three proinflammatory cytokine upon stimulation relative to the control-stimulated group. Similar results were also found by Costanzo et al., who reported that the mRNA expression of IL-1β, IL-6 and IL-10 was decreased after administration with krill oil in C57BL/6 mice-induced colitis with dextran sodium sulphate (DSS) [45]. Ozonated krill oil also showed a reduction effect on the expression of the proinflammatory cytokines IL-1β and IL-6 at the mRNA expression level in LPS-induced RAW264.7 [32]. However, these studies did not provide the chemical compositions of the krill oil used. In this study, among the three krill lipid fraction-treated cells, KLF-E with the higher phospholipid, EPA and DHA contents showed significantly higher inhibition of mRNA expression of TNF-α, IL-1β and IL-6 compared to those of the other krill lipid fractions. These results were basically consistent with the inhibitory secretion of the three proinflammatory cytokines by the diverse krill lipid fractions.

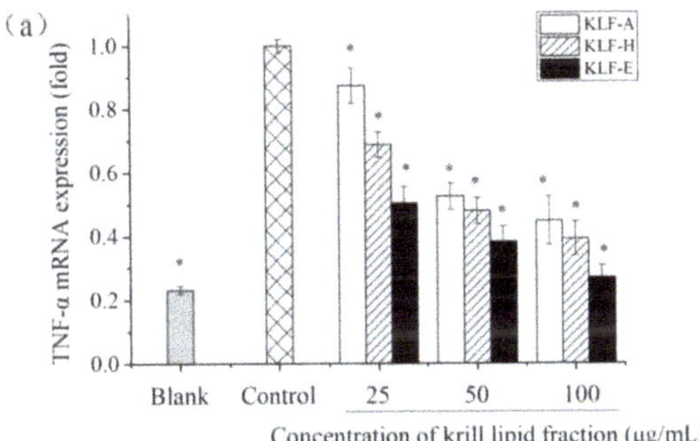

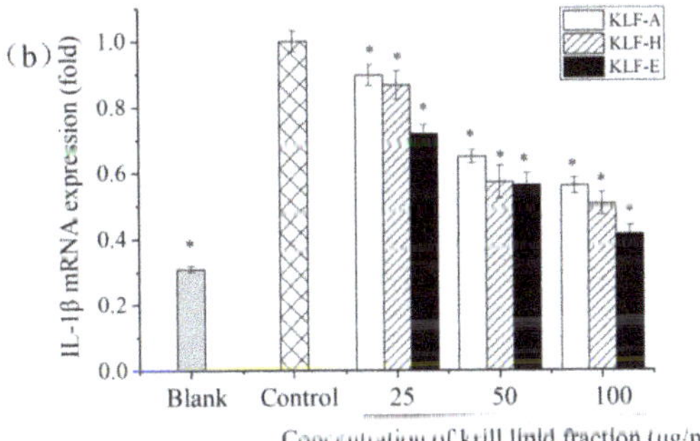

Figure 4. *Cont.*

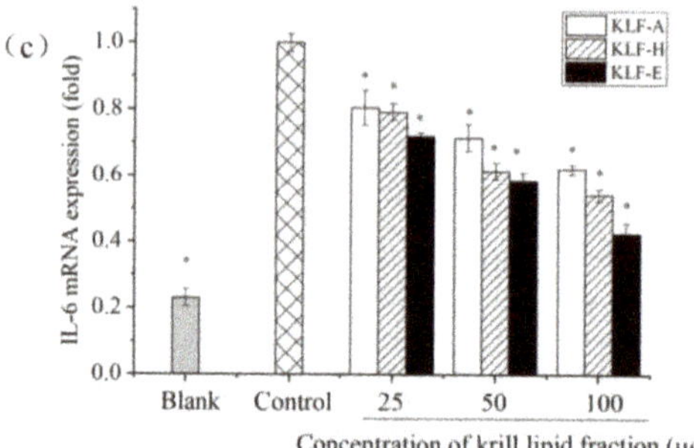

Figure 4. Effect of three krill lipid fractions on mRNA expression of cytokine (**a**) TNF-α, (**b**) IL-1β and (**c**) IL-6 in RAW264.7 cells with LPS stimulation. Blank group referred to the untreated cells; cells treated with LPS alone were considered as the control group. * $p < 0.05$ versus the control group; one-way analysis of variance (ANOVA) combined with Duncan's multiple-range test was used to analyze the significance. Abbreviations: LPS, lipopolysaccharide; KLF-A, the krill lipid fraction extracted with acetone in the first step; KLF-H, the krill lipid fraction extracted with hexane in the second step; KLF-E, the krill lipid fraction extracted with ethanol in the third step.

3.5. Effect of Three Krill Lipid Fractions on Genes Expression of iNOS and COX-2 in RAW264.7 Cells

To further investigate the impact of three krill lipid fractions on proinflammatory cytokines, the mRNA expression levels of two important inducible enzymes in the inflammatory response, iNOS and COX-2, were also determined by q-PCR. The mRNA expression of both iNOS and COX-2 is limited under normal conditions. As previously mentioned, iNOS is highly expressed in the infected or inflammatory situation because a series of intracellular signaling pathways can be initiated and large amounts of NO are released at inflammatory sites. Additionally, as another important inducible enzyme, the transcription of COX-2 can be initiated in inflammatory response [46,47].

Figure 5 demonstrates the q-PCR results of iNOS and COX-2 mRNA expression. It can be seen that, compared with the blank group, the mRNA expression levels of iNOS and COX-2 were significantly increased after LPS stimulation ($p < 0.05$). After the intervention of krill lipid fraction, the gene expression of the two inducible enzymes decreased, and the downregulation effects were in a concentration-dependent manner for a certain krill lipid fraction. Additionally, a similar dose–response profile was observed for the three tested concentrations among the three lipid fractions. At the maximum administration concentration (100 μg/mL), the expression levels of iNOS in the cells treated with KLF-A, KLF-H and KLF-E were only 0.54, 0.44 and 0.33 times as much as those in the control group, respectively; the gene expression levels of COX-2 were 0.54, 0.47 and 0.43 times as much as those in the control group, respectively. These results indicated an effective downregulation effect of krill lipid fractions on the mRNA of iNOS and COX-2. Similar to other inflammatory markers, the higher the amount of phospholipids in the krill lipid, the more significant downregulation effect on the mRNA expression of iNOS and COX-2.

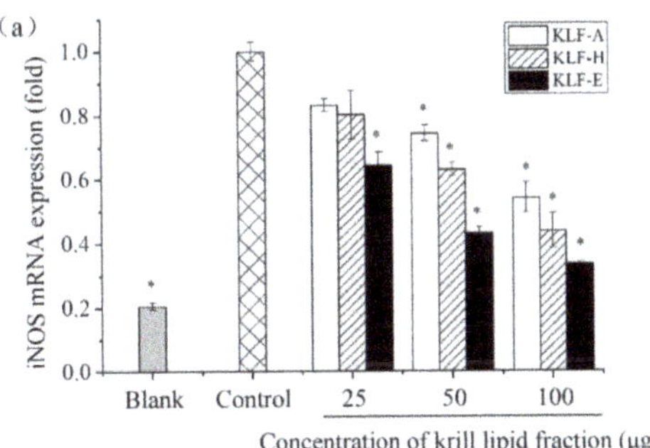

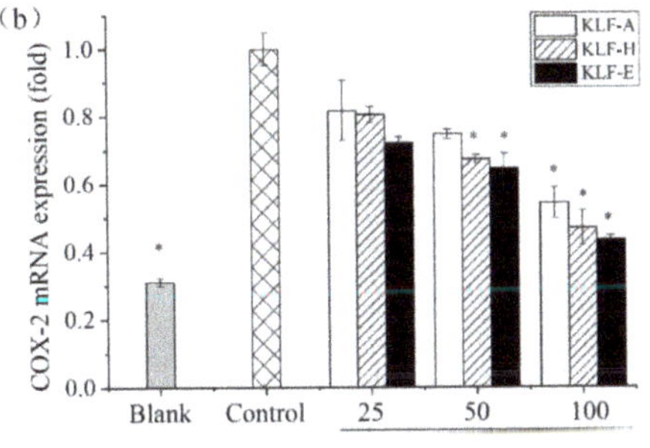

Figure 5. Effect of three krill lipid fractions on mRNA expression of inducible enzymes (**a**) iNOS and (**b**) COX-2 in RAW264.7 cells with LPS stimulation. Blank group referred to the untreated cells; cells treated with LPS alone were considered as the control group. * $p < 0.05$ versus the control group; one-way analysis of variance (ANOVA) combined with the Games–Howell test was used to analyze the significance. Abbreviations: LPS, lipopolysaccharide; KLF-A, the krill lipid fraction extracted with acetone in the first step; KLF-H, the krill lipid fraction extracted with hexane in the second step; KLF-E, the krill lipid fraction extracted with ethanol in the third step.

Unlike common edible oils that consist mainly of triacylglycerols, krill oil is characterized by a high phospholipids content and is rich in EPA and DHA. Moreover, some active lipid accompaniments, such as astaxanthin and tocopherol, are also present in krill [6], which further improves the nutrition value of krill oil. However, diverse processing methods have resulted in krill oil products with different compositions, and this limits our understanding of the association between composition and functionalities of krill oil. The present study was first undertaken to elucidate and compare the anti-inflammatory potential of different krill lipid fractions: KLF-A, KLF-H and KLF-E. These fractions mainly differed in the phospholipids content and minor components including tocopherols and astaxanthin, as well as the EPA and DHA contents. Indeed, n-3 PUFAs have demonstrated potent anti-inflammatory properties [15,16,48,49]. There has been some evidence showing that n-3 PUFAs exhibit anti-inflammatory properties by inhibiting both the IL-1, 2, 6 synthesis and the protein kinase C signaling pathway [50–52]. In our study, although to varying degrees, all the three krill lipid fractions containing n-3 PUFA-modulated proinflammatory cytokines IL-1β and IL-6 both at the protein expression and mRNA expression level in LPS-induced RAW264.7. Furthermore, some studies have reported that n-3 PUFA bound to phospholipids may be more bioactive and more bioavailable than n-3 PUFA as triacylglycerides [51,52]. It has been proven that EPA and DHA in krill lipids were mostly combined with phospholipids [6]. Thus, it is reasonable to hypothesize that lipid fractions with a high content of EPA and DHA as phospholipids would exert higher anti-inflammatory activity. The present study confirmed that the three krill lipid fractions with different compositions showed significant differences in the anti-inflammatory properties. Generally, KLF-E, which is the most abundant in EPA and DHA, was more favorable to reduce the production of NO, the secretion of inflammatory cytokines TNF-α, IL-1β and IL-6 and the expression of inflammatory-related genes in inflammatory cells. Especially at the highest concentration (100 µg/mL), KLF-E showed a more obvious superiority in anti-inflammatory responses compared to the other fraction lipids. The delivery efficiency of neutral oil into cell is limited because it is hard to dissolve in the culture and across membrane transport. Therefore, an oil-in-water emulsion has been considered as an effective delivery system of liposoluble components in cell experiments, and phospholipids were the commonly used emulsifiers for preparing emulsions [47,53]. In the present study, KLF-H and KLF-E contained a high content of phospholipids (35.02 and 62.79 g/100 g, respectively). In addition, EPA and DHA in krill lipids were mostly combined with phospholipids, which was beneficial to transport anti-inflammatory active factors EPA and DHA into cells. For KLF-A, although it contains high concentrations of the anti-inflammatory factors astaxanthin and tocopherol, it only contains 2.39 g/100 g of phospholipids. The low content of phospholipids indicates a low content of EPA and DHA, and this might also limit the transporting efficiency of other anti-inflammatory active components to cells, such as astaxanthin and tocopherol. In addition, there were slightly smaller differences between KLF-A, KLF-H and KLF-E in regulating NO production, inflammatory factor secretion and gene expression at the lowest administration concentration used in our study. Saw et al. found a synergistic anti-inflammatory effect of DHA/EPA with curcumin at low dosage in treating LPS-induced RAW264.7 cells [54]. In this study, although the contents of EPA and DHA in KLF-H were less than those in KLF-E, it contained a small amount of astaxanthin (30.03 mg/kg), while the astaxanthin content in KLF-E was very little (9.50 mg/kg). As a physiological active component with anti-inflammatory activity [55], the small quantity of astaxanthin and EPA/DHA in krill oil may also have a synergistic anti-inflammatory effect. However, when the dosage is increased to 50 and 100 µg/mL, this synergistic effect might be weakened by extremely high concentration of EPA/DHA. This could explain why KLF-E showed more significant anti-inflammatory activity than KLF-H and KLF-A at higher dosages (50 and 100 µg/mL), but less significant at 25 µg/mL. In addition, although there has been no direct evidence to support the synergistic anti-inflammation effect of n-3 PUFA and tocopherol, or tocopherol and astaxanthin, this is an interesting research topic to further illustrate the krill lipid anti-inflammation mechanism. Thus, more intensive study should be focused on the synergistic effect of different lipid components to support our results in future.

4. Conclusions

In this study, the LPS-stimulated RAW 264.7 macrophages model was used to evaluate the anti-inflammatory activities of the three krill lipid fractions with different compositions. The results clearly indicate that treatment of all the three krill lipid fractions decreased the production of LPS-induced proinflammatory mediators NO, TNF-α, IL-1β and IL-6, as well as the gene expression of proinflammatory cytokines and two inducible enzymes, iNOS and COX-2, in inflammatory cells. However, diverse krill lipid fractions differentially reduced LPS-induced inflammatory markers in RAW264.7 macrophages. Generally, KLF-E, with the highest phospholipids, EPA and DHA contents, showed the highest anti-inflammatory effect among the three krill lipid fractions. As most of EPA and DHA are associated with phospholipids, the results indicated that krill oil manufacturers should optimize technology to increase the phospholipids content if they want to produce products with tailor-made anti-inflammatory activity. However, the synergistic effect of n-3 PUFA and astaxanthin in anti-inflammatory activity still needs to be explored in future.

Author Contributions: Conceptualization, D.X. and J.J.; Methodology, F.H.; Software, J.J.; Validation, D.X., Q.J. and X.W. (Xiaosan Wang); Formal Analysis, X.W. (Xiaosan Wang); Resources, D.X. (Xingguo Wang); Data Curation, J.J.; Writing—Original Draft Preparation, D.X.; Writing—Review and Editing, F.H. and J.J; Visualization, J.J.; Supervision, J.J. and Q.J.; Project Administration, D.X.; Funding Acquisition, D.X. All authors have read and agreed to the published version of the manuscript.

Funding: This research was funded by Natural Science Foundation of China, grant number 32102046, Natural Science Foundation of Anhui Province, grant number 2008085QC142 and Scientific Research Foundation of Anhui Polytechnic University, grant number 2019YQQ017; Xjky2020063.

Conflicts of Interest: The authors declare no conflict of interest.

References

1. Barnes, P.J.; Karin, M. Nuclear factor-κB—A pivotal transcription factor in chronic inflammatory diseases. *N. Engl. J. Med.* **1997**, *336*, 1066–1071. [CrossRef]
2. Monteiro, R.; Azevedo, I. Chronic inflammation in obesity and the metabolic syndrome. *Mediat. Inflamm.* **2010**, 289645. [CrossRef]
3. Nathan, C. Points of control in inflammation. *Nature* **2002**, *420*, 846–852. [CrossRef]
4. Gautam, R.; Jachak, S.M. Recent developments in anti-inflammatory natural products. *Med. Res. Rev.* **2009**, *29*, 767–820. [CrossRef]
5. Haddad, P.S.; Azar, G.A.; Groom, S.; Boivin, M. Natural health products, modulation of immune function and prevention of chronic diseases. *Evid. Based Complement Alternat. Med.* **2005**, *2*, 513–520. [CrossRef]
6. Xie, D.; Gong, M.; Wei, W.; Jin, J.; Wang, X.; Wang, X.; Jin, Q. Antarctic krill (*Euphausia superba*) oil: A comprehensive review of chemical composition, extraction technologies, health benefits, and current applications. *Compr. Rev. Food Sci. F* **2019**, *18*, 514–534. [CrossRef]
7. Deutsch, L. Evaluation of the effect of neptune krill oil on chronic inflammation and arthritic symptoms. *J. Am. Coll. Nutr.* **2007**, *26*, 39–48. [CrossRef]
8. Batetta, B.; Griinari, M.; Carta, G.; Murru, E.; Ligresti, A.; Cordeddu, L.; Giordano, E.; Sanna, F.; Bisogno, T.; Uda, S.; et al. Endocannabinoids may mediate the ability of (n-3) fatty acids to reduce ectopic fat and inflammatory mediators in obese zucker rats. *J. Nutr.* **2009**, *139*, 1495–1501. [CrossRef]
9. Bonaterra, A.G.; Driscoll, D.; Schwarzbach, H.; Kinscherf, R. Krill oil-in-water emulsion protects against lipopolysaccharide-induced proinflammatory activation of macrophages in vitro. *Mar. Drugs* **2017**, *15*, 74. [CrossRef]
10. Costanzo, M.; Cesi, V.; Prete, E.; Negroni, A.; Palone, F.; Cucchiara, S.; Oliva, S.; Leter, B.; Stronati, L. Krill oil reduces intestinal inflammation by improving epithelial integrity and impairing adherent-invasive *Escherichia coli* pathogenicity. *Diges. Liver Dis.* **2016**, *48*, 34–42. [CrossRef]
11. Ierna, M.; Kerr, A.; Scales, H.; Berge, K.; Griinari, M. Supplementation of diet with krill oil protects against experimental rheumatoid arthritis. *BMC Musculoskel Dis.* **2010**, *11*, 1–11. [CrossRef] [PubMed]
12. Grimstad, T.; Bjørndal, B.; Cacabelos, D.; Aasprong, O.G.; Janssen, E.A.M.; Omdal, R.; Svardal, A.; Hausken, T.; Bohov, P.; Portero-Otin, M.; et al. Dietary supplementation of krill oil attenuates inflammation and oxidative stress in experimental ulcerative colitis in rats. *Scand J. Gastroenterolo.* **2012**, *47*, 49–58. [CrossRef] [PubMed]
13. Cicero, A.F.; Rosticci, M.; Morbini, M.; Cagnati, M.; Grandi, E.; Parini, A.; Borghi, C. Lipid-lowering and anti-inflammatory effects of omega 3 ethyl esters and krill oil: A randomized, cross-over, clinical trial. *Arch. Med. Sci.* **2016**, *12*, 507–512. [CrossRef] [PubMed]
14. Skarpańska-Stejnborn, A.; Pilaczyńska-Szcześniak, Ł.; Basta, P.; Foriasz, J.; Arlet, J. Effects of supplementation with neptune krill oil (euphasia superba) on selected redox parameters and pro-Inflammatory markers in athletes during exhaustive exercise. *J. Hum. Kinet.* **2010**, *25*, 49. [CrossRef]
15. Cabre, E.; Manosa, M.; Gassull, M.A. Omega-3 fatty acids and inflammatory bowel diseases—A systematic review. *Br. J. Nutr.* **2012**, *107* (Suppl. 2), S240–S252. [CrossRef]
16. Chapkin, R.S.; Kim, W.; Lupton, J.R.; McMurray, D.N. Dietary docosahexaenoic and eicosapentaenoic acid: Emerging mediators of inflammation. *Prostag. Leukot. Ess.* **2009**, *81*, 187–191. [CrossRef]
17. Lordan, R.; Redfern, S.; Tsoupras, A.; Zabetakis, I. Inflammation and cardiovascular disease: Are marine phospholipids the answer? *Food Func.* **2020**, *11*, 2861–2885. [CrossRef]
18. Ambati, R.R.; Phang, S.M.; Ravi, S.; Aswathanarayana, R.G. Astaxanthin: Sources, extraction, stability, biological activities and its commercial applications—A review. *Mar. Drugs* **2014**, *12*, 128–152. [CrossRef]
19. Grammas, P.; Hamdheydari, L.; Benaksas, E.J.; Mou, S.; Pye, Q.N.; Wechter, W.J.; Floyd, R.A.; Stewart, C.; Hensley, K. Anti-inflammatory effects of tocopherol metabolites. *Biochem. Biophys. Res. Commun.* **2004**, *319*, 1047–1052. [CrossRef]
20. Kimble, L.; Mathison, B.; Chew, B.P. Astaxanthin mediates inflammatory biomarkers associated with arthritis in human chondrosarcoma cells induced with interleukin. *FASEB J.* **2013**, *27*, 37–51. [CrossRef]
21. Akeson, A.; Woods, C.; Mosher, L.; Thomas, C.; Jackson, R. Inhibition of IL-1β expression in THP-1 cells by probucol and tocopherol. *Atherosclerosis* **1991**, *86*, 261–270. [CrossRef]
22. Fox, E.S.; Brower, J.S.; Bellezzo, J.M.; Leingang, K.A. N-Acetylcysteine and α-tocopherol reverse the inflammatory response in activated rat Kupffer cells. *J. Immunol.* **1997**, *158*, 5418–5423.
23. Wu, S.J.; Liu, P.L.; Ng, L.T. Tocotrienol-rich fraction of palm oil exhibits anti-inflammatory property by suppressing the expression of inflammatory mediators in human monocytic cells. *Mol. Nutr. Food Res.* **2008**, *52*, 921–929. [CrossRef]
24. Sun, D.; Cao, C.; Li, B.; Chen, H.; Li, J.; Cao, P.; Liu, Y. Antarctic krill lipid extracted by subcritical n-butane and comparison with supercritical CO_2 and conventional solvent extraction. *LWT-Food Sci. Technol.* **2018**, *94*, 1–7. [CrossRef]

25. Xie, D.; Jin, J.; Sun, J.; Liang, L.; Wang, X.; Zhang, W.; Wang, X.; Jin, Q. Comparison of solvents for extraction of krill oil from krill meal: Lipid yield, phospholipids content, fatty acids composition and minor components. *Food Chem.* **2017**, *33*, 434–441. [CrossRef]

26. Yamaguchi, K.; Murakami, M.; Nakano, H.; Konosu, S.; Kokura, T.; Yamamoto, H.; Kosaka, M.; Hata, K. Supercritical carbon dioxide extraction of oils from Antarctic krill. *J. Agric. Food Chem.* **1986**, *34*, 904–907. [CrossRef]

27. Xie, D.; Mu, H.; Tang, T.; Wang, X.; Wei, W.; Jin, J.; Wang, X.; Jin, Q. Production of three types of krill oils from krill meal by a three-step solvent extraction procedure. *Food Chem.* **2018**, *248*, 279–286. [CrossRef]

28. Li, D.M.; Zhou, D.Y.; Zhu, B.W.; Chi, Y.L.; Sun, L.M.; Dong, X.P.; Qin, L.; Qiao, W.Z.; Murata, Y. Effects of krill oil intake on plasma cholesterol and glucose levels in rats fed a high-cholesterol diet. *J. Sci. Food Agric.* **2013**, *93*, 2669–2675. [CrossRef]

29. Jung, H.A.; Jin, S.E.; Ahn, B.R.; Lee, C.M.; Choi, J.S. Anti-inflammatory activity of edible brown alga eisenia bicyclis and its constituents fucosterol and phlorotannins in LPS-stimulated RAW264.7 macrophages. *Food Chem. Toxicol.* **2013**, *59*, 199–206. [CrossRef]

30. Moro, C.; Palacios, I.; Lozano, M.; D'Arrigo, M.; Guillamón, E.; Villares, A.; Martínez, J.A.; García-Lafuente, A. Anti-inflammatory activity of methanolic extracts from edible mushrooms in LPS activated RAW 264.7 macrophages. *Food Chem.* **2012**, *130*, 350–355. [CrossRef]

31. Fricke, H.; Gercken, G.; Schreiber, W.; Oehlenschläger, J. Lipid, sterol and fatty acid composition of antarctic krill (*Euphausia superba* Dana). *Lipids* **1984**, *19*, 821–827. [CrossRef]

32. Kim, H.D.; Lee, S.B.; Ko, S.C.; Jung, W.K.; Kim, Y.M.; Kim, S.B. Anti-inflammatory effect of ozonated krill (*Euphausia superba*) oil in lipopolysaccharide-stimulated RAW 264.7 macrophages. *Fish. Aquat. Sci.* **2018**, *21*, 15. [CrossRef]

33. Mosmann, T. Rapid colorimetric assay for cellular growth and survival: Application to proliferation and cytotoxicity assays. *J. Immunol. Methods* **1983**, *65*, 55–63. [CrossRef]

34. Frede, K.; Henze, A.; Khalil, M.; Baldermann, S.; Schweigert, F.J.; Rawel, H. Stability and cellular uptake of lutein-loaded emulsions. *J. Funct. Foods* **2014**, *8*, 118–127. [CrossRef]

35. Sessa, M.; Balestrieri, M.L.; Ferrari, G.; Servillo, L.; Castaldo, D.; D'Onofrio, N.; Donsì, F.; Tsao, R. Bioavailability of encapsulated resveratrol into nanoemulsion-based delivery systems. *Food Chem.* **2014**, *147*, 42–50. [CrossRef]

36. Sharma, J.N.; Al-Omran, A.; Parvathy, S.S. Role of nitric oxide in inflammatory diseases. *Inflammopharmacology* **2008**, *15*, 252–259. [CrossRef]

37. Bian, K.; Ke, Y.; Kamisaki, Y.; Murad, F. Proteomic modification by nitric oxide. *J. Pharmacol. Sci.* **2006**, *101*, 271–279. [CrossRef]

38. Guzik, T.J.; Korbut, R.; Adamek-Guzik, T. Nitric oxide and superoxide in inflammation and immune regulation. *J. Physiol. Pharmacol.* **2003**, *54*, 469–487. [PubMed]

39. Lawrence, T.; Willoughby, D.A.; Gilroy, D.W. Anti-inflammatory lipid mediators and insights into the resolution of inflammation. *Nat. Rev. Immunol.* **2002**, *2*, 787–795. [CrossRef]

40. Tayal, V.; Kalra, B.S. Cytokines and anti-cytokines as therapeutics—An update. *Eur. J. Pharmacol.* **2008**, *579*, 1–12. [CrossRef] [PubMed]

41. Ko, S.C.; Jeon, Y.J. Anti-inflammatory effect of enzymatic hydrolysates from *Styela clava* flesh tissue in lipopolysaccharide-stimulated RAW 264.7 macrophages and in vivo zebrafish model. *Nutr. Res. Prac.* **2015**, *9*, 219–226. [CrossRef] [PubMed]

42. Feghali, C.A.; Wright, T.M. Cytokines in acute a- chronic inflammation. *Front. Biosci.* **1997**, *2*, d12–d26. [PubMed]

43. Jaffer, U.; Wade, R.G.; Gourlay, T. Cytokines in the systemic inflammatory response syndrome: A review. *HSR Proc. Intensive Care Cardiovasc. Anesth.* **2010**, *2*, 161–175. [PubMed]

44. McCarty, M.F. Interleukin-6 as a central mediator of cardiovascular risk associated with chronic inflammation, smoking, diabetes, and visceral obesity: Down-regulation with essential fatty acids, ethanol and pentoxifylline. *Med. Hypotheses* **1999**, *52*, 465–477. [CrossRef]

45. Costanzo, M.; Cesi, V.; Palone, F.; Pierdomenico, M.; Colantoni, E.; Leter, B.; Vitali, R.; Negroni, A.; Cucchiara, S.; Stronati, L. Krill oil, vitamin D and *Lactobacillus reuteri* cooperate to reduce gut inflammation. *Benef. Microbes* **2018**, *9*, 389–399. [CrossRef]

46. Kim, J.Y.; Park, S.J.; Yun, K.J.; Cho, Y.W.; Park, H.J.; Lee, K.T. Isoliquiritigenin isolated from the roots of Glycyrrhiza uralensis inhibits LPS-induced iNOS and COX-2 expression via the attenuation of NF-kappaB in RAW 264.7 macrophages. *Eur. J. Pharmacol.* **2008**, *584*, 175–184. [CrossRef]

47. Chang, M.; Qiu, F.; Lan, N.; Zhang, T.; Guo, X.; Jin, Q.; Liu, R.; Wang, X. Analysis of phytochemical composition of camellia oleifera oil and evaluation of its anti-Inflammatory effect in lipopolysaccharide-stimulated RAW 264.7 macrophages. *Lipids* **2020**, *55*, 353–363. [CrossRef]

48. Ibrahim, A.; Mbodji, K.; Hassan, A. Anti-inflammatory and anti-angiogenic effect of long chain n-3 polyunsaturated fatty acids in intestinal microvascular endothelium. *Clin. Nutr.* **2011**, *30*, 678–687. [CrossRef]

49. Yang, B.; Li, R.; Greenlief, C.M.; Fritsche, K.L.; Gu, Z.; Cui, J.; Lee, J.C.; Beversdorf, D.Q.; Sun, G.Y. Unveiling anti-oxidative and anti-inflammatory effects of docosahexaenoic acid and its lipid peroxidation production lipopolysaccharide-stimulated BV-2 microglial cells. *J. Neuroinflamm.* **2018**, *9*, 202. [CrossRef]

50. Kim, W.; Khan, N.A.; McMurray, D.N.; Prior, I.A.; Wang, N.; Chapkin, R.S. Regulatory activity of polyunsaturated fatty acids in T-cell signaling. *Prog. Lipid Res.* **2010**, *49*, 250–261. [CrossRef]

51. Lapointe, L.; Harvey, S.; Aziz, H.; Jordan, R.; Hegele, A.; Lemieux, P. A single-dose, comparative bioavailability study of a formulation containing OM3 as phospholipid and free fatty acid to an ethyl ester formulation in the fasting and fed states. *Clin. Ther.* **2019**, *41*, 426–444. [CrossRef]
52. Awada, M.; Meynier, A.; Soulage, C.O. n-3 PUFA added to high-fat diets affect differently adiposity and inflammation when carried by phospholipids or triacylglycerols in mice. *Nutr. Metab.* **2013**, *10*, 23. [CrossRef]
53. Lu, M.; Zhang, T.; Jiang, Z.; Guo, Y.; Qiu, F.; Liu, R.; Zhang, L.; Chang, M.; Liu, R.; Jin, Q.; et al. Physical properties and cellular antioxidant activity of vegetable oil emulsions with different chain lengths and saturation of triglycerides. *LWT-Food Sci. Technol.* **2020**, *121*, 108948. [CrossRef]
54. Saw, C.; Huang, Y.; Kong, A.N. Synergistic anti-inflammatory effects of low doses of curcumin in combination with polyunsaturated fatty acids: Docosahexaenoic acid or eicosapentaenoic acid. *Biochem. Pharmacol.* **2010**, *3*, 421–430. [CrossRef]
55. Ohgami, K.; Shiratori, K.; Kotake, S.; Nishida, T.; Mizuki, N.; Yazawa, K.; Ohno, S. Effects of astaxanthin on lipopolysaccharide-Induced inflammation in vitro and in vivo. *Invest. Opth. Vis. Sci.* **2003**, *44*, 2694. [CrossRef]

Review

Recent Advances in Pickering Double Emulsions and Potential Applications in Functional Foods: A Perspective Paper

Junjia Zhang, Jieyu Zhu, Yujia Cheng and Qingrong Huang *

Department of Food Science, Rutgers University, 65 Dudley Road, New Brunswick, NJ 08901, USA
* Correspondence: qhuang@sebs.rutgers.edu; Tel.: +1-848-932-5514; Fax: +1-732-932-6776

Abstract: Double emulsions are complex emulsion systems with a wide range of applications across different fields, such as pharmaceutics, food and beverage, materials sciences, personal care, and dietary supplements. Conventionally, surfactants are required for the stabilization of double emulsions. However, due to the emerging need for more robust emulsion systems and the growing trends for biocompatible and biodegradable materials, Pickering double emulsions have attracted increasing interest. In comparison to double emulsions stabilized solely by surfactants, Pickering double emulsions possess enhanced stability due to the irreversible adsorption of colloidal particles at the oil/water interface, while adopting desired environmental-friendly properties. Such advantages have made Pickering double emulsions rigid templates for the preparation of various hierarchical structures and as potential encapsulation systems for the delivery of bioactive compounds. This article aims to provide an evaluation of the recent advances in Pickering double emulsions, with a special focus on the colloidal particles employed and the corresponding stabilization strategies. Emphasis is then devoted to the applications of Pickering double emulsions, from encapsulation and co-encapsulation of a wide range of active compounds to templates for the fabrication of hierarchical structures. The tailorable properties and the proposed applications of such hierarchical structures are also discussed. It is hoped that this perspective paper will serve as a useful reference on Pickering double emulsions and will provide insights toward future studies in the fabrication and applications of Pickering double emulsions.

Keywords: pickering emulsions; double emulsions; colloidal particles; encapsulation; controlled release

Citation: Zhang, J.; Zhu, J.; Cheng, Y.; Huang, Q. Recent Advances in Pickering Double Emulsions and Potential Applications in Functional Foods: A Perspective Paper. *Foods* **2023**, *12*, 992. https://doi.org/10.3390/foods12050992

Academic Editors: Qianchun Deng, Ruijie Liu and Xin Xu

Received: 24 December 2022
Revised: 13 February 2023
Accepted: 24 February 2023
Published: 26 February 2023

1. Introduction

Double emulsions are generally defined as emulsions of emulsions, as they are emulsion systems in which a primary emulsion is subsequently dispersed into the continuous phase of the secondary emulsion. Double emulsions, composed of three phases, are considered to be multiple emulsions with ternary structures [1], though the terms "double emulsion" and "multiple emulsion" are used somewhat interchangeably in practice [2]. Double emulsions are commonly classified into either W/O/W (water-in-oil-in-water) or O/W/O (oil-in-water-in-oil) emulsions. The W/O/W emulsion is composed of a continuous aqueous phase with a dispersed oil phase, which is also the continuous phase encapsulating the innermost aqueous dispersed phase, and vice versa for the O/W/O double emulsion. Double emulsions' complex structure and multiple interfaces can be denoted as $W_1/O/W_2$ and/or $O_1/W/O_2$ [3] to distinguish their phases and interfaces. However, it is worth mentioning that there have been recent studies exploring the fabrications of water-in-water-in-water (W/W/W) [4,5] and oil in oil in oil (O/O/O) [6] double emulsions, through methods such as phase inversions, transitions, and utilizations of microfluidic devices.

Conventionally, double emulsions are prepared with surfactants as stabilizing agents, while two kinds of surfactants are usually required for the contrasting interfaces [3]. Hence, surfactants of different surface hydrophobicity are used in combination, depending on the surface characteristic (W/O or O/W). However, such double emulsions stabilized solely by surfactants usually lacked kinetic stability [7] and are prone to destabilization because the low molecular weight surfactants can not provide robust interfacial adsorption and can further lead to surfactant migrations between the interfaces. Ultimately, coalescence occurs and induces the collapse of the multiple interfaces, which results in the formation of a simple emulsion [8]. The primary destabilization routes [9] for conventional double emulsions are illustrated in Figure 1. However, other routes of destabilization, such as Ostwald ripening [10], can also threaten the overall stability, especially for polydisperse double emulsions. In addition to the potential instability of double emulsions stabilized solely by surfactants, irritancy [11], environmental concerns [12], and unpleasant sensory effects [2] associated with surfactants further limited their potential towards widespread commercial applications.

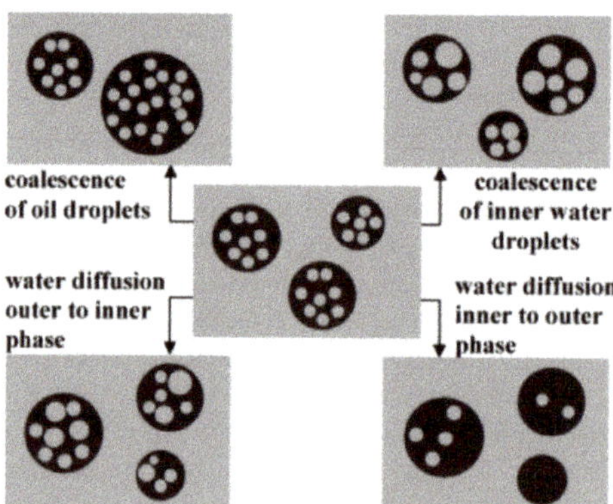

Figure 1. Schematic representation of destabilization routes of conventional double emulsions. Reprinted with permission from [9].

Thus, considering the above limitations, double emulsions stabilized by dispersed colloidal particles have gained increased interest. Stabilization of emulsions realized by colloidal particles, also known as Pickering emulsions, was formally acknowledged since the publication of Pickering [13]. Compared to conventional surfactants, interface-adsorbed colloidal particles can establish desorption energy significantly greater than the thermal energy of Brownian motion [14] and can achieve the irreversible adsorption that provides substantial stabilization for double emulsions against coalescence. For instance, the tunable wetting specificity of the colloidal particles enables the stabilization of either the O/W or the W/O interface by adjusting the surface chemistry and/or the hydrophilicity of the particles. The double emulsion is stabilized solely by silica nanoparticles, with tailored particle hydrophobicity to accommodate the multiple interfaces [15]. Moreover, the unique interfacial adsorption properties of the colloidal particles make such double emulsions to be excellent and rigid templates for the fabrication of microspheres [16], microcapsules [17], colloidosomes [18], and other supra-colloidal structures. From now on, this article will refer to double emulsions stabilized by colloidal particles as Pickering double emulsion (PDE). Though it is worth mentioning that the phrases Pickering double emulsion and double Pickering emulsion are used interchangeably among publications. Ideally, the phrase Pickering double emulsion and double Pickering emulsion can be distinguished by double emulsion stabilized by a combination of colloidal particles and co-surfactants, and double emulsion is stabilized solely by colloidal particles. Nonetheless, PDE will be adopted in this article and will include both double emulsions stabilized by a combination of colloidal particles and co-surfactant and by colloidal particles only.

Oza and Frank [19] were recognized as the first to fabricate PDE, where stable W/O/W double emulsion was successfully prepared through a combination of colloidal microcrystalline cellulose and Span surfactant. Since then, clay [20], silica [15,21], metal [22–25], polymeric [26–28], and other colloidal particles have been studied for their potential to stabilize double emulsions. Moreover, colloidal particles obtained from naturally derived substances were also investigated in their potential to stabilize PDE due to the increasing trend for biocompatible and biodegradable materials.

Because of the appealing properties of PDE, the utilization of colloidal particles in the fabrication of double emulsions and their subsequent applications have received considerable interest. Thus, in the hope of providing insights toward future studies in relevant topics, the aim of this paper is to provide perspectives toward advances in PDE over the last decade. The paper will lead with stabilization strategies of PDE and the corresponding colloidal particles. Emphasis is then placed on the proposed and investigated applications of PDE, and the paper concludes with the challenges, limitations, and outlook of PDE.

2. Stabilization Strategy

It is worth mentioning that there are two major variations in the number of step(s) of emulsification during the assembly of PDE: the one-step and two-step methods. The two-step method remains the more commonly employed method. It consists of emulsification of a primary emulsion followed by subsequent emulsification into the external dispersed phase to form a secondary emulsion. In contrast, for the one-step method, the emulsification of PDE is achieved in a single homogenization step. Thus, coupled with the two-step method, the combination of hydrophobic and hydrophilic particles with/without co-surfactants is the most prevalent stabilization strategy for the preparation of PDE. As a hydrophilic/hydrophobic particle with/without co-surfactant will be incorporated for stabilization of the primary emulsion for the first step, and another hydrophobic/hydrophilic particle with/without co-surfactant will be utilized for stabilization of the secondary emulsion leading to the formation of the PDE. However, other stabilization strategies were also applied to achieve one-step and/or two-step emulsification, for example, the incorporation of particles with intermediate wettability, environmentally responsive particles, in situ modified particles, and more. The following sections start with an overview of the more conventional stabilization strategy of the combination of hydrophobic and hydrophilic particles for PDE, then move forward to the elaboration of other stabilization strategies. Meanwhile, a summary of the stabilization strategy, the corresponding emulsification steps, colloidal particles utilized, and applications are presented in Table 1.

Table 1. Summary of Stabilizing Strategy, colloidal particles, and applications of PDE.

Emulsification Step	Stabilization Strategy	Colloidal Particle (and Co-Surfactant)	Application	Ref
Two-Step	Hydrophobic and hydrophilic particles (co-surfactant)	O_1/W: Hydrophilic N20 silica W/O_2: Hydrophobic H30 silica	Synthesis of tunable capsule clusters	[17]
Two-Step		O_1/W: Hydrophilic silica Ludox HS-40 W/O_2: Hydrophobic silica particle HDK H2000		[21]
Two Step		O_1/W: Hydrophilic silica particle SLM 1466 W/O_2: Hydrophobic silica particle SLM1472	Proposed application as an entrapping reservoir for active ingredients	[15]
Two Step		O_1/W: Hydrophilic G37-H60-B30 block copolymer worms O_1/W/O_2: Hydrophobic L16-B37 Block copolymer worms		[26]
Two-Step		W_1/O: Hydrophobic H30 silica nanoparticle O/W_2: Hydrophilic mesoporous silica nanoparticles (MSN)	Fabrication of tunable Janus microspheres with magnetism and dual anisotropy of porosity	[29]

Table 1. *Cont.*

Emulsification Step	Stabilization Strategy	Colloidal Particle (and Co-Surfactant)	Application	Ref
Two-Step		W_1/O: Lecithin and hydrophobic AEROSIL R974 silica O/W_2: Hydrophilic HKD N20 silica nanoparticle	Preparation of high internal phase Pickering double emulsions	[30]
Two-Step		W_1/O: Span 80 and hydrophobic SiO_2 O/W_2: Hydrophilic SiO_2	Fabrication of multi-hollow microspheres	[31]
Two-Step		O_1/W: Laponite RD clay W/O_2: Hydrophobic modified H30 silica particle	Suspension polymerization template for synthesis of multi-hollow polymer microspheres	[16]
Two-Step		O_1/W: Fe_2O_3 nanoparticle and Laponite RD clay W/O_2: H30 silica particle and oleic acid coated Fe_2O_3 nanoparticle	Proposed application as polymerization vessel to fabricate nanocomposite polymer microspheres	[32]
Two-Step		W_1/O: Hydrophilic H30 silica nanoparticle O/W_2: P(NIPAm-co-MAA) microgels	Stimulus-responsive (pH/temp) emulsion for controlled release.	[33]
One-Step		W_1/O/W_2: Hydrophilic SiO_2 nanoparticle and PLGA	Fabrication of microporous microsphere	[34]
Two-Step		W_1/O: Modified hydrophobic SiO_2 particle O/W_2: PVA	Fabrication of microsphere with aqueous core	[35]
Two-Step		O_1/W: Modified A380 silica and CTAB-functionalized silica W/O_2: Down Corning surfactant DC3225C	Fabrication of wax–water–SiO_2 microcapsule with thermo-stimulable release property	[36]
Two-Step		W_1/O: Hydrophobic silica/PEI hybrid nanoparticles O/W_2: hydrophilic silica/PEI hybrid nanoparticles		[37]
Two-Step		W_1/O: Graphene oxide (GO) nanoparticle O/W_2: Hydroxyapatite (HA) nanoparticle	Fabrication of multi-drug containing composite microcapsule for controlled release	[38]
Two-Step		O_1/W: PEI and Fe_3O_4 nanoparticle W/O_2: SM-$CaCO_3$ nanoparticle	Template for polymerization-induced synthesis of magnetic Fe_3O_4/polyamine hybrid microsphere	[25]
Two-Step		W_1/O: Fe_3O_4 nanoparticle O/W_2: Cellulose nanocrystals	Fabrication of multihollow magnetic imprinted microspheres	[39]
One-Step		W_1/O/W_2: PVA and iron oxide nanoparticles	Synthesis of multi-drug encapsulated nanocapsules with magnetism	[40]
Two-Step		O_1/W: APTMS coated Fe_3O_4 nanoparticle and pluronic F68 W/O_2: PGPR	Synthesis of magnetic porous microspheres for absorption	[41]
Two-Step		O_1/W: Au-TEG nanoparticle W/O_2: CdSe QDs		[22]
Two-Step		W_1/O: PGPR O/W_2: Modified quinoa starch		[42]
Two-Step		W_1/O: PGPR O/W_2: Waxy starch	Encapsulation of sucrose to enhance sweetness perception	[43]
Two-Step		W_1/O: PGPR O/W_2: Octenylsuccinate quinoa starch	Encapsulation of anthocyanin	[44]
Two-Step		W_1/O: PGPR O/W_2: Kafirin	Encapsulation of anthocyanin	[45]
Two-Step		W_1/O: Lecithin O/W_2: Zein nanoparticles		[46]
Two-Step		W_1/O: PGPR O/W_2: Bacterial cellulose	Co-encapsulation and controlled release	[47,48]
Two-Step		W_1/O: Span-80 /O/W_2: β-cyclodextrin	Encapsulation of *Lactobacillus dellbrueckii*	[49]
Two-Step		W_1/O: PGPR O/W_2: Sugar beet pectin–bovine serum albumin nanoparticles	Co-encapsulation of betanin and curcumin	[50]

Table 1. *Cont.*

Emulsification Step	Stabilization Strategy	Colloidal Particle (and Co-Surfactant)	Application	Ref
Two-Step		O_1/W: Soluble WPC-GA complex W/O_2: Hydrophobic particles obtained from insoluble WPC-GA complexes		[51]
Two-Step		O_1/W: Nano-fibrillated cellulose (NFC)/Sulfated cellulose nanocrystals (CNC) W/O_2: Modified NFC/modified CNC	Proposed applications in the fields of food, pharmaceuticals, and cosmetics	[52]
Two-Step		W_1/O: lipophilic lignin O/W_2: hydrophilic lignin	Fabrication of molecularly imprinted multi-hollow microspheres	[53]
Two-Step		W_1/O: fat crystals O/W_2: Sodium caseinate		[54]
Two-Step		O_1/W: Gelatin and xanthan gum W/O_2: Vegetable fat crystal	Proposed for low-fat formulation for margarines and spreads	[55]
One-Step		O_1/W/O_2: Carnauba wax		[56]
Two-step		O_1/W: Cyclodextrin W/O_2: Candelilla wax		[57]
One-Step		O_1/W/O_2: Microbowls	Template to obtain supracolloidal systems	[58]
One-Step		O_1/W/O_2: Modified Boehmite alumina particles	Controlled release for pharmaceutical field	[23]
One-Step		W_1/O/W_2: PDEA microgel particle		[59]
One-Step		O_1/W/O_2: Palmitoyl chloride-modified diatomite particles	Synthesis of porous polyacrylamide particles	[60]
One-Step	Environmentally responsive particles	O_1/W/O_2: Terpolymer-grafted silica nanoparticles	pH-dependent controlled release	[11]
One-Step		O_1/W/O_2: PDAA nanoparticle		[28]
Two-Step	In situ modified particles	W_1/O: In situ modified cross-linked starch nanoparticles O/W_2: Hydroxyethyl cellulose		[61]
Two-Step		W_1/O: in situ modified PEI/silica particles O/W_2: PEI/silica particles	Fabrication of colloidosomes-in-colloidosomes structure	[62]
Two-Step		W_1/O: In situ modified amphiphilic silica O/W_2: In situ modified amphiphilic silica	Fabrication of hollowed microspheres	[63]
One-Step		W_1/O/W_2: In situ modified corn-peptide-functionalized calcium phosphate	Free fatty acid scavenging and lipid oxidation retardation	[64]

2.1. Combination of Hydrophobic and Hydrophilic Particles with/without Co-Surfactants

2.1.1. Silica Particles

Silica particles are one of the most typical interfacial stabilizers in the preparation of PDE due to their simplicity, wide availability, and versatility. It is found that PDE prepared with silica particles of extreme hydrophilicity and/or hydrophobicity exhibit poor stability against coalescence [15], as colloidal particles should be partially wetted by both the continuous and dispersed silica particles with intermediate hydrophilicity, and hydrophobicity yields improved stability. Moreover, by adsorbing Poly(ethylene imine) (PEI) onto the surface of fumed silica particles, Williams et al. prepared a "hybrid" PEI/silica particle that could stabilize both the W/O and O/W interfaces of PDE through adjustment of the PEI/silica mass ratio [37]

A combination of hydrophilic and hydrophobic silica particles is usually required to stabilize the two contrasting interfaces present in PDE [17,29]. In some conditions, hydrophobic/hydrophilic silica particles are used in combination with co-surfactants [30,31], magnetic nanoparticles [16,32,33,65], and polymeric particles [34,35].

2.1.2. Polymeric Particles

Besides silica particles, polymeric particles are also widely used in PDE. Graphene oxide [38], with its high surface-to-volume ratio, has been explored for stabilization of the W_1/O interface of a $W_1/O/W_2$ PDE by tuning the hydrophilic functional group content through thermal reduction. Hydrophilic and hydrophobic graphene oxide quantum dots [66] were fabricated and employed for the stabilization of $W_1/O/W_2$ PDE. In addition, Thompson et al. [26] prepared hydrophilic and hydrophobic polymer worms through reversible addition–fragmentation chain transfer (RAFT)—mediated polymerization-induced self-assembly and investigated their capability in stabilizing the two interfaces of PDE. Furthermore, Lei et al. [59] has fabricated a high internal phase W/O/W PDE through one-step emulsification by using poly(2-(diethylamino)ethyl methacrylate) (PDEA) microgel particles as a Pickering stabilizer. The team synthesized PDEA microgel particles from 2-(diethylamino)ethyl methacrylate (DEA) monomers and expected the microgel particles to stabilize only the O/W interface, but W/O/W PDEs were achieved in one-step emulsification instead. It was hypothesized that some DEA monomers remained and contributed to the W/O interface stabilization. The combined effect of DEA and PDEA together contributed to the formation of W/O/W PDE.

2.1.3. Metallic Particles

To endow specific functionality such as electrical, catalytic, and magnetic properties, metallic nanoparticles such as Fe_2O_3 [33,67], Fe_3O_4 [25,32,39–41], and tetra(ethylene glycol), functionalized Au [22] has been utilized for the preparation of PDE. Meanwhile, these colloidal particles are preferentially wetted by the aqueous phase, making them capable of stabilizing the O/W interface and, thus, incorporating additional hydrophobic particles with/without co-surfactant was needed to stabilize the W/O interface of the PDE. Through hydrophobic modification, metallic particles such as oleic acid Fe_2O_3 coated particles can also serve as a stabilizer for the W/O interface [33].

2.1.4. Naturally Derived Particles

Due to the increased interest in applications of PDE among pharmaceutical, cosmetic, and food sciences, recent trends in fabricating PDE with naturally derived colloidal particles continue to grow steadily. In the fabrication of $W_1/O/W_2$ PDE, naturally derived colloidal particles such as oligosaccharide particles (e.g., cyclodextrin [49]), polysaccharide particles (e.g., starch and modified starch [42–44]); water-insoluble protein particles (e.g., kafirin [45] and zein [46]); fat (e.g., wax [36,56] and mono-and triglyceride crystal [54]); and protein-polysaccharide conjugates (e.g., sugar beet pectin-bovine serum albumin [50] particles) have been used. However, as these particles are preferentially wetted by the aqueous phase, they are prone to stabilizing the O/W_2 interface of the PDE. Co-surfactants such as PGPR, Span, and lecithin are usually incorporated as well for the stabilization of the W_1/O interface.

The surface hydrophilicity of some naturally-derived particles can also be easily tuned. For example, soluble complexes and solid particles dried from insoluble complexes [51], obtained from interactions between whey protein concentrate (WPC) and gum Arabic (GA), have been used to stabilize the O_1/W and W/O_2 interfaces of $O_1/W/O_2$ PDE. By adjusting the drying temperature during solid particle formation from insoluble WPC-GA complexes, the interface adsorption characteristics of such particles can be further tailored to satisfy the target need [51]. For instance, the fabrication of $O_1/W/O_2$ PDE was also achieved with native and modified nanocellulose particles [52]. The unmodified nanocellulose particles with slight hydrophilicity can stabilize the O_1/W interface and form a three-dimensional nanofibrils network to protect the emulsion droplets against creaming, while modified nanocellulose particles were able to stabilize the W/O_2 interface due to the enhanced hydrophobicity introduced from chemical modification with lauroyl chloride [52]. Similarly, hydrophobic and oleic acid-modified lignin, and hydrophilic and unmodified lignin [53] have been combined to stabilize surfactant-free $W_1/O/W_2$ PDE.

Besides proteins and polysaccharides, fats have also been investigated for stabilizing Pickering double emulsions due to their unique capability of crystallization-induced network stabilization and interfacial adsorption as colloidal particles. $W_1/O/W_2$ PDE was prepared by the incorporation of mono- and triglyceride in the primary W_1/O emulsion and subsequently emulsified into the external sodium caseinate containing the W_2 phase [54]. It was observed that the fat crystals were capable of adsorbing at the W_1/O interface by the formation of smooth "shells." A combination of biopolymers, such as a mixture of gelatin and xanthan gum, and crystallized solid fat was applied to successfully prepare the $O_1/W/O_2$ emulsion [55]. It was postulated that the synergistic effect of interfacial adsorption stabilizes the W/O_2 interface by crystallized individual fat particles and the physical entrapment of droplets in the network formed by bulk crystallization [55]. Similarly, networked lamellar crystals and a solid layer of adsorbed crystal from carnauba wax [56] have achieved simultaneous stabilization of the O_1/W and the W/O_2 interfaces.

Under the aspect of stabilization strategies incorporating naturally derived particles for PDE, especially considering potential applications for edible purposes, gelation of corresponding phases of PDE have also been studied, not only for improving overall stability, but also for lowering the concentration of co-surfactants needed, such as PGPR, which has been reported for irritancy as well as a negative impact on sensory qualities at high concentrations [2]. It has been found that when the external W_2 phase of the $W_1/O/W_2$ PDE is gelled with an appropriate level of alginate, the storage and encapsulation stability of the PDE is greatly improved [47]. Meanwhile, $W_1/O/W_2$ PDE with beeswax-induced gelation [47,48] of the intermediate oil phase has also demonstrated enhanced stability against processing conditions such as the freeze–thaw cycle and variations in osmotic pressure.

2.2. Particles with Intermediate Wettability

Besides the hydrophilic and hydrophobic particles discussed above, particles possessing intermediate wettability were also applied in the stabilization of PDE. Heterogeneities in surface properties could be attributed to distinctive particle morphology, as presented in Nonomura, Kobayashi, and Nakagawa's work, which reported the successful fabrication of PDE stabilized by microbowls [58]. According to the authors' definition, microbowls are hollowed silica resin particles with holes on the surface. Different PDEs are achieved by combining variously shaped microbowls with oil of different compositions. It is postulated that microbowls demonstrate large contact angle hysteresis due to their distinctive particle shape (Figure 2); the different contact angles caused by microbowls' morphology enables them to perform not only as a high HLB surfactant, but also as a low HLB surfactant that further allows them to stabilize both outer and inner drop surfaces of the multiple emulsions.

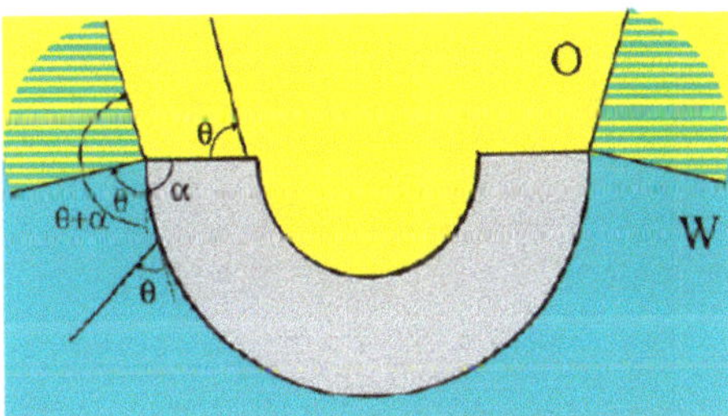

Figure 2. Schematic Illustration of stabilization mechanism of the environmentally responsive terpolymer grafted silica colloidal particles. Reprinted with permission from [11].

Aside from the influence of particle morphology, various degrees of chemical modification can also generate particles with heterogeneous surface properties. Chen et al. found OS2 particles, a C10-C13 alkylbenzene sulfonic acid hydrophobic-treated boehmite alumina particle with contact angle of about 90°, can establish polydisperse O/W/O PDE in a one-step emulsification process [23]. While claiming OS2 particles as the sole stabilizer of the O/W/O PDEs, the authors also postulated that the OS2 may undergo different degrees of modification and, as a result, exhibit different levels of hydrophobicity: some OS2 particles showed a contact angle slightly larger than 90°, while others have a contact angle equal or less than 90°. Thus, they attributed the formation of PDEs to the stabilization provided by OS2 particles with heterogeneous hydrophobicity. Another similar finding was reported in the work of Bai et al. [60], using surface-modified diatomite particles as the sole stabilizer. Hydrophobic groups were grafted onto diatomite particles by reacting with palmitoyl chloride. It has been speculated that the one-step PDE formation may be attributed to the particle's heterogeneity in surface wettability, as the surface modification may yield particles with varied amounts of hydrophobic groups and, thus, varied hydrophobicity [60].

2.3. Particles with Environmentally Responsive Property

Surface modifications have enabled the tailoring of particles that can adjust their surface wettability in response to certain environmental stimuli, such as pH and/or dispersed phase components. These particles, which were regarded as environmentally responsive particles, were also applied in the stabilization of PDE. Due to their flexible surface wettability, such particles can adapt to either the aqueous or oil phases. Therefore, a single kind of particle would be sufficient to accommodate the multiple interfaces of the PDE, which means that it can stabilize either the O/W or W/O interface. For example, amphiphilic silica particles are produced by rafting terpolymers consisting of hydrophilic poly(ethylene glycol) PEG, and hydrophobic polystyrene (PS), and anchoring the block poly[(3-triisopropyloxysilyl) propyl methacrylate] (PIPSMA) onto the particle surface [11]. The terpolymer-grafted silica particles can stabilize either the W/O or the O/W interfaces based on the solvent environment, as if the nanoparticle is dispersed in oil prior to emulsification, the nanoparticle with active hydrophobic PS chains can stabilize the W/O interface and vice versa (Figure 3) [11]. Environmentally responsive particles can also respond to changes in environmental pH. Zhu et al. [28] has synthesized poly(dodecylacrylate-co-acrylic acid) (PDAA) nanoparticles with hydrophobic as well as hydrophilic regions on particle surfaces that can exhibit responsive surface characteristics toward environmental pH. Such polymeric particles can be wetted by both the aqueous and the oil phase under controlled pH value and are, thus, capable of stabilization for both O/W and W/O interfaces.

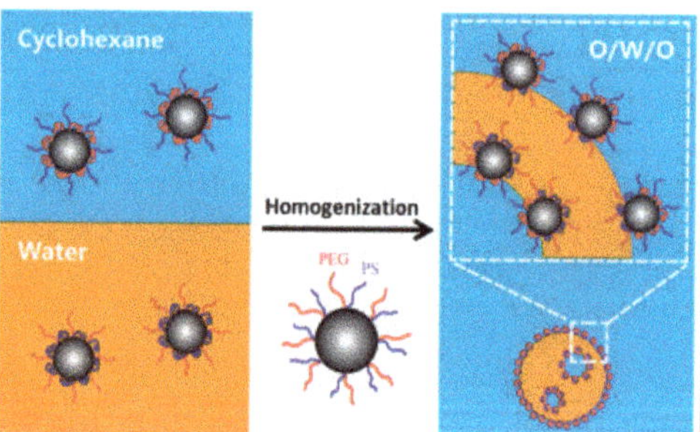

Figure 3. Schematic Illustration of contact angle hysteresis at the surface of microbowl particles. Reprinted with permission from [58].

2.4. In Situ Modified Particles

In situ modification is another form of surface modification that can alter the surface properties of a particle. However, in situ modification specifies that particles are modified during the homogenization process by interacting with other components presented in the emulsion precursor system, and the modification site being the interface. Stable PDEs can also be achieved by using colloidal particles that are in situ modified. For example, cross-linked starch nanoparticles (CSTN) were exposed to poly (styrene-co-maleic anhydride) (SMA) which reacted with the hydroxyl group on the surface of CSTN, and consequently rendered the hydrophilic CSTN to be moderately hydrophobic. The in situ modified CSTN can therefore stabilize the W/O interface satisfactorily and can generate stable W/O/W PDEs along with hydroxyethyl cellulose that stabilize the O/W interface [61]. Similarly, the surface wettability of PEI/silica hybrid particles is adjusted by in situ modification to stabilize PDEs [62]. While unmodified PEI/silica particles arre able to stabilize the O/W interface, PEI/silica particles modified by 1-undecanal exhibit enhanced hydrophobicity. Having the W_1/O interface stabilized by the in situ modified PEI/silica particles and the O/W_2 interface stabilized by the unmodified PEI/silica particles, stable W/O/W PDEs have been achieved. Zhang et al. [63] also used in situ modified amphiphilic silica particles to achieve W/O/W PDEs through a two-step emulsification process. Primary W/O emulsions were first prepared by using the mixture of styrene (St), tetrathoxysilane (TEOS), hexadecane, and g-(trimethoxysilyl) propylmethacrylate (MPS) as the oil phase, and aqueous triethylamine (TEA) solution as the inner water phase. By the hydrolysis-condensation of TEOS under basic conditions, silica particles were formed and modified by MPS at the O/W interface. Then W/O/W emulsions were fabricated by adding water as the outer phase. The partially modified silica nanoparticles were able to stabilize both the inner and outer droplets of the double emulsions. Aside from adjusting the Pickering particle's surface wettability, in situ modification can also render stabilizing particles multifunctional. In the work of Ruan et al., corn-peptide-functionalized calcium phosphate (CP-CaP) particles were modified in situ by free fatty acids that presented in the oil phase during the emulsification process [64]. Such modification not only made the CP-CaP particles hydrophobic but also reduced the free fatty acid content in the oil phase, which can contribute to lipid oxidation retardation. The in situ modified CP-CaP particles were able to stabilize both the O/W and W/O interfaces and generate W/O/W PDEs in a sign emulsification step.

3. Applications of PDE

3.1. Encapsulation of Drugs and/or Nutraceuticals

Pickering emulsion systems have been studied extensively for encapsulation [68] to protect the encapsulant from degradation or destabilization through direct contact with undesirable environments, such as harsh pH conditions, light, heat, oxidation, and more. In addition, the droplet-in-droplet structure of PDE for enhanced protection had raised promising potentials for them in encapsulation. However, due to the complex nature of PDE, such as its multiple interfaces and the potential diffusion and interactions of encapsulants between phases [57], careful design and preliminary studies are crucial for successful encapsulation and release.

3.1.1. Lipophilic Compounds

Encapsulation of lipophilic compound within the O_1 phase of O_1/W/O_2 PDE are not as prevalent comparing to encapsulation of water-soluble compound in the W_1 phase of W_1/O/W_2 PDE, Partly due to the reason that oil is not as suitable as water when incorporated as the continuous phase for oral delivery. However, one way to overcome this issue is to remove the outermost O_2 phase and to use the hierarchical structure with O_1/W morphology as the encapsulation medium for lipophilic compounds. Thermo-responsive wax-in-water microcapsule fabricated from O/W/O PDE [36] as the shell material, would be broken by expansion of the melting core at 44 °C, has been proposed with the potential of encapsulation of lipophilic compound and heat-stimulated release. However, the thermally induced release could disrupt the heat-sensitive encapsulant prior to target delivery, thus, the physical and chemical properties of the encapsulant should also be evaluated prior to encapsulation into the emulsion matrix.

3.1.2. Hydrophilic Compound

With carmine encapsulation [42,69] in the W_1 phase, followed by emulsification, storage, centrifugation release, and determination of carmine present in the W_2 phase, high encapsulation efficiency (EE) and encapsulation stability (ES) of the Pickering double emulsion has been determined. Similarly, through measurement of the released encapsulant from the innermost to the outermost phase right after emulsification and after a set storage period, $W_1/O/W_2$ PDE also yielded satisfactory EE and ES for encapsulant such as anthocyanin [44,45] and sucrose [43]. The high EE and ES of these PDEs are partially contributed by the robust protection provided by the irreversibly adsorbed colloidal particles, which also prevent coalescence between droplets and diffusion of droplets between phases [69], though other factors, such as osmotic pressure gradient [45] should also be considered for comprehensive assessment in long-term stability.

3.1.3. Co-Encapsulation of Lipophilic and Hydrophilic Compounds

The unique three-phase structure of PDE also gives rise to the potential of co-encapsulation [70]: in the case of $W_1/O/W_2$ PDE, simultaneous encapsulation of hydrophilic and hydrophobic active ingredients in the W_1 and the O phase has been achieved. With aggregated gliadin nanoparticles adsorbed at the interface and gelatin-induced immobilization of the W_1 phase [70], enhanced encapsulation stability and 2- and 4-fold increases in the bioaccessibility of W_1 encapsulated EGCG (epigallocatechin-3-gallate) and O encapsulated quercetin have been obtained. Similarly, betanin has been successfully encapsulated in a gel-like W_1 phase and curcumin in the oil phase of medium-chain triglyceride, having an encapsulation efficiency of 65.3% and 84.1%, respectively. This PDE co-stabilized by PGPR and sugar beet pectin-bovine serum albumin nanoparticles (SBNPs) was able to prolong the storage stability and bioaccessibility of betanin and curcumin [50].

In addition to EE and ES, the digestion and the release profile of the encapsulant are also crucial parameters to be assessed. The digestion profile of PDE could be investigated by simulated digestion coupled with microscopic analysis of the emulsion morphology before and after digestion [46]. For instance, the destabilization mechanism of the colloidal particles and the emulsion matrix and the subsequent release property under digestion are also crucial factors for delivery. As in the investigation of the in vitro digestion profile of a kafirin stabilized $W_1/O/W_2$ Pickering double emulsion [45], though the emulsion system exhibits stable EE and ES during storage for anthocyanin encapsulated in the W_1 phase, kafirin nanoparticles undergo flocculation and structural collapse during the gastric digestion phase due to the enzymatic digestible property of kafirin. The subsequent emulsion destabilization and release of the droplet contents has been proposed as an outside capsule for target release during the intestinal phase. While $W_1/O/W_2$ Pickering double emulsion stabilized by OSA-modified starch particles [44] has demonstrated the controlled release of W_1 encapsulated anthocyanin aimed at the intestinal phase, the OSA modified starch-endowed resistance to the acidic and enzymatic condition during the simulated gastric digestion can maintain emulsion integrity for passage to the intestinal digestion.

Regarding the digestion profile of PDE as encapsulation and co-encapsulation delivery systems, recent studies have discovered insightful relationships between tuning the structural properties of the PDE and the corresponding release of the encapsulants. As of PDE's multi-compartment nature, it was found that when coupled with gelation of selected W_1, O, and/or W_2 interfaces of $W_1/O/W_2$, the release profile and the bioaccessibility of the encapsulants can be tuned [47]. More specifically, the release profile and the bioaccessibility of the encapsulant could be manipulated through different combinations of gelation interface(s), as well as the gelation strength. Similarly, incorporations of crystallizable emulsifiers into PDE and the induced crystallization at different interfaces have also been investigated [71]; in which it was found that the site of crystallization played important roles in the rate of structural degradation and lipid digestion of the PDE delivery system, and ultimately offered potentials for tunable release of the encapsulants.

On top of the contributions toward encapsulation and delivery, manipulations of the structural properties of the PDE could also serve insightful purposes on its sensory profile when considering food and beverage applications. For example, studies [47,48] have found that by gelation of individual phases and/or combination of phases within PDE, as well as controlled gelation strength, the tribological properties of the PDE can be manipulated and result in tunable oral sensation under the aspect of in-mouth smoothness.

3.1.4. Microbes

PDE had also been adapted for probiotic encapsulation to achieve improved cell viability upon release. Compared to surfactant stabilized double emulsion, the presence of solid particles in the PDE was able to prolong the viability of entrapped cells because the interface-adsorbed particles and the droplet-within-droplet structure of the double emulsion protected the encapsulated microbes from direct contact with the acidic environment during digestion, thereby preventing the rapid loss of viability due to poor tolerance to acidic medium. For instance, β-cyclodextrin-stabilized $W_1/O/W_2$ emulsion had significantly enhanced the cell viability of *Lactobacillus delbrueckii* encapsulated in the W_1 phase upon release when compared to the surfactant stabilized counterparts [49]. Analogously, *Lactobacillus acidophilus* (LA) encapsulated in the W_1 phase of a $W_1/O/W_2$ emulsion fabricated with β-cyclodextrin particles yielded improved viability during 14 days storage, and high survival (84%) rate after simulated gastrointestinal digestion in comparison to the free and unencapsulated LA [72]. Furthermore, a 3-fold increase in colon-adhesion efficiency was determined for the $W_1/O/W_2$ encapsulated LA compared to the unencapsulated LA, due to factors such as facilitated adhesion through micelle formation from free fatty acids after oil phase digestion and the gelling effect of sodium alginate incorporated in the W_1 phase that immobilized the LA [72].

3.2. PDE-Templated Hierarchical Structure

3.2.1. Microsphere

Microspheres are spherical particles of diameters within the micrometer range, normally from 1 μm to 1000 μm [73]. Microspheres have a wide range of applications, such as drug encapsulation and delivery [74,75], controlled release [76], catalysis [67], enzyme immobilization [77], sensor [78], and adsorption [41]. Typically, microspheres can be fabricated by solvent evaporation coupled with multiple emulsion templates [79], coacervation methods [80], spray drying [81], polymerization techniques [82], and/or a combination of the above techniques [73,82]. Among the multiple emulsion technique mentioned above, PDE has become an attractive template for the fabrication of microspheres, in contrast to surfactant stabilized double emulsions; because compared to conventional low-molecular-weight surfactants, Pickering stabilizers can assemble at the interfaces without the problems of diffusion between the internal and the external phases, thereby providing a robust template. Moreover, the tunable droplet characteristic allows the fabrication of either monodisperse [27] or polydisperse [16] PDE, in which monodisperse PDE consists of a primary emulsion with a single dispersed droplet, while the polydisperse PDE is made up of a primary emulsion with more than one dispersed droplets. Thus, combined with the advantages mentioned above, PDE is a controllable and versatile template for microsphere fabrication.

The droplet-within-droplet characteristic and the multiple interfaces of PDE also make it ideal for the preparation of microspheres when coupled with polymerization, which is an important and prevalent technique in the fabrication of microspheres. When coupling with the polymerization technique, monomers and initiators can be incorporated into either the O or W phase for $W_1/O/W_2$ [30] or $O_1/W/O_2$ [16] PDE accordingly. Followed by heat, cross-linking, or photo-induced polymerization, the subsequent removal of the external continuous phase will yield the fabricated microsphere. For instance, a hollow microsphere can also be made by the additional removal of the innermost dispersed phase.

One of the advantages of using PDEs as the template for the fabrication of microspheres is that the structure of microspheres can be tailored by adjusting the structure and the property of the PDE. It is found that by varying the volume ratio of the aqueous and the oil phase of the PDE, the pore size and the pore structure can be adjusted for the fabricated porous microsphere (Figure 4) [32,35]. Hu et al. also found that the pore structure of the porous PLGA microsphere can be controlled by variation of the initial PLGA concentration presented in the oil phase of the $W_1/O/W_2$ template PDE [34]. Regulation of the inner structure of the microsphere, from closed-celled to hollowed and to interconnected, can also be achieved by tuning the concentration of the colloidal particles with/without co-surfactants [31]. On the other hand, there has been increased interest in the fabrication of molecularly imprinted microspheres from templating PDE [39,53,66]; for the synthesis of molecularly imprinted microspheres, template molecule and functional monomers are usually incorporated in the O phase of $W_1/O/W_2$ PDE, followed by polymerization and the subsequent removal of the template molecule. Advantages, such as good tunability leading to tailored porosity and the potential of a multi-hollow morphological structure for improved adsorption, have made PDE an attractive template for molecularly imprinted microspheres with versatility.

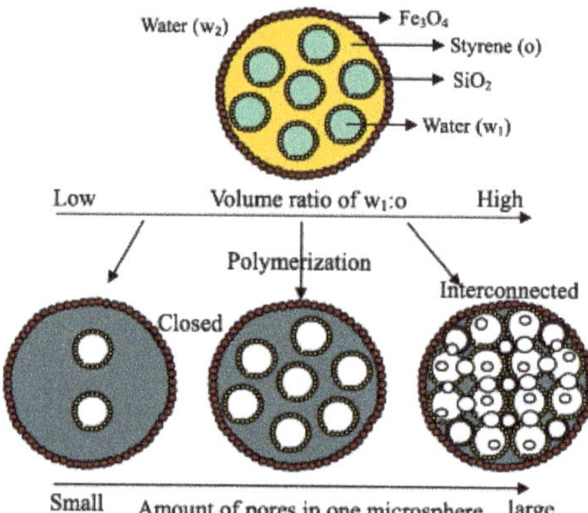

Figure 4. Preparation of multihollow microspheres with tunable pore structures by PDE templating. Reprinted with permission from [32].

Moreover, colloidal particles presented in PDE can not only serve as an interfacial stabilizer but also endows specific functionalities to microspheres. Magnetic microspheres have been fabricated from PDE by incorporation of interface-adsorbed Fe_3O_4 colloidal particles [25,32,41], which provided the microsphere responsiveness to the magnetic stimulus (Figure 5) [39] and has been proposed with applications in the field of carbon fixation, catalysis, heavy metal removal, and wastewater treatment. Ning et al. have further prepared Janus microspheres with dual anisotropy of porosity and magnetism from PDE, with the strategy of polymerization under a magnetic field, as the primary W_1/O emulsion droplets stabilized by Fe_3O_4 particles are concentrated within one side toward the magnetic field [29].

3.2.2. Microcapsule

Microencapsulation is the process of encapsulating micron-sized solid, liquid, or gas particles in a shell to protect the encapsulant from the environment [83], in which the microcapsule is one of the vehicles for microencapsulation [84]. Similar to microspheres, microcapsules were defined within the micrometer ranges, and the phrases microcapsule and microsphere were used somewhat analogously in the literature [85]. However, microcapsules and microspheres have different internal structures and morphologies (Figure 6) [84,86]. Different from microspheres, microcapsules possess a distinct core-shell structure with encapsulants containing a core and a surrounding shell built by a layer of polymer(s) and/or solid particles [86]. Because of their distinct core-shell structures, microcapsules are excellent candidates for encapsulation and controlled release of bioactive components [84], masking undesirable attributes such as taste, odor, gastric irritation [87], and more [88]. Techniques to fabricate microcapsules include coating [88], coacervation and phase separation [89], spray drying [90], emulsion and solvent evaporation [84], polymerization [91], and/or a combination of these techniques.

Figure 5. Photograph of magnetic microsphere particles suspended in water and under the presence of an externally placed magnet. Reprinted with permission from [39].

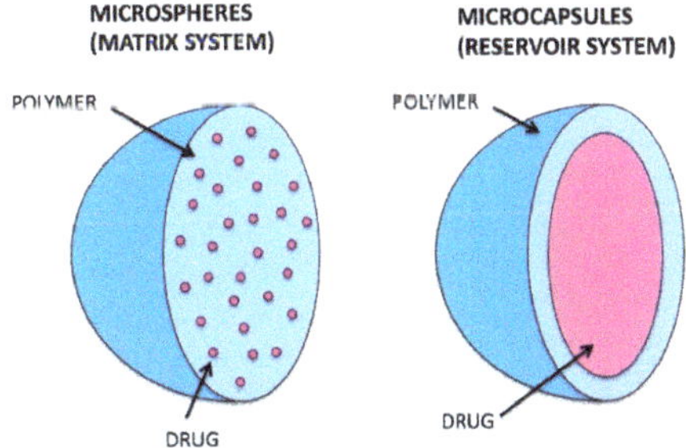

Figure 6. Structure of microspheres (left) and microcapsules(right). Reprinted with permission from [86].

Thermal responsive microcapsules consisting of a polymeric and colloidal crystal shell, with proposed applications in molecular labeling and biochemical sensors, were synthesized from microfluidic achieved monodispersed O/W/O PDEs via photo-initiated polymerization (Figure 7) [27]. Moreover, wax-in-water microcapsules [36] were assembled from monodispersed $O_1/W/O_2$ PDE, with the silica shell of the microcapsules fabricated from mineralization of surface adsorbed silica particles induced by the mineralization agent tetraethoxy-orthosilane (TEOS). Aside from monodispersed PDE, multi-compartment microcapsules, which enable the co-encapsulation of incompatible compounds for synergistic effects, can be templated from polydisperse PDEs. Multi-compartment microcapsules with a capsule-in-capsule structure [17] were prepared from the templated polydisperse $O_1/W/O_2$ PDE, by in situ polymerizations at both the O_1/W and the W/O_2 interfaces coupled with cross-link reaction between the interface adsorbed silica nanoparticles and the shell forming polymers. Two potent chemotherapy drugs, doxorubicin and paclitaxel were loaded in the W_1 and O phase of a $W_1/O/W_2$ PDE, which was consequently synthesized into core-shell nanocapsule after polymerization with poly(vinyl alcohol) originally present in the O phase [40]. The PVA served the dual purpose of a surfactant as well as shell constituent, whereas the magnetic iron oxide acted as both the shell stabilizer and the trigger for remote drug release under high-frequency magnetic fields. The resulting multi-drug-containing nanocapsules were biocompatible and were proposed for magneto-chemotherapy application [40]. Similarly, W/O/W PDE templating Graphene oxide (GO)@polylactic acid (PLA)@hydroxyapatite (HA) composite microcapsule [38], possessing biocompatible, biodegradable, and pH-sensitive properties has been proposed with potential to load hydrophilic and hydrophobic active compounds simultaneously.

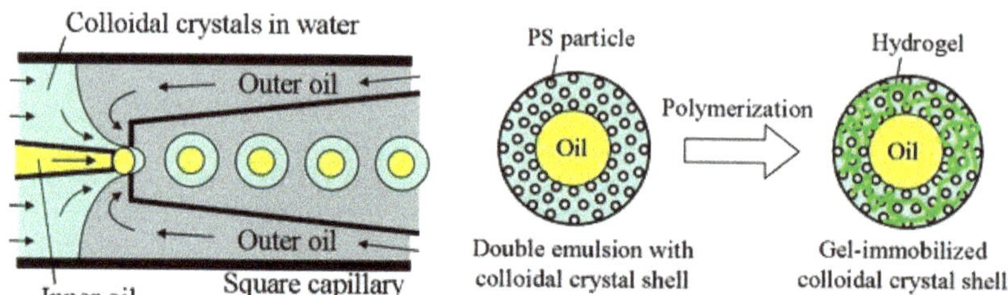

Figure 7. Fabrication of microcapsules with gel-immobilized colloidal crystal shells using capillary microfluidics and photopolymerization. Reprinted with permission from [27].

Microcapsules whose shells consist of densely packed colloidal particles are called colloidosomes [18]. Similar to other microcapsules, the sizes and physical properties, such as permeability, mechanical strength, and biocompatibility, of colloidosomes can be precisely tuned through the proper choice of colloids and preparation conditions for their assembly. The high degree of control over their physical properties makes colloidosomes attractive structures for encapsulation and controlled release of active ingredients.

Conventionally, simple W/O Pickering emulsions were commonly adapted to template the colloidal shell structures. However, this approach required subsequent transferring of the colloidosome into a continuous aqueous phase either by centrifugation or repeated washing, which would often induce extra damage to the colloidosome [18]. Compared to simple Pickering emulsions, PDEs have recently been reported to be better templates for colloidosome fabrication, as colloidosomes generated from PDE templates require no further phase transferring and exhibit a narrower range of size distribution. Furthermore, the utilization of PDE as a colloidosome fabrication template also enabled the control over the thickness of the colloidal shell, a vital factor that further affects the permeability and mechanical strength of the colloidosomes by changing the dimension of the PDE templates.

Lee and Weitz were the first to prepare monodisperse semipermeable nanoparticle colloidosomes from W/O/W double emulsion templates generated from glass capillary microfluidic devices (Figure 8a) [18]. Hydrophobic silica (SiO_2) nanoparticles were dispersed in the oil phase and were reported to adsorb onto the W_1/O and O/W_2 interfaces upon the formation of W/O/W PDE, forming colloidal shell structures. The colloidosomes were subsequently fabricated upon removal of the oil phase (Figure 8b) [18]. The colloidosomes-in-colloidosomes structure can also be fabricated from W/O/W PDEs [62]. Hydrophobic and hydrophilic PEI/Silica hybrid particles, obtained by varying PEI surface concentrations of silica nanoparticles, are used to stabilize the W_1/O interface and O/W_2 interface, respectively. Cross-linkers were also loaded into the system prior to the emulsification step to provide covalent stabilization, and colloidosomes-in-colloidosomes structures can be achieved through either aqueous phase cross-linking or oil droplet cross-linking, depending on the cross-linker of choice [62].

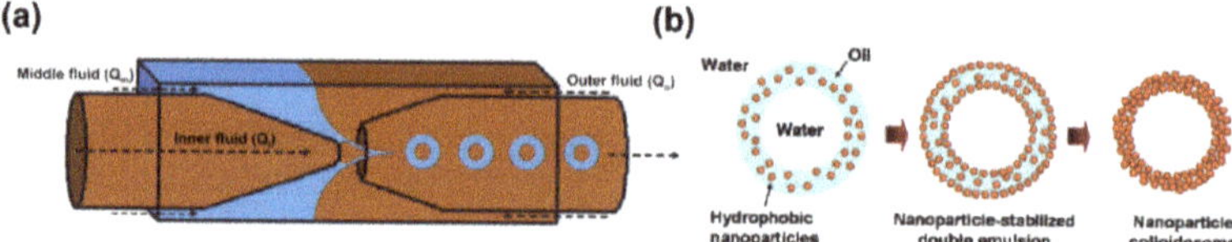

Figure 8. (**a**) Fabrication of PDE via microcapillary device. (**b**) Formation of nanoparticle colloidosomes from W/O/W PDE. Reprinted with permission from [18].

4. Perspectives: Challenges and Outlooks of PDE

Despite the expanding literature on the exploration of colloidal particles for stabilization of the W/O and O/W interfaces of PDE, a majority of the literature has relied on stabilization strategies through a combination of hydrophilic and hydrophobic colloidal particles. However, the incorporation of two kinds of colloidal particles usually requires two separate emulsification steps. Potential destabilization of the primary emulsion might be introduced during the second emulsification step. Furthermore, compared to single-step emulsification, two-step emulsification is more complex, which could limit its scale-up potential for industrial applications. Thus, a potential future research trend lies in developing colloidal particles that could simultaneously stabilize the contrasting interfaces of PDE while achieving emulsification in one step for facile and robust preparation of PDE.

In response to the consumers' growing demand for organic and clean-labeled products, naturally derived substances have been utilized as a material for food and drug manufacturing. This trend is also reflected in the preparation of PDE, as almost all PDEs stabilized by naturally derived particles were reported in the last ten years. However, PDEs stabilized solely by naturally derived particles were sparse, as co-surfactants are usually required for stabilization of the W/O interface, with PGPR being the most popular option. Despite being excellent stabilizers, most synthetic surfactants have low biocompatibility, are strictly regulated by international regulatory authorities, and can introduce undesirable sensory attributes. Thus, the utilization of hydrophobic naturally derived particles in replacement of synthetic surfactant is preferable in the formulation of PDEs.

In the future, several potential routes can be investigated to achieve the application potentials of PDE, especially in the field of food, beverage, and nutraceutical sciences:

(1) Research towards naturally derived particles that can stabilize simple W_1/O emulsion.

(2) To tune the surface wettability of hydrophilic naturally derived particles via surface modification methods. Moreover, the current application of PDE in oral delivery focuses primarily on the fabrication of the W/O/W delivery system, while exploration for O/W/O delivery is lacking. This is in alignment with the board range of products that are based on aqueous continuous systems. However, despite being limited, there are popular food products with continuous oil phases, such as butter, migraine, and spreads. More investigations in the preparation of O/W/O PDE for oral delivery could shed light on new formulations of the existing lipid-based food products.

(3) PDEs are regarded as suitable delivery media for many nutraceuticals and phytochemicals, though PDEs are still rarely adapted as delivery systems for active ingredients due to their poor and uncontrollable stability when exposed to complex human digestive environments. Recent studies are starting to explore the precise tuning of the digestion and release of the encapsulants by engineering the numerous tunable properties of the PDEs. However, more controlled release solutions, such as pH-triggered release, salt-induced release, and temperature-controlled release, could be explored for the PDE as a crucial delivery system to better suit the human digestive system.

(4) As described in previous sections, hierarchical structures such as microspheres and microcapsules, templated after PDEs, possess desirable characteristics and potentials for applications in terms of controlled release. However, their application in food products is rather limited as non-biocompatible polymers are often required for the polymerization step, which is a critical step to the formation of microstructure. PDE formulated with generally regarded as safe (GRAS) material could potentially give rise to microstructure with biocompatibility. Moreover, the microstructures' reliance on synthetic polymers could be negated by researching the topic of biopolymerization, which is the polymerization of biopolymers, such as cellulose and wax. If a microcapsule could be templated after GRAS PDE, via bio-polymerization, then its food application could be granted.

Meanwhile, in many studies, the morphology and configuration of the templating PDE, resulting from adjusting the parameters during preparation of the PDE, have been determined to play critical roles in the microstructure and properties of the fabricated hierarchical structures. However, rather few studies have examined the correlation between variations in the microstructure of the hierarchical structures resulting from tuning the configuration of the templating PDE, and the corresponding changes in the performance of such hierarchical structures in their proposed field of application. Efforts should be made to clarify such correlations and to provide insights for increasing the competency of PDE as a template for the fabrication of hierarchical structures with optimized and tailored performances. For instance, several studies have demonstrated the dual-function role of colloidal particles in PDE, for the cases of using iron oxides as interface adsorbed particles to achieve stabilization of the PDE while providing magnetic property to the fabricated hierarchical structure. As such, the development of nanoparticles responsive to environmental stimuli, such as pH, ionic strength, temperature, mechanical stress, and others, could be another emerging topic seeking to strengthen the versatility of PDE as a template material for hierarchical structures with special purposes. It would then be worthwhile for future studies to venture into exploring colloidal particles that possess dual-function roles and/or environmental responsive properties to provide improved efficacy or increased versatility of the fabricated hierarchical structures based on the templating PDE.

5. Conclusions

PDE represents a promising alternative to single emulsion and conventional double emulsion stabilized solely by surfactants, with many unique advantages. Compared to a single emulsion, the multi-phases and the consequent multi-compartment properties of PDE allow for more complex tunability and applications, for example, manipulation of the properties of three phases versus two, as well as the potential for co-encapsulation and delivery. While compared to conventional double emulsions, PDE offered enhanced stability and environmental friendliness, meeting satisfaction toward practical purposes. Meanwhile, the rigidity and particle-adsorbed interfaces opened room for PDE to serve as a synthesis template for other hierarchical structures. Consequently, different kinds of colloidal particles and stabilization strategies have been studied extensively. For instance, a broad application prospect has been proposed for PDE; the applications of PDE were outlined in the illustrative Figure 9, from a synthesis template for microspheres, microcapsules, and colloidosomes, to encapsulation with extra layer of protection and co-encapsulation of lipid-soluble and/or water-soluble compounds, which possess promising applications across various fields, such as pharmaceuticals, food and beverage, material sciences, personal care, and dietary supplements. However, it is likely that the potential of PDE has yet to be fully achieved.

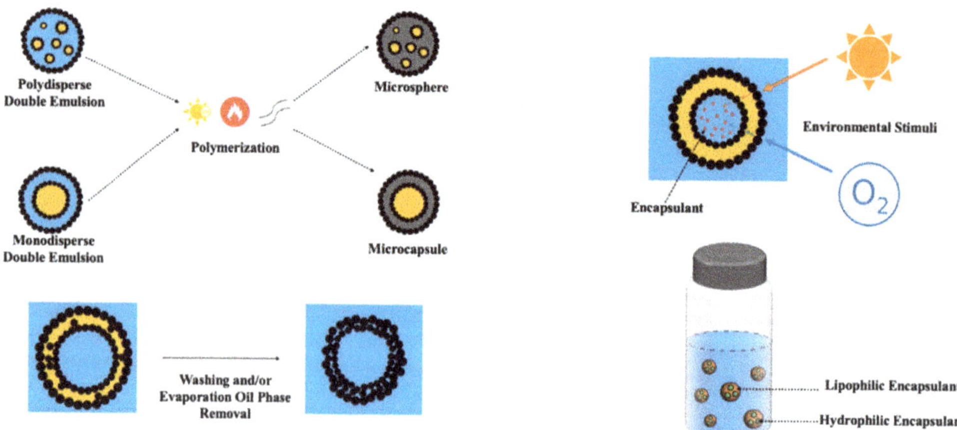

Figure 9. Applications of PDE.

Author Contributions: J.Z. (Junjia Zhang): conceptualization, literature review, writing—original draft. J.Z. (Jieyu Zhu): conceptualization, writing—editing. Y.C.: literature review, writing—editing. Q.H.: supervision, conceptualization, funding acquisition, writing-review and editing. All authors have read and agreed to the published version of the manuscript.

Funding: This work was supported by the United States Department of Agriculture, the National Institute of Food and Agriculture (2019-67017-29176).

Data Availability Statement: Data is contained within the article.

Conflicts of Interest: The authors declare no conflict of interest.

References

1. Aserin, A. *Multiple Emulsion: Technology and Applications*; John Wiley & Sons: Hoboken, NJ, USA, 2008; Volume 1.
2. Muschiolik, G.; Dickinson, E. Double Emulsions Relevant to Food Systems: Preparation, Stability, and Applications. *Compr. Rev. Food Sci. Food Saf.* **2017**, *16*, 532–555. [CrossRef] [PubMed]
3. Dickinson, E. Double Emulsions Stabilized by Food Biopolymers. *Food Biophys.* **2010**, *6*, 1–11. [CrossRef]
4. Beldengrün, Y.; Dallaris, V.; Jaén, C.; Protat, R.; Miras, J.; Calvo, M.; García-Celma, M.; Esquena, J. Formation and stabilization of multiple water-in-water-in-water (W/W/W) emulsions. *Food Hydrocoll.* **2019**, *102*, 105588. [CrossRef]
5. Song, Y.; Shum, H.C. Monodisperse w/w/w Double Emulsion Induced by Phase Separation. *Langmuir* **2012**, *28*, 12054–12059. [CrossRef] [PubMed]
6. Mohamed, L.A.; Dyab, A.K.F.; Taha, F. Non-aqueous castor oil-in-glycerin-in-castor oil double (o/o/o) Pickering emulsions: Physico-chemical characterization and in vitro release study. *J. Dispers. Sci. Technol.* **2019**, *41*, 102–116. [CrossRef]
7. Florence, A.; Whitehill, D. Some features of breakdown in water-in-oil-in-water multiple emulsions. *J. Colloid Interface Sci.* **1981**, *79*, 243–256. [CrossRef]
8. Aveyard, R.; Binks, B.P.; Clint, J.H. Emulsions stabilised solely by colloidal particles. *Adv. Colloid Interface Sci.* **2003**, *100-102*, 503–546. [CrossRef]
9. Mezzenga, R.; Folmer, B.M.; Hughes, E. Design of Double Emulsions by Osmotic Pressure Tailoring. *Langmuir* **2004**, *20*, 3574–3582. [CrossRef] [PubMed]
10. Taylor, P. Ostwald ripening in emulsions. *Adv. Colloid Interface Sci.* **1998**, *75*, 107–163. [CrossRef]
11. Liu, M.; Chen, X.; Yang, Z.; Xu, Z.; Hong, L.; Ngai, T. Tunable Pickering Emulsions with Environmentally Responsive Hairy Silica Nanoparticles. *ACS Appl. Mater. Interfaces* **2016**, *8*, 32250–32258. [CrossRef] [PubMed]
12. Cserháti, T.; Forgács, E.; Oros, G. Biological activity and environmental impact of anionic surfactants. *Environ. Int.* **2002**, *28*, 337–348. [CrossRef]
13. Pickering, S.U. CXCVI.—Emulsions. *J. Chem. Soc. Trans.* **1907**, *91*, 2001–2021. [CrossRef]
14. Xiao, J.; Li, Y.; Huang, Q. Recent advances on food-grade particles stabilized Pickering emulsions: Fabrication, characterization and research trends. *Trends Food Sci. Technol.* **2016**, *55*, 48–60. [CrossRef]
15. Binks, P.; Dyab, A.; Fletcher, P. Multiple emulsions stabilised solely by nanoparticles. In Proceedings of the 3rd World Congress on Emulsions, Lyon, Paris, 1 September 2002.
16. Gao, Q.; Wang, C.; Liu, H.; Chen, Y.; Tong, Z. Dual nanocomposite multihollow polymer microspheres prepared by suspension polymerization based on a multiple pickering emulsion. *Polym. Chem.* **2009**, *1*, 75–77. [CrossRef]
17. Yang, Y.; Wang, C.; Tong, Z. Facile, controlled, large scale fabrication of novel capsule clusters. *RSC Adv.* **2013**, *3*, 4514–4517. [CrossRef]
18. Lee, D.; Weitz, D.A. Double Emulsion-Templated Nanoparticle Colloidosomes with Selective Permeability. *Adv. Mater.* **2008**, *20*, 3498–3503. [CrossRef]
19. Oza, K.P.; Frank, S.G. Multiple emulsions stabilized by colloidal microgrystalline cellulose. *J. Dispers. Sci. Technol.* **1989**, *10*, 163–185. [CrossRef]
20. Sekine, T.; Yoshida, K.; Matsuzaki, F.; Yanaki, T.; Yamaguchi, M. A novel method for preparing oil-in-water-in-oil type multiple emulsions using organophilic montmorillonite clay mineral. *J. Surfactants Deterg.* **1999**, *2*, 309–315. [CrossRef]
21. Midmore, B.; Herrington, T. Silica-Stabilised Multiple Emulsions. In *Trends in Colloid and Interface Science XIII*; Springer: Berlin/Heidelberg, Germany, 1999; pp. 115–120.
22. Miesch, C.; Kosif, I.; Lee, E.; Kim, J.-K.; Russell, T.P.; Hayward, R.C.; Emrick, T. Nanoparticle-Stabilized Double Emulsions and Compressed Droplets. *Angew. Chem. Int. Ed.* **2011**, *51*, 145–149. [CrossRef] [PubMed]
23. Chen, J.; Vogel, R.; Werner, S.; Heinrich, G.; Clausse, D.; Dutschk, V. Influence of the particle type on the rheological behavior of Pickering emulsions. *Colloids Surfaces A: Physicochem. Eng. Asp.* **2011**, *382*, 238–245. [CrossRef]
24. Zhu, Y.; Hua, Y.; Zhang, S.; Wang, Y.; Chen, J. Open-cell macroporous bead: A novel polymeric support for heterogeneous photocatalytic reactions. *J. Polym. Res.* **2015**, *22*, 57. [CrossRef]
25. Wang, F.; Zhang, X.; Shao, L.; Cui, Z.; Nie, T. Synthesis of magnetic Fe_3O_4/polyamine hybrid microsphere using O/W/O Pickering emulsion droplet as the polymerization micro-reactor. *RSC Adv.* **2015**, *5*, 22188–22198. [CrossRef]

26. Thompson, K.L.; Mable, C.J.; Lane, J.A.; Derry, M.J.; Fielding, L.A.; Armes, S.P. Preparation of Pickering Double Emulsions Using Block Copolymer Worms. *Langmuir* **2015**, *31*, 4137–4144. [CrossRef] [PubMed]
27. Kanai, T.; Lee, D.; Shum, H.C.; Shah, R.K.; Weitz, D.A. Gel-Immobilized Colloidal Crystal Shell with Enhanced Thermal Sensitivity at Photonic Wavelengths. *Adv. Mater.* **2010**, *22*, 4998–5002. [CrossRef] [PubMed]
28. Zhu, Y.; Sun, J.; Yi, C.; Wei, W.; Liu, X. One-step formation of multiple Pickering emulsions stabilized by self-assembled poly(dodecyl acrylate-co-acrylic acid) nanoparticles. *Soft Matter* **2016**, *12*, 7577–7584. [CrossRef]
29. Ning, Y.; Wang, C.; Ngai, T.; Tong, Z. Fabrication of Tunable Janus Microspheres with Dual Anisotropy of Porosity and Magnetism. *Langmuir* **2013**, *29*, 5138–5144. [CrossRef] [PubMed]
30. Guan, X.; Ngai, T. pH-Sensitive W/O Pickering High Internal Phase Emulsions and W/O/W High Internal Water-Phase Double Emulsions with Tailored Microstructures Costabilized by Lecithin and Silica Inorganic Particles. *Langmuir* **2021**, *37*, 2843–2854. [CrossRef]
31. Yin, D.; Li, B.; Liu, J.; Zhang, Q. Structural diversity of multi-hollow microspheres via multiple Pickering emulsion co-stabilized by surfactant. *Colloid Polym. Sci.* **2014**, *293*, 341–347. [CrossRef]
32. Zou, S.; Liu, H.; Yang, Y.; Wei, Z.; Wang, C. Multihollow nanocomposite microspheres with tunable pore structures by templating Pickering double emulsions. *React. Funct. Polym.* **2013**, *73*, 1231–1241. [CrossRef]
33. Zou, S.; Wang, C.; Gao, Q.; Tong, Z. Surfactant-Free Multiple Pickering Emulsions Stabilized by Combining Hydrophobic and Hydrophilic Nanoparticles. *J. Dispers. Sci. Technol.* **2013**, *34*, 173–181. [CrossRef]
34. Hu, Y.; Zou, S.; Yang, Y.; Tong, Z.; Wang, C. Facile Fabrication of Macroporous PLGA Microspheres via Double-Pickering Emulsion Templates. *Macromol. Chem. Phys.* **2015**, *216*, 714–720. [CrossRef]
35. Zhou, H.; Shi, T.; Zhou, X. Aqueous core polystyrene microspheres fabricated via suspension polymerization basing on a multiple pickering emulsion. *J. Appl. Polym. Sci.* **2013**, *131*. [CrossRef]
36. Depardieu, M.; Nollet, M.; Destribats, M.; Schmitt, V.; Backov, R. Thermo-Stimulable Wax@ Water@ SiO$_2$ Multicore-Shell Capsules. *Part. Part. Syst. Charact.* **2013**, *30*, 185–192. [CrossRef]
37. Williams, M.; Warren, N.J.; Fielding, L.A.; Armes, S.P.; Verstraete, P.; Smets, J. Preparation of Double Emulsions using Hybrid Polymer/Silica Particles: New Pickering Emulsifiers with Adjustable Surface Wettability. *ACS Appl. Mater. Interfaces* **2014**, *6*, 20919–20927. [CrossRef]
38. Guo, H.; Wang, Y.; Huang, Y.; Huang, F.; Li, S.; Shen, Y.; Zhu, M.; Xie, A. A GO@PLA@HA Composite Microcapsule: Its Preparation and Multistage and Controlled Drug Release. *Eur. J. Inorg. Chem.* **2017**, *2017*, 3312–3321. [CrossRef]
39. Zhu, W.; Ma, W.; Li, C.; Pan, J.; Dai, X. Well-designed multihollow magnetic imprinted microspheres based on cellulose nanocrystals (CNCs) stabilized Pickering double emulsion polymerization for selective adsorption of bifenthrin. *Chem. Eng. J.* **2015**, *276*, 249–260. [CrossRef]
40. Hu, S.-H.; Liao, B.-J.; Chiang, C.-S.; Chen, P.-J.; Chen, I.-W.; Chen, S.-Y. Core-Shell Nanocapsules Stabilized by Single-Component Polymer and Nanoparticles for Magneto-Chemotherapy/Hyperthermia with Multiple Drugs. *Adv. Mater.* **2012**, *24*, 3627–3632. [CrossRef] [PubMed]
41. Zhu, Y.; Zheng, Y.; Wang, F.; Wang, A. Fabrication of magnetic porous microspheres via (O$_1$/W)/O$_2$ double emulsion for fast removal of Cu^{2+} and Pb^{2+}. *J. Taiwan Inst. Chem. Eng.* **2016**, *67*, 505–510. [CrossRef]
42. Marefati, A.; Sjöö, M.; Timgren, A.; Dejmek, P.; Rayner, M. Fabrication of encapsulated oil powders from starch granule stabilized W/O/W Pickering emulsions by freeze-drying. *Food Hydrocoll.* **2015**, *51*, 261–271. [CrossRef]
43. Al Nuumani, R.; Vladisavljević, G.T.; Kasprzak, M.; Wolf, B. In-vitro oral digestion of microfluidically produced monodispersed W/O/W food emulsions loaded with concentrated sucrose solution designed to enhance sweetness perception. *J. Food Eng.* **2020**, *267*, 109701. [CrossRef]
44. Lin, X.; Li, S.; Yin, J.; Chang, F.; Wang, C.; He, X.; Huang, Q.; Zhang, B. Anthocyanin-loaded double Pickering emulsion stabilized by octenylsuccinate quinoa starch: Preparation, stability and in vitro gastrointestinal digestion. *Int. J. Biol. Macromol.* **2019**, *152*, 1233–1241. [CrossRef]
45. Xiao, J.; Lu, X.; Huang, Q. Double emulsion derived from kafirin nanoparticles stabilized Pickering emulsion: Fabrication, microstructure, stability and in vitro digestion profile. *Food Hydrocoll.* **2017**, *62*, 230–238. [CrossRef]
46. Jiang, H.; Zhang, T.; Smits, J.; Huang, X.; Maas, M.; Yin, S.; Ngai, T. Edible high internal phase Pickering emulsion with double-emulsion morphology. *Food Hydrocoll.* **2020**, *111*, 106405. [CrossRef]
47. Chen, M.; Li, W.; Wang, W.; Cao, Y.; Lan, Y.; Huang, Q.; Xiao, J. Effects of gelation on the stability, tribological properties and time-delayed release profile of double emulsions. *Food Hydrocoll.* **2022**, *131*, 107753. [CrossRef]
48. Chen, M.; Abdullah; Wang, W.; Xiao, J. Regulation Effects of Beeswax in the Intermediate Oil Phase on the Stability, Oral Sensation and Flavor Release Properties of Pickering Double Emulsions. *Foods* **2022**, *11*, 1039. [CrossRef] [PubMed]
49. Eslami, P.; Davarpanah, L.; Vahabzadeh, F. Encapsulating role of β-cyclodextrin in formation of pickering water-in-oil-in-water (W$_1$/O/W$_2$) double emulsions containing Lactobacillus dellbrueckii. *Food Hydrocoll.* **2017**, *64*, 133–148. [CrossRef]
50. Tang, X.-Y.; Wang, Z.-M.; Meng, H.-C.; Lin, J.-W.; Guo, X.-M.; Zhang, T.; Chen, H.-L.; Lei, C.-Y.; Yu, S.-J. Robust W/O/W Emulsion Stabilized by Genipin-Cross-Linked Sugar Beet Pectin-Bovine Serum Albumin Nanoparticles: Co-encapsulation of Betanin and Curcumin. *J. Agric. Food Chem.* **2021**, *69*, 1318–1328. [CrossRef] [PubMed]

51. Estrada-Fernández, A.; Román-Guerrero, A.; Jiménez-Alvarado, R.; Lobato-Calleros, C.; Alvarez-Ramirez, J.; Vernon-Carter, E. Stabilization of oil-in-water-in-oil (O_1/W/O_2) Pickering double emulsions by soluble and insoluble whey protein concentrate-gum Arabic complexes used as inner and outer interfaces. *J. Food Eng.* **2018**, *221*, 35–44. [CrossRef]

52. Cunha, A.G.; Mougel, J.-B.; Cathala, B.; Berglund, L.A.; Capron, I. Preparation of Double Pickering Emulsions Stabilized by Chemically Tailored Nanocelluloses. *Langmuir* **2014**, *30*, 9327–9335. [CrossRef]

53. Pan, J.; Yin, Y.; Gan, M.; Meng, M.; Dai, X.; Wu, R.; Shi, W.; Yan, Y. Fabrication and evaluation of molecularly imprinted multi-hollow microspheres adsorbents with tunable inner pore structures derived from templating Pickering double emulsions. *Chem. Eng. J.* **2015**, *266*, 299–308. [CrossRef]

54. Frasch-Melnik, S.; Spyropoulos, F.; Norton, I.T. W1/O/W2 double emulsions stabilised by fat crystals–formulation, stability and salt release. *J. Colloid Interface Sci.* **2010**, *350*, 178–185. [CrossRef] [PubMed]

55. Patel, A.R. Surfactant-free oil-in-water-in-oil emulsions stabilized solely by natural components-biopolymers and vegetable fat crystals. *MRS Adv.* **2017**, *2*, 1095–1102. [CrossRef]

56. Szumała, P.; Luty, N. Effect of different crystalline structures on W/O and O/W/O wax emulsion stability. *Colloids Surfaces A: Physicochem. Eng. Asp.* **2016**, *499*, 131–140. [CrossRef]

57. Yu, S.-C.; Bochot, A.; Le Bas, G.; Chéron, M.; Mahuteau, J.; Grossiord, J.-L.; Seiller, M.; Duchêne, D. Effect of camphor/cyclodextrin complexation on the stability of O/W/O multiple emulsions. *Int. J. Pharm.* **2003**, *261*, 1–8. [CrossRef] [PubMed]

58. Nonomura, Y.; Kobayashi, N.; Nakagawa, N. Multiple Pickering Emulsions Stabilized by Microbowls. *Langmuir* **2011**, *27*, 4557–4562. [CrossRef] [PubMed]

59. Lei, L.; Zhang, Q.; Shi, S.; Zhu, S. High internal phase emulsion with double emulsion morphology and their templated porous polymer systems. *J. Colloid Interface Sci.* **2016**, *483*, 232–240. [CrossRef]

60. Bai, Y.; Pei, X.; Zhao, B.; Xu, K.; Zhai, K.; Wang, C.; Zhang, F.; Tan, Y.; Zhang, B.; Wang, Y.; et al. Multiple pickering high internal phase emulsions stabilized by modified diatomite particles via one-step emulsification process. *Chem. Eng. Sci.* **2020**, *212*, 115341. [CrossRef]

61. Nikfarjam, N.; Qazvini, N.T.; Deng, Y. Cross-linked starch nanoparticles stabilized Pickering emulsion polymerization of styrene in w/o/w system. *Colloid Polym. Sci.* **2013**, *292*, 599–612. [CrossRef]

62. Williams, M.; Armes, S.P.; Verstraete, P.; Smets, J. Double Emulsions and Colloidosomes-in-Colloidosomes Using Silica-Based Pickering Emulsifiers. *Langmuir* **2014**, *30*, 2703–2711. [CrossRef] [PubMed]

63. Zhang, J.; Ge, X.; Wang, M.; Yang, J.; Wu, Q.; Wu, M.; Liu, N.; Jin, Z. Hybrid hollow microspheres templated from double Pickering emulsions. *Chem. Commun.* **2010**, *46*, 4318–4320. [CrossRef] [PubMed]

64. Ruan, Q.; Zeng, L.; Ren, J.; Yang, X. One-step formation of a double Pickering emulsion via modulation of the oil phase composition. *Food Funct.* **2018**, *9*, 4508–4517. [CrossRef]

65. Dyab, A.K.F. Macroporous Polymer Beads and Monoliths From Pickering Simple, Double, and Triple Emulsions. *Macromol. Chem. Phys.* **2012**, *213*, 1815–1832. [CrossRef]

66. Hosseinzadeh, B.; Nikfarjam, N.; Kazemi, S.H. Hollow molecularly imprinted microspheres made by w/o/w double Pickering emulsion polymerization stabilized by graphene oxide quantum dots targeted for determination of l-cysteine concentration. *Colloids Surfaces A Physicochem. Eng. Asp.* **2020**, *612*, 125978. [CrossRef]

67. Guo, C.; Ge, M.; Liu, L.; Gao, G.; Feng, Y.; Wang, Y. Directed Synthesis of Mesoporous TiO_2 Microspheres: Catalysts and Their Photocatalysis for Bisphenol A Degradation. *Environ. Sci. Technol.* **2009**, *44*, 419–425. [CrossRef]

68. Albert, C.; Beladjine, M.; Tsapis, N.; Fattal, E.; Agnely, F.; Huang, N. Pickering emulsions: Preparation processes, key parameters governing their properties and potential for pharmaceutical applications. *J. Control. Release* **2019**, *309*, 302–332. [CrossRef] [PubMed]

69. Matos, M.; Timgren, A.; Sjöö, M.; Dejmek, P.; Rayner, M. Preparation and encapsulation properties of double Pickering emulsions stabilized by quinoa starch granules. *Colloids Surfaces A Physicochem. Eng. Asp.* **2013**, *423*, 147–153. [CrossRef]

70. Chen, X.; McClements, D.J.; Wang, J.; Zou, L.; Deng, S.; Liu, W.; Yan, C.; Zhu, Y.; Cheng, C.; Liu, C. Coencapsulation of (−)-Epigallocatechin-3-gallate and Quercetin in Particle-Stabilized W/O/W Emulsion Gels: Controlled Release and Bioaccessibility. *J. Agric. Food Chem.* **2018**, *66*, 3691–3699. [CrossRef] [PubMed]

71. Li, W.; Wang, W.; Yong, C.; Lan, Y.; Huang, Q.; Xiao, J. Effects of the Distribution Site of Crystallizable Emulsifiers on the Gastrointestinal Digestion Behavior of Double Emulsions. *J. Agric. Food Chem.* **2022**, *70*, 5115–5125. [CrossRef] [PubMed]

72. Wang, L.; Song, M.; Zhao, Z.; Chen, X.; Cai, J.; Cao, Y.; Xiao, J. Lactobacillus acidophilus loaded pickering double emulsion with enhanced viability and colon-adhesion efficiency. *LWT* **2019**, *121*, 108928. [CrossRef]

73. Sahil, K.; Akanksha, M.; Premjeet, S.; Bilandi, A.; Kapoor, B. Microsphere: A review. *Int. J. Res. Pharm. Chem.* **2011**, *1*, 1184–1198.

74. Floyd, J.A.; Galperin, A.; Ratner, B.D. Drug encapsulated polymeric microspheres for intracranial tumor therapy: A review of the literature. *Adv. Drug Deliv. Rev.* **2015**, *91*, 23–37. [CrossRef] [PubMed]

75. Cho, C.-S.; Islam, M.A.; Firdous, J.; Choi, Y.-J.; Yun, C.-H. Design and application of chitosan microspheres as oral and nasal vaccine carriers: An updated review. *Int. J. Nanomed.* **2012**, *7*, 6077–6093. [CrossRef] [PubMed]

76. Freiberg, S.; Zhu, X.X. Polymer microspheres for controlled drug release. *Int. J. Pharm.* **2004**, *282*, 1–18. [CrossRef] [PubMed]

77. Jiang, D.-S.; Long, S.-Y.; Huang, J.; Xiao, H.-Y.; Zhou, J.-Y. Immobilization of Pycnoporus sanguineus laccase on magnetic chitosan microspheres. *Biochem. Eng. J.* **2005**, *25*, 15–23. [CrossRef]

78. Zhao, B.; Liu, Z.; Liu, Z.; Liu, G.; Li, Z.; Wang, J.; Dong, X. Silver microspheres for application as hydrogen peroxide sensor. *Electrochem. Commun.* **2009**, *11*, 1707–1710. [CrossRef]
79. O'Donnell, P.B.; McGinity, J.W. Preparation of microspheres by the solvent evaporation technique. *Adv. Drug Deliv. Rev.* **1997**, *28*, 25–42. [CrossRef]
80. Arshady, R. Microspheres and microcapsules, a survey of manufacturing techniques Part II: Coacervation. *Polym. Eng. Sci.* **1990**, *30*, 905–914. [CrossRef]
81. He, P.; Davis, S.S.; Illum, L. Chitosan microspheres prepared by spray drying. *Int. J. Pharm.* **1999**, *187*, 53–65. [CrossRef]
82. Ganesan, P.; John, A.J.D.; Sabapathy, L.; Duraikannu, A. Review on microsphere. *Am. J. Drug Discov. Dev.* **2014**, *4*, 153–179. [CrossRef]
83. Ghosh, S.K. *Functional Coatings and Microencapsulation: A General Perspective*; Wiley: Hoboken, NJ, USA, 2006; pp. 1–28.
84. Jyothi, N.V.N.; Prasanna, P.M.; Sakarkar, S.N.; Prabha, K.S.; Ramaiah, P.S.; Srawan, G. Microencapsulation techniques, factors influencing encapsulation efficiency. *J. Microencapsul.* **2010**, *27*, 187–197. [CrossRef]
85. Paulo, F.; Santos, L. Design of experiments for microencapsulation applications: A review. *Mater. Sci. Eng. C* **2017**, *77*, 1327–1340. [CrossRef] [PubMed]
86. Herrero-Vanrell, R.; Bravo-Osuna, I.; Andrés-Guerrero, V.; Vicario-De-La-Torre, M.; Martínez, I.T.M. The potential of using biodegradable microspheres in retinal diseases and other intraocular pathologies. *Prog. Retin. Eye Res.* **2014**, *42*, 27–43. [CrossRef]
87. Jyothi, S.S.; Seethadevi, A.; Prabha, K.S.; Muthuprasanna, P.; Pavitra, P. Microencapsulation: A review. *Int. J. Pharm. Biol. Sci.* **2012**, *3*, 509–531.
88. Bansode, S.S.; Banarjee, S.K.; Gaikwad, D.D.; Jadhav, S.L.; Thorat, R.M. Microencapsulation: A review. *Int. J. Pharm. Sci. Rev. Res.* **2010**, *1*, 38–43.
89. Gouin, S. Microencapsulation: Industrial appraisal of existing technologies and trends. *Trends Food Sci. Technol.* **2004**, *15*, 330–347. [CrossRef]
90. Baranauskienė, R.; Venskutonis, P.R.; Dewettinck, K.; Verhé, R. Properties of oregano (*Origanum vulgare* L.), citronella (*Cymbopogon nardus* G.) and marjoram (*Majorana hortensis* L.) flavors encapsulated into milk protein-based matrices. *Food Res. Int.* **2006**, *39*, 413–425. [CrossRef]
91. Yow, H.N.; Routh, A.F. Formation of liquid core–polymer shell microcapsules. *Soft Matter* **2006**, *2*, 940–949. [CrossRef] [PubMed]

MDPI

St. Alban-Anlage 66

4052 Basel

Switzerland

Tel. +41 61 683 77 34

Fax +41 61 302 89 18

www.mdpi.com

Foods Editorial Office

E-mail: foods@mdpi.com

www.mdpi.com/journal/foods